Y0-BYB-155

# MONOGRAPHIAE BIOLOGICAE

EDITOR

P. VAN OYE
Gent

VOLUMEN XVI

DR. W. JUNK PUBLISHERS - THE HAGUE - 1966

# SALINITY AND ARIDITY

## New Approaches to Old Problems

edited by

HUGO BOYKO

DR. W. JUNK PUBLISHERS - THE HAGUE - 1966

Printed in the Netherlands
Zuid-Nederlandsche Drukkerij N.V. — 's-Hertogenbosch

## CONTENTS

## PART III: STUDIES ON PLANT AND ANIMAL LIFE IN A BRINE

# PREFACE

When I accepted the invitation to write a book on my own experiences in productivizing deserts and in plantgrowing by irrigation with sea-water, I immediately proposed to broaden the basis and to include also other new approaches to these problems of global impact. From the start, the main stress was to be laid on the methods, and not only on the results of experiments — and with regard to the new principles and natural laws not only on their formulation and explanation, but also on their applicability.

The verification in other countries of the first successful experiments of direct irrigation with sea-water soon led to a rapid growth of interest in the development of these new methods. Accordingly, further steps were sponsored by the highest international bodies like UNESCO and the World Academy of Art and Science (WAAS). Particular thanks are due to both international bodies for their active sponsorship of research in these new fields.

Two famous biologists, both Fellows of the World Academy, were kind enough to read some of the manuscripts and to render their most constructive advice, the plant ecologist Pièrre DANSERAU and the microbiologist Stuart MUDD.

The entire work on the actual experiments, of writing our own parts of the book and editing the other contributions would not have been complete without the horticulturist Dr. Elisabeth BOYKO, working on my side as my wife and coworker since more than 40 years. The linguistic capacities of my Assistant, Mr. Zeev AVNI, decisively helped to make several chapters more palatable for the reader.

Last but not least, this book ows much of its existence to the late Professor Walter W. WEISBACH and to Mrs. WEISBACH, who were its main initiators, and to the excellent editorial and technical advice of Mr. Klaus J. PLASTERK of the Publisher's firm, Dr. W. Junk.

When the results of the first experiments were made known by a few scientific papers, the more imaginative part of the scientific world saw immediately that ways were opened by these new principles and methods to conquest vast areas of sandy deserts on the one hand, and on the other to use for this purpose, on an economic basis, the saline waters of their underground and even those of the sea itself.

Solutions could be anticipated, which deeply involve the welfare of developed and developing countries alike, but after the theoretical findings, the main condition for practical success was a concentric international and national assistance to further research.

The international UNESCO — WAAS — Italy Symposium on this subject (Rome, Sept. 5—9, 1965) paved the way for these concentric efforts and this book, together with the complementary Volume containing the Proceedings of that Symposium, are destined to serve as the basic material for future work in this line and with it in many new fields of biology and geophysics, in so far as they are as closely connected with the phenomena of salinity and aridity as botany, zoology, agriculture, physiology, microbiology, biochemistry, biophysics, hydrology and soil science.

Rehovot, March 1966 HUGO BOYKO

Remark: Dr. HUGO BOYKO is an authority of world renown on desert ecology. Retired as Ecological Adviser in the Government of Israel, he is Honorary Consultant of UNESCO and served since 1947 as Chairman of the International Commission on Applied Ecology of IUBS. He was elected President of the World Academy of Art and Science (WAAS) at its IIIrd. Plenary Meeting in Rome, September 1965, as successor to the first President, Lord BOYD ORR.

The Publishers

Barren gravel hills in the absolute desert area of Eilat just before planting in October 1949.
The barrels are filled with water brought by command car from a saline well, 18 km to the north in Wadi Araba. A coastal salt swamp (sabha) and the Red Sea (Gulf of Eilat or Gulf of Aqaba) is to be seen in the depression.

This part of the "Desert Garden of Eilat" shows the same spot as above, ten years later (1959).
Irrigation water during the first seven years had a fluctuating Total Dilution of Salts (T.D.S.) of 2000 to 6000 mg/litre.
Details, including the plant list, are to be found in the Proceedings of the International Symposium on Highly Saline and Seawater Irrigation with and without Desalination (Vol. IV of the WAAS-Series-publications of the World Academy of Art and Science).

# PART I:

# GENERAL PART

# SALINITY AND ARIDITY
# NEW APPROACHES TO OLD PROBLEMS
## (An Introduction, a Summary and an Outlook)

BY

HUGO BOYKO

(with 4 figs.)

It is certainly not new that the rate of human reproduction has become a dangerous problem. Overpopulation has always led groups and whole peoples into risky wanderings and local wars. What is new, however, is that even a regional reproduction of too high a rate has become a dangerous world problem; dangerous not only for the whole world of humanity simultaneously, but even for all living beings and perhaps for life itself on our globe.

It is generally assumed that before the year 2000 the very unwise species homo sapiens will have doubled its population number by uncontrolled reproduction and will populate the planet "Earth" with 6,000 Millions, or even more. If we shall not find, before this date, the solution for feeding them properly, then each of them, single-handed, in small groups and grouped into major parts of the whole population will fight the others with all means at their disposal — and the means are by far more horrifying than those of a mere cannibalism of former times, or those of the Second World War, including the Hiroshima Bomb, which appear like a child's play in comparison.

A span of about 35 years only separates us from this foreseeable future, and this is a very short time in the history of mankind. It is the time when the children of today will have reached their best creative age.

On the other hand, the rapid growth of science and technology in quantity as well as in quality during the past few decades of the 20th century has led to developments which are decisively changing our ways of life as individuals as well as as groups, and beyond this also our lines of thinking.

One cannot help being surprised ever anew in considering the revolutionary changes in our scientific knowledge, in our technological progress and in our social structure during a present single life span.

Since the end of the 19th century, we jumped from the candle or the petrol-lamp to Neon-light and from the ancient Ikarus dream to the jet plane and the space ship. Though we are still looking through the good old microscope magnifying several hundred times, we have also the electron microscope at our disposal, magnifying several hundred thousand times and opening to our astonished eyes

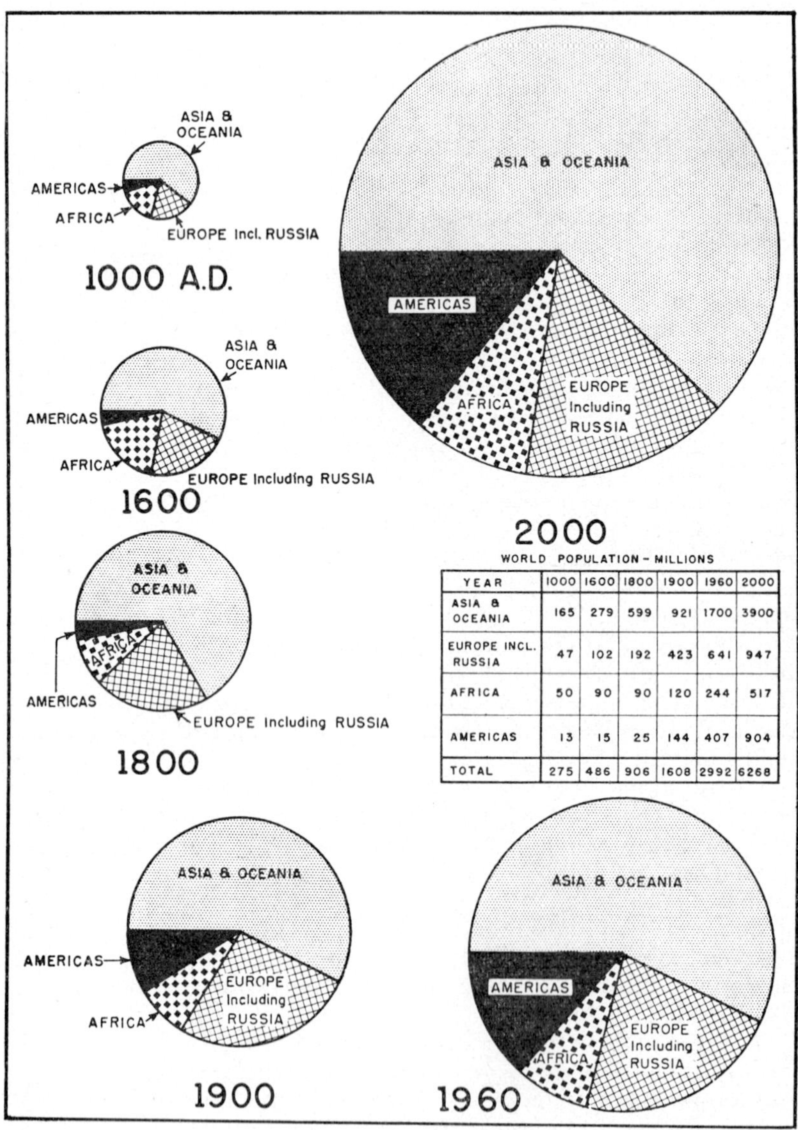


| YEAR | 1000 | 1600 | 1800 | 1900 | 1960 | 2000 |
|---|---|---|---|---|---|---|
| ASIA & OCEANIA | 165 | 279 | 599 | 921 | 1700 | 3900 |
| EUROPE INCL. RUSSIA | 47 | 102 | 192 | 423 | 641 | 947 |
| AFRICA | 50 | 90 | 90 | 120 | 244 | 517 |
| AMERICAS | 13 | 15 | 25 | 144 | 407 | 904 |
| TOTAL | 275 | 486 | 906 | 1608 | 2992 | 6268 |

Fig. 1. A Thousand Years of World Population Growth. In 1000 A.D., Asia accounted for 60% of the world's population, Europe, including Russia, for about 17%, Africa, 18% and the Americas, 4%. By 1960, Asia's percentage had declined to somewhat under 60, that of Europe and the USSR had increased to 22% and the Americas, to 14%. Africa's portion declined to 8%. By 2000, Asia may comprise about 65% of the total, Europe and the USSR, 15%, the Americas, 15% and Africa, 8%. Russia includes Asiatic and European Russia.
Reproduced from: ANNABELLE DESMOND "How Many People Have Ever Lived On Earth?", Population Bulletin, Vol. XVIII, no. 1, February 1962, by courtesy of the publishers.

new worlds of infinitesimal smallness, in which the virus is already a giant.

We made the surprising experience that much of what we have learned in our childhood as axiomatic truth, turned out to be nonsense, that for instance elements can be changed, by nature as well as artificially, and that the atom is not only not indestructible and not physically and chemically indivisible, but contains the greatest power known to us in nature, and that we are able to obtain and to use this immense energy just by its very destruction.

More new branches of science sprang up during this short time of about seventy years than in the 7000 or more years of science's history. Dramatic results rapidly followed one another, in Physics, in Chemistry and particularly in the new border sciences of Ecology, Bio-physics and Biochemistry. We are penetrating deeper and deeper into the secrets of life and it is no longer a completely utopian thought that we shall in a not very distant future be able to call "Halt" to death as an effective order.

Today already life expectancy is steadily rising in almost all

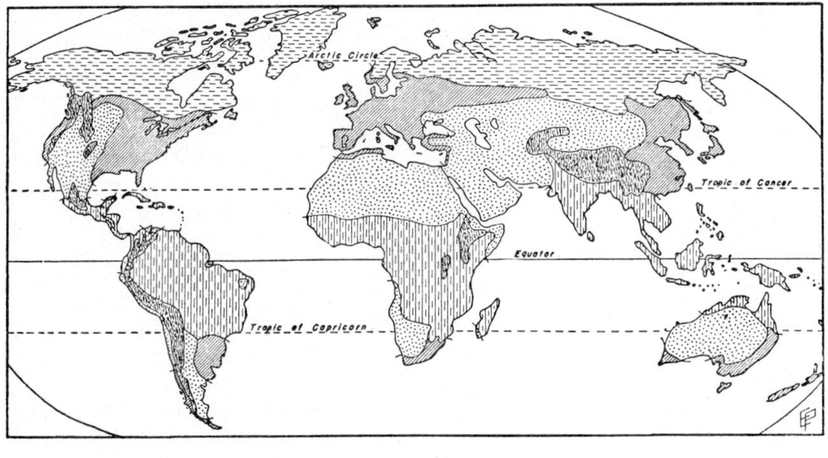

Fig. 2. Areas of the world hospitable and hostile to man's occupance. Reproduced from R. H. Fifield & G. S. Pearcy, *Geopolitics in Principle and Practice*. Boston: Ginn and Company, by courtesy of the publishers.

countries as a consequence of modern hygiene and the rapid progress of medical research, whereas birth-rates remain more or less unchanged. The resulting population growth mentioned above and compared with that in former times may be seen from the instructive graph (Fig. 1) drawn by ANNABELLE DESMOND (Mrs. ROBERT COOK) (1962).

In his survey of the present distribution of the human population in relation to the geographical factors, WHITE (1964) gives the following figures: "About one-half of the world's people are concentrated on a mere 5% of the earth's surface, whereas 57% of the land supports only 5% of the people".

The reason is to be seen from the map (Fig. 2) of "areas hospitable and hostile to man's occupancy" presented by FIFIELD & PEARCY in their book "Geopolitics in Principle and Practice". This map shows clearly enough that the arid zones with their deserts and semi-deserts play a major rôle among the "hostile areas". A more detailed distribution of these deserts and semi-deserts is presented in the map of Fig. 3 and by Table I (p. 8).

The connection of these introductory words with the topic of this book in general and of this article in particular is easily to be seen. The threatening overpopulation on the one hand, and the finding of new ways to convert areas hostile up to now into hospitable ones and to productivize desert areas in the size of a whole continent on the other hand, are too closely connected to be overlooked in their correlation. There are not many ways to keep the present equilibrium of the human race as a whole. Planned parenthood is one way. To raise the productivity potential of our globe is another way, and this book deals with a major problem of the latter.

In 30 to 40 years, our earth will have to feed twice as large a population as today, and even today food production and its distribution is not sufficient for large parts of the world population.

If we human beings, and in the foreline we, the scientists, do not find in time the means to keep the equilibrium as it is now, or perhaps even to better it, then far bigger powers than ours will step in and — whether we like it or not — an equilibrium will be re-achieved far removed from the ideals of mankind and may be even removed from the minimum requirements of our very existence.

As high and as atonishing as the achievements of the human race may be, the "Basic Law of Universal Balance" is stronger than the energies of all living beings together and certainly stronger than that of one single species. This basic law*, to be dealt with in more detail in this book by means of the example of the "Global

* The Basic Law of Universal Balance reads: "Wherever and whenever the natural equilibrium is disturbed, re-adjustment takes place until a state of balance is re-achieved" (BOYKO, 1960).

Salt Circulation" (Part II, page 180) leaves no doubt that we have to use all our knowledge and all our power in order not to lose the equilibrium in which we are living, and — as a matter of course — not to destroy it ourselves.

There is only one way out of mankind's present dilemma and to prevent conflict of a catastrophic scale: Science and Technology, in close cooperation with farseeing Statesmanship have to mobilize all their strength and co-operative spirit in order to find the necessary transnational solution, outside of all group interests, and to convince all political leaders to put these solutions into effect by combined effort.

Fortunately, these considerations no longer sound like a lonely voice in the desert. The rise of a bright light is more and more discernible above the misty horizon.

These considerations are the underlying motive of most of the activities of the Specialized Agencies of UNO and they were the main reason for the foundation of the World Academy of Art and Science*.

The Specialized Agencies, particularly UNESCO and the Food and Agricultural Organization of UNO (FAO) are in the forefront to support research on natural resources and to raise food production all over the earth. The World Academy, with its primarily advisory capacity, supports on the one hand actions with regard to planned parenthood in overpopulated countries and on the other hand specific research tasks in the fields of raising the production capacity of our globe, preferring in its support particularly new fields of research and border sciences, for these are, as history has taught us, always the most handicapped ones. All these efforts are directed to achieve a more or less lasting contribution to an equilibrium and to Human Welfare in general*.

In these strenuous efforts, two broad fields of scientific research are rather new, and therefore still lagging behind, generally, although both are probably of decisive importance in our struggle to overcome the difficulties of the next few generations.

One is the conquest of the oceans for their practically inexhaustible mining possibilities including their freshwater potential, and for their immense food potential. (Fresh water is obtainable by desalination, the food content is to be exploited by scientific

---

* According to the statutes of WAAS (page 3)
"The two main purposes are:
(a) to gradually build up a TRANSNATIONAL forum in which the vital problems of mankind can be responsibly discussed and thouroughly studied by the best minds of our generation and of the generations to follow, from an objective, scientific and global point of view;
(b) to act as an objective advisory body for the leading international organizations and for mankind as a whole, OUTSIDE OF ALL GROUP-INTERESTS."

breeding instead of the present mere hunting in them). The other field is the conquest of the deserts as new agricultural areas by utilization of saline waters including the waters of the sea.

As mentioned above, if we are speaking in this book of utilizing sea-water, we are not including here the methods of desalination. In this latter field, about 100 Institutions are already working on all continents, with remarkable results. The costs are, however, very high and will probably remain so for long, if not for ever, even for domestic and industrial use. They are certainly, at present at least, prohibitive for almost all agricultural purposes. Exceptions are only a few luxury crops of horticulture with a particularly high market value, e.g. orchids and other expensive flowers, certain seeds, primeurs etc. The importance of desalination of sea-water lies more in the fact that it can save considerable amounts of natural fresh water now used in cities and industries of arid regions for agricultural purposes there.

This book, however, deals mainly with new approaches to the solution of the steadily growing water problem, and new methods

**Table I.**

Desert areas.

| Approximate: | square kilometres | square miles | Per-centage |
|---|---|---|---|
| Total Land area | 145,040,000 | 56,000,000 | 100 |
| Inland deserts and semi-deserts | 21,756,000 | 8,400,000 | 15 |
| Sand dune areas including coastal dunes | 12,950,000 | 5,000,000 | 9 |
| *Deserts* | | | |
| Sahara | 9,065,000 | 3,500,000 | 41.7 |
| Arabian deserts | 2,590,000 | 1,000,000 | 11.9 |
| Iranian deserts | 389,000 | 150,000 | 1.8 |
| Turkestan desert | 1,942,000 | 750,000 | 8.9 |
| Sind desert (India, Pakistan) | 596,000 | 230,000 | 2.7 |
| Taklan Makan desert | 777,000 | 300,000 | 3.6 |
| North American deserts (part of the Great Basin, Mojave, Sonoran, Chihuahuan desert) | 1,294,000 | 500,000 | 5.9 |
| Kalahari desert | 570,000 | 220,000 | 2.6 |
| Australian desert | 3,367,000 | 1,300,000 | 15.5 |
| Atacama-Peruvian desert | 363,000 | 140,000 | 1.7 |
| Argentine (Patagonian) (mainly semi-) desert | 673,000 | 260,000 | 3.1 |
| Various smaller deserts, e.g. of Sinaï, Israel, Somaliland etc. | 129,000 | 50,000 | 0.6 |
| | 21,756,000 | 8,400,000 | 100 |

* These rough figures are based partly on vegetation maps and partly on various other sources. Cold deserts are not included.

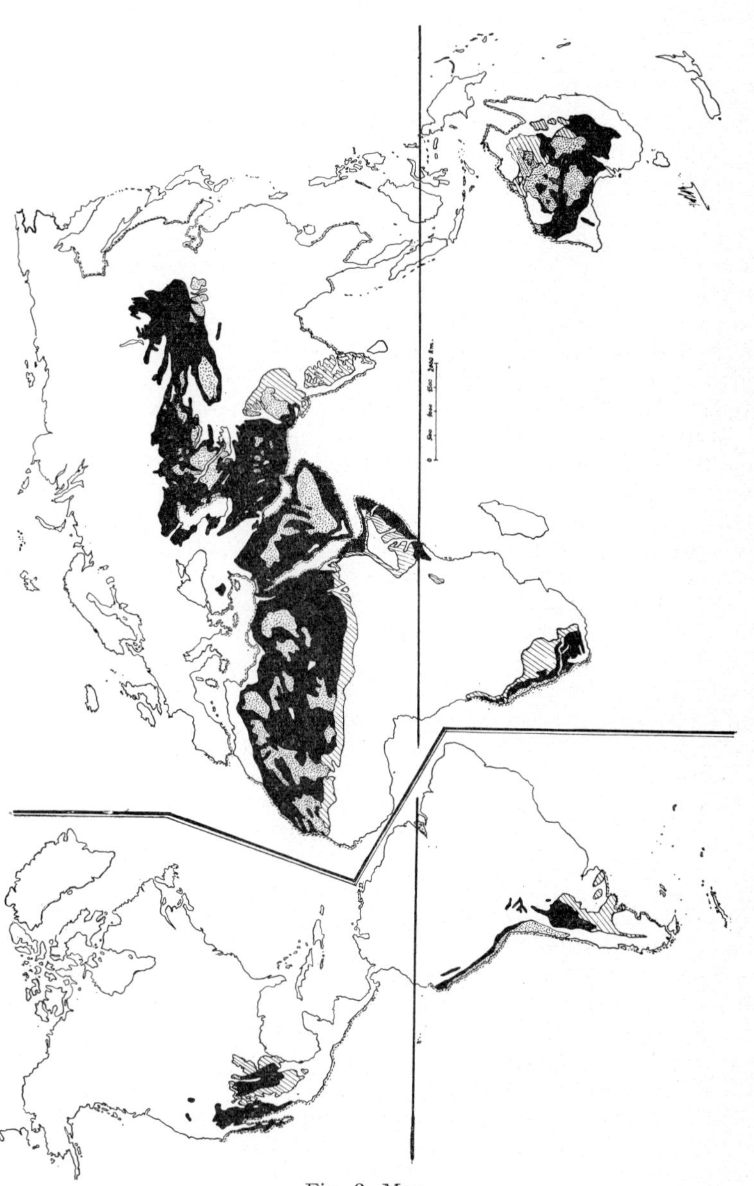

Fig. 3. Map.

a) Deserts of Sand Dunes
b) Deserts without large, continuous areas of sand dunes
c) Semi – Deserts
d) Coastal areas of the warm and temperate zones with a dry period of more than three months.

Remark : Not included in this map are;
1.) the coastal dunes, except if connected with inland dune deserts;
2.) the dune areas of islands);
3.) the large and frequently dune building sand areas inside the steppe regions.

The main contures of this map are taken from the Oxford Atlas, 1961; deviations of details from regional maps and own observations.

and results of recent experiments in various countries are presented here.

Two of the oldest enemies of agriculture, salinity and aridity, are dealt with here, both mostly in causal correlation, particularly in those regions where, many thousands of years ago, stood the cradle of agriculture.

The map (fig. 3) shows that arid and semi-arid areas cover a whole third of the terra firma on our earth's surface, and as we can see from Table I, deserts and semi-deserts alone cover almost 22 million square km (8.4 million sq. miles).

Wherever we are confronted with an arid climate, we frequently find as a consequence also saline soil and saline waters. But the problem of salinity is not restricted to arid zones. Broad coastal strips along all oceans and along the many salt lakes of the earth have to endure a continuous saltspray from the sea and only those living beings adapted to this kind of salinity can populate these salt-infested habitats.

The map (fig. 3) shows that about two thirds of the coastal areas are situated in more or less humid zones and one third in semi-arid or arid ones. A great part of these areas, in humid and arid zones alike, are covered by sand dunes, endangering fields, forests and settlements in their vicinity. If the windfactor, as the sand-moving and thereby destructive climatic element, is outbalanced by precipitation or irrigation, then the permeability of the soil plays an eminent rôle for all plant life in this struggle for existence, simply by preventing salt accumulation.

The experiments described in this book show that the salt tolerance of most plant species is raised several times if the soil is not a soil in the agricultural sense but dune sand. Salt water of high concentration and in some cases even sea-water can be used to productivize vast areas of shifting dunes and other sandcovered areas. The experiments (in chronological order) of H. and E. Boyko in Israel, of I. E. Gomez in Spain, of M. P. D. Meyering in Western Germany and of the team of Indian scientists at the Central Salt-station in Bhavnagar, India under the direction of Dr. Datar, are proof of it. On the other hand, the experiments in Southern Italy (see article of G. Lopez, p. 294) and those in Sweden by G. Hellgren (see Table V) show that in regions with more effective rainfalls even normal agricultural soils can be irrigated with highly saline water. These facts find their expression in the general rules of saline irrigation formulated on page 145. The application of the new principles as described in Part II and the new research field of plantgrowing under highly saline conditions, or extremely arid ones or in an environment combining both, are very promising in regard to food production and the production of plant raw materials for industry in the future.

Basic and applied research are likewise involved. This is made clear already in the general survey of V. J. CHAPMAN with regard to halophytic vegetation and of P. C. RAHEJA about soil and land use problems. It is for this reason that this book has to deal not only with knowledge accumulated already, but with many new principles and theories and with new experiments and observations as presented in Parts II and III. On the basis of these new approaches to these old problems quite a number of theories prevailing up to now seem to become obsolete. Thus it could be shown that:

(1) the opinion is wrong that a salinity of more than 600 mg/litre chlorine in agricultural irrigation water is a priori destructive for soil and plants. Experiments with more than 10,000 mg/l Cl had surprising success with quite a number of economic plants, if carried out on dune sand or on a substrate with a similarly high permeability and not on usual agricultural soil;

(2) the osmotic pressure as such is not the most decisive factor. Most of the experiments indicate that many if not most plants can adapt themselves to relatively high osmotic pressure and to fluctuations of it;

(3) not the absolute quantity of ions is determining the growth and life limits. H. HEIMANN's article gives an example that the relative quantity of the components in the composition of the solution is the most decisive one;

(4) the opinion that plant vitality is diminished by high salinity does not concur with the facts. Extensive experiments and measurements proved exactly the contrary and led to the formulation of the new principle of "raised vitality" (see page 175);

(5) not in all cases salt accumulation in soil will take place in the long run and prohibit further cultivation. The experiments in Israel, and the chemical analyses after years of daily sea- or saline water irrigation made it clear that no salt accumulation is to be feared if sand is taken as substratum. This most important result was obtained even in the arid climate of the Negev in Israel. BOYKO's new theories of "Partial Root-contact" connected with "Subterranean Dew", the "Viscosity Principle" and that of the "Global Salt Circulation" give the necessary scientific explanation to the experimental facts and to the "Biological Desalination" (p. 165);

(6) even in normal agricultural soils of much lower permeability, irrigation waters with a high Total Salt Concentration (T.S.C.) can successfully be used, if the specific climatic or soil conditions are taken into account, as it is being done in the experiments in Italy (p. 294), Sweden (HELLGREN, 1959) and Tunis (UNESCO, 1964).

Completely new vistas are opened in the biochemical field: The Indian scientists in Bhavnagar found a higher amount of alkaloids in certain plants when irrigated with sea-water (see page 323);

in similar experiments a new growth-stimulating ferment could be isolated in La Jolla at the Oceanographical Institute of the University of California (see p. 178); in a "Dying Experiment", the first of its kind, much stronger drought resistance could be achieved in all plants experimented with, the higher the concentration of the irrigation water was, from fresh water up to oceanic concentration (see p. 246), indicating biochemical processes still completely unknown to us. All these revolutionary results were already found in the very first steps of research in this direction, undertaken independently from one another and in three different countries and climates. They are promising enough to instigate deeper-going experiments on this line wherever the conditions may render them possible.

Even the two general articles, that on halophile vegetation problems (p. 23) and that on saline soils and their reclamation (p. 43) deviate much from the usual ones by new points of view in addition to the mere fact that the problems are treated by scientists living in the Southern Hemisphere (V. J. Chapman, New Zealand) or, respectively, in the Tropics (P. C. Raheja, India). In consequence, the greater part of examples is taken from Australia and Oceania for the first, and from the Subcontinent India for the second one, both unusual among the prevailing textbooks on such subjects, complementing them, therefore, in a valuable way.

Part II with its new theories, methods and experiments dealing with direct irrigation with sea-water and other highly saline waters opens not only new and diversified fields for scientific research, but also highly promising ways for agriculture, planning and land use. They justify the hope that large parts of the earth's surface, unused up to now and thought of as unusable, can be brought to intensive productivity. There can be no doubt that the success of these experiments on a world-wide scale would be a major victory of science for Human Welfare and peace, simply because it could be a decisive victory in the fight for food. Besides this it would offer new water resources for agriculture in addition to the limited freshwater resources at present at our disposal.

Shifting dunes were always seen as arch-enemies of man. He knew very well the dangers of the slowly encroaching mountains of sand, covering fields and forests, houses and cities on their slow but steady way. Witnesses of their immense destructive power are to be found in humid and arid climates alike; the ghost trees and ruined houses emerging slowly on the opposite side of the dunes, after having been buried for decades or centuries, can be seen in the deserts of the Sahara or in the Negev, in the Gobi desert as well as in the coastal dunes of the cool, temperate and humid climate of Lithuania or elsewhere. Sea-water and saline water are known since prehistoric times as enemies of quite another kind. The damage caused to agricultural crops by flooding from the sea was always felt as catastrophic, and well known as likewise destructive was, in ancient times already, the slower process of salination brought about by constant irrigation with saline water.

To the history of the last mentioned process, excavations in Southern Irak gave some elucidating examples.

JACOBSEN & ADAMS (1958) described the archeological findings with regard to salination in this region. Pottery dating from about 3500 B.C. suggests that at that time the proportion of wheat to barley was nearly equal. A little more than 1000 years later, the less salt-tolerant wheat accounted for only one sixth of the crop — and in 2100 B.C. for even less than 2%. By 1700 B.C., the cultivation of wheat had been completely abandoned. The average yield, laboriously recorded by the priests, had shrunk from 2537 litres per hectare in about 2400 B.C. to an average of only 897 litres per hectare about 700 years later (1700 B.C.).

In those times, salination of the once fertile soils was, in itself, enough to cause increasing poverty and depopulation, not to speak of the numerous wars to win new arable lands. This information supports also the theory that there have been no considerable climatic changes during the last 4000 years in these desert and semi-desert parts of South-West Asia, where at present, from the Negev in Israel to the Rajastan desert of Pakistan and India, considerable parts are in process of reclamation.

The principles and methods described here and in the following chapters may prove to play an important rôle in this reclamation work, although their application on an economic scale has started only recently, after having finally overcome the various administrative obstacles.

However, it is no wonder that after thousands of years of these unfortunate experiences with saline irrigation, the idea of using salt water and even sea-water for crop irrigation sounded insane to most of the usually conservatively thinking officials and scientists, particularly if these new methods were to be used in just those areas of the barren sand dunes where, even without such dangerous irrigation, almost nothing could grow. In order to combine these two dangers for finding positive results, a synoptic and synthetic approach was essential, such as ecology has adopted during the last decades.

The magnitude of the problem and of the possible economic and social impact of the results of these experiments cannot be exaggerated, even not from a global point of view.

A great part of the land masses of the earth has at present not sufficient precipitation or other fresh water at its disposal to support a more intensive agriculture and allows a poor pasture economy only, just sufficient for a nomadic way of life. Other vast areas are covered by so desolate deserts that they do not allow even this. But a great part of these waste lands have potentially very fertile soils, even according to the prevailing standards, if only the water problem could be solved.

On the other hand, a surprisingly great part of our earth is covered by waste lands of a different kind, depending in their origin more on the wind climate than on precipitation, namely by sand dunes. The total area of all dune regions is according to WHITFIELD & BROWN (1948) 3,200 million acres (i.e. about 13 million square kilometres). This is an area about twice as large as the United States of America and seven times as large as their agricultural area.

It is hardly surprising that orthodox thinking, hampered by inertia, could not quickly adjust to the global impact of the new principles and that, therefore, the first and successful experiments of direct irrigation with water of 10,000 mg/l T.S.C. up to sea-water from the ocean met with scepticism and outright opposition supported by vested interests. The idea to satisfy the thirst of sandy deserts by highly saline water and sea-water requires an adaptability of thought as well as freedom from preconceived notions.

Such an adaptability is nowhere easily to be found, and this fact is true for the entire history of Science and Technology, which is full of examples where the search for new ways was not only opposed by all means, but in many cases severely punished.

A scientific truth can, however, neither be suppressed nor can it be hidden in the long run. From 1954 on, similar experiments with brackish water were started in other places and in various countries at the request of the author, and with the support of UNESCO also the results of the first experiments with sea-water of oceanic concentration were verified.

Today, a chain of experimental stations along a climatic profile from India to Sweden is carrying out the work, and preparations are made to expand it in various directions. Several dozens of economic species have successfully been experimented with under direct irrigation with sea-water up to oceanic concentration and several hundred species under irrigation with highly saline water of various composition.

A few examples taken from the many to be described in the following chapters show lightninglike the possible economic and social influence of these experiments. Prospective results have been achieved, for instance, with wheat in India; fibre plants for textile and cellulose raw materials, various vegetables, fodder plants and woody species in Israel; a valuable fodder plant of high nutrition value in Western Germany; cereals, vegetables and flowers in Italy and Spain; numerous pasture plants and grasses in Sweden, etc.

To this economic aspect, some aspects of basic science, and

certainly of no less importance, are added. It is generally assumed that the present forms of life or at least most of them had their origin in the waters of the ocean. This assumption seems to receive certain support by the results of the experiments described in this book and particularly also by the investigations of S. FLOWERS, the zoologist and F. R. EVANS, the botanist, on plant and animal life in the saturated brine of the Great Salt Lake in Utah. 270,000 mg/l (milligram per litre) or 27% Total Salt Content (T.S.C.) is the highest known more or less constant concentration of any of the greater salt lakes. Only flat salt lakes drying out periodically surpass this concentration during this process, becoming gradually a wet salt layer and finally a crust of dry salt covering the lake's bottom until the next basin filling water influx. In spite of the enormous osmotic pressure of this brine in the Great Salt Lake of Utah, the two authors could find about 30 different species, from unicellular

**Table II.**

Chemical Composition of the Ocean, the East Mediterranean Sea, the Great Salt Lake and the Dead Sea.

| | Ocean<br>% | East Mediterranean Sea<br>% | Great Salt Lake<br>% | Dead Sea<br>% |
|---|---|---|---|---|
| Total Salt Content (T.S.C.) | 3.44 | 4.08—4.36 | 26.40 | 27.48 |
| % of T.S.C.: | | | | |
| NaCl | 78.198 | 72.706 | 87.908 | 29.986 |
| $MgCl_2$ | 9.302 | 8.096 | 3.524 | 51.820 |
| KCl | 1.744 | 1.995 | | 4.294 |
| $CaCl_2$ | | | | 12.008 |
| $MgBr_2$ | | | | 1.419 |
| $Na_2SO_4$ | 6.395 | | none | |
| $MgSO_4$ | | 3.929 | 7.942 | |
| $K_2SO_4$ | | | | |
| $CaSO_4$ | 4.070 | 6.800 | 0.095 | 0.473 |
| $CaCO_3$ | | | 0.530 | |
| $Li_2SO_4$ | | | | |
| $SiO_2$ | | | | |
| $Fe_2O_3, Al_2O_3$ | | | trace | |
| various | 0.291 | 6.473 | | |
| | 100% | 100% | 100% | 100% |

The sources of these analyses are:

| | |
|---|---|
| Ocean: | HOWE, 1956. |
| East Mediterranean Sea: | Analyses of the Chemical Laboratory of the Agricultural Research Station, Rehovot, (1957). |
| Great Salt Lake: | FLOWERS & EVANS (p. 371 of this book). |
| Dead Sea: | Laboratories of Palestine Potash Ltd. Jerusalem (after NOVOMEYSKY, 1936). |

**Table III.**

Number of species living in the Sea compared with that in the Great Salt Lake and the Dead Sea.

| Species of | Ocean | East. Medit. Sea | Great Salt Lake | Dead Sea |
|---|---|---|---|---|
| Bacteria and Algae to Mammals | ×.10,000 | ×.1,000 | | |
| Bacteria and Algae to Arthropoda | | | >27* | |
| Flagellates (one or two) | | | | 1—2 (plus Bacteria) |

* see Table IV

**Table IV.**

Number of known species living in the main body of the Great Salt Lake: (after FLOWERS & EVANS (see p. 372).

| | |
|---|---|
| Bacteria | 12 |
| Cyanophyta | 2 |
| Chlorophyta | 2 |
| Flagellates | several |
| Protozoa | several |
| Ciliates | 7 |
| Amoeba | 2 |
| Arthropoda | 2 |
| larvae of several Insects | ? |
| | >27 |

ones to such highly organized ones as Arthropoda (see p. 373), some of them in relatively abundant populations.

The Dead Sea, for comparison, has a similarly high Total Salt Content of 27% and constitutes therefore too an almost saturated brine at certain times, but in spite of the some times lower Total Salt Concentration there, no life is to be found except for one or two Flagellates and a few Bacteria (VOLCANI, 1940).

The reason for this discrepancy between two extremely saline inland lakes as life sustaining media is to be seen in the different composition of their salt content. Both have quantitatively a similarly high concentration of near saturation, but whereas the Great Salt Lake has qualitatively a composition similar to that of the Oceans (i.e. a similar percentual relation of the various components to one another), the analysis of the Dead Sea shows in comparison great differences. Table II elucidates this in more detail.

At the first glance we can see for instance that in the Dead Sea analysis there are more than 50% magnesium chloride against less than 10% in the three other analyses. Just the contrary relation is to be seen with sodium chloride: whereas the Dead Sea contains only about 30%, we have in the three true sea-water types between 70 and 90%. All three are also similar in their content of sulphates (8 to 11%) against 0.5% only in the Dead Sea. Another striking contrast is to be seen in its high content of calcium chloride and magnesium bromide as against apparently none in the three other analyses, and in the more than twice as high percentage of potassium chloride. Other differences of possibly decisive influence on plant and animal life may also lie in the existence or non-sufficient content of trace elements not yet investigated in this connection.

This too seems to prove the theory that the relative composition of the salt content as it exists in the waters of the ocean is more decisive for the sustenance of life than the quantity of concentration. In a letter to the author, the Nobel Laureate for Chemistry and Member of the World Academy of Art and Science, HAROLD C. UREY, expresses his opinion about the experiments on direct irrigation with sea-water with some basic thoughts on the history of life:

"I am very glad that the ideas in regard to the sea-water irrigation project are going along so well. It does seem to me that it is an important line of research. It seems odd that such an enormous amount of the plant material growing in the sea should not prove to be useful as a source of food, or that the wayward ones that have wandered out of the sea to the land should not be returned to the sea and produce food."

Such views open vast research fields.

The principle of "The Balance of the Ionic Environment" for instance has been taken as the working basis by H. HEIMANN in his studies, two of which are described in this book as an example. His experiments are carried out with water of relatively low brackish concentration, but on normal agricultural soils, whereas most of the sea-water experiments described here by various authors were carried out on dune sand (see Table V). The countries where experiments of this latter kind (i.e. on sand) were made, are, in chronological order (up to the end of 1964):

Israel, USSR, Spain, Western Germany and India; Experiments on usual agricultural soils with the brackish water from the Baltic Sea (6000 mg/l T.S.C.) in Sweden; Underground water with a relatively high T.S.C. but of other composition of the salt content than sea-water have been used in the experiments (on usual agricultural soils) in Southern Italy with a fluctuating chlorine content of the irrigation water up to 9000 mg/l and of a lower chlorine content but with more sulphates and a T.S.C. of 2,000 to 6,000 (fluctuating up to 10,000 mg/l) near the Red Sea in Israel. Similar scientific work is now being done as a project of the Special Fund of UNO in Tunis, where practical experience since centuries shows that the specific soils and waters there can be used for agricultural purposes in spite of a T.S.C. of about 5000 mg/l.

A "Chronological Table" (Table V) may, in a condensed form, show the development of this line of research in various places.

The twelve stations enumerated range from the semi-arid tropics of India and the hot arid desert of Israel to the cool temperate and humid coast of Sweden and from the atlantic climate of Western Europe to the continental climate of Central Asia east of the Caspian Sea. All these experiments and all these new approaches to the old problems of salinity and aridity are very promising. They are of particular importance in the present vital fight of science for human welfare and for a sound future of our race.

The future, seen by so many competent authorities as highly alarming, would have a much more hopeful aspect if science could concentrate its activities more intensively than now in the direction of achieving the necessary equilibrium between population growth and natural resources. The line of research dealt with in these chapters is only a part of such a concentric attack. If sponsored and actively supported by the various administrations, governmental and institutional ones, it may well become a major part of it*.

The experiments, observations and new principles described in this book, present however an outlook far beyond their own limited fields. A transnational teamwork with such aims in mind irradiates rays of hope for a better cooperation between the various groups of mankind and for a better future for all.

Seen from a global point of view, the scale of the experiments up to now is very small. From the budgetary point of view, the sums at their disposal were a few thousand dollars only, which is an infinitesimal fraction of the sums used for instance for desalination experiments. Both lines of research are aiming at the same goal and, most important, both lines are complementing each other. Whereas desalination is working towards sufficient water for urban and industrial needs, supplying thus the immediate water needs of a great part of the whole population, the experiments dealt with here are aiming at the conquest of new agricultural soils, at converting waste lands and deserts into productive agricultural areas, and

* It is for this reason that the World Academy of Art and Science in cooperation with UNESCO and the Italian Authorities convened an International Conference dealing with this subject. The main aim of this Conference was to achieve an international team of Institutions and scientists working on it in close cooperation, on a world-wide and transnational scale (see H. Osvald, p. 357).

at the utilization of large amounts of water at very low cost from sources unused up to now.

In some cases the best economic solution may be found in a combination of both methods. One of the reasons is that the energy needed for electrolytic desalination of oceanic water from a concentration of 3.5% to one of 0.5% T.S.C. is only about one tenth of the energy needed to desalinate such a brackish water of 0.5% to 0.05% T.S.C.

The experiments described in Part II show that for many agricultural crops, grown according to these methods on sand, we do not need a lower concentration than 0.5% T.S.C. We would, therefore, by such a combination of both methods, need only a tenth of the energy normally necessary for the desalination of oceanic water to a concentration suitable for domestic and industrial purposes, namely to about 0.05% T.S.C.

A graphical representation of these relations shows approximately the following curve (fig. 4):

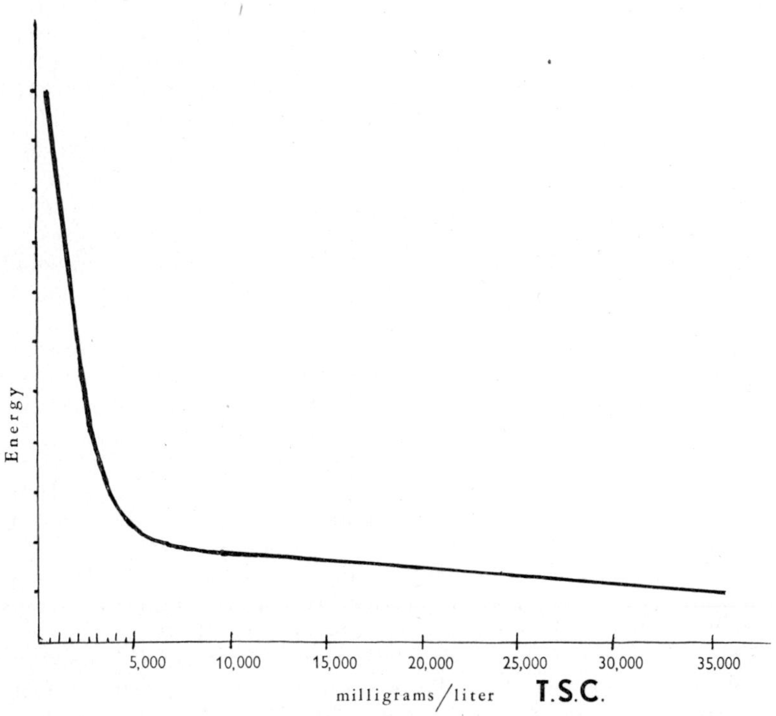


Fig. 4. Relation between Energy and Total Salt Content (T.S.C.) in Desalination by the Electrolytic Process.

There is also an other additional economic and social outlook: Fresh food on the spot, greeneries and growing industrial plant raw materials by saline irrigation combined with an additional amount of desalinated water for the necessary potwater and for certain industries or mining purposes opens new possibilities for densely populating those vast land reserves constituting only desolate deserts today and, at least on some places, potential and particularly healthy recreation centres of tomorrow.

The pictures on the frontispiece indicate that this is not an utopian vision any more. It shows the first garden of Eilat, near the shore of the Gulf of Eilat (Red Sea), laid out and planted by Elisabeth Boyko (see p. 220) on the completely barren gravel hills there. The irrigation water at disposal was a desert underground water with a Total Salt Content of 2,000 to 6,000 mg/l, fluctuating sometimes up to 10,000 mg/l, in an absolute hot desert climate of less than 25 mm average yearly rainfall.

Without this achievement, the simultaneous (1949) foundation of the harbour town Eilat would never have succeeded (March 1949 no inhabitants, 300 when the garden was planted in October-November 1949, 7000 in December 1962 and 12,000 in December 1964). For these greeneries and trees on this formerly desolate place provided not only shadow and a place of recreation, but gave the pioneering population the necessary trust in their future.

For science itself, all these new principles and experiments, and likewise the unique observations on plant and animal life in the Great Salt Lake of Utah, have led until now already to surprising discoveries in the ecological, physiological, agronomical, geophysical and biochemical field, opening new vistas and even new branches of science.

Some of these new views are:

1. A better understanding of plant reaction to osmotic pressure, of the practical value of electrical conductivity for plant physiology and agronomy, of the actual rôle of specific iones, e.g. of sodium, potassium, chlorine, etc.;

2. Saline waters of sea-water composition have a much more favourable effect than other salt compositions of the same concentration;

3. Vitality and drought resistance are raised with higher concentrations of the irrigation water;

4. The viscosity theory in connection with the principle of partial root contact and subterranean dew is throwing new light on various problems of soil science and root physiology (p. 147 ff);

5. the theory of "Global Salt Circulation" elucidates an important geophysical aspect and explains some of the surprising results (p. 180); it presents also a striking example of the "Basic Law of Universal Balance" (p. 195), and includes also the new principle of "Biological Desalination" (p. 165).

6. The article of FLOWERS & EVANS (p. 367) reveals the ecologically astonishing fact that a relatively great number of plant and animal species could find their living possibilities in an almost saturated brine, i.e. that of the Great Salt Lake, because the relative composition is similar to that of sea-water, and provided thereby also a better understanding of the necessary "Balance of the Ionic Environment";

7. The experiments show that a surprisingly high plant adaptability to a sudden raise or fall of osmotic pressure is widely spread. Successful growing of many species could be shown under such conditions, of species which are generally known as non salt tolerant, among them many valuable economic plants;

8. The experiments open an outlook on almost unlimited new discoveries in the field of bio-chemistry;

9. and last but not least, also on the easy applicability of their basic results to applied research. They open new ways of practical application in agriculture, horticulture and forestry on a major scale.

We must, however, not forget that all the work done up to now can be seen only as a few first steps in this new and most prospective direction.

If these lines and the following presentations of new principles, methods and facts will stimulate ever more and more research in this direction and on the basis of the results as many new agronomical trials of application as possible, then this book will have achieved its aim.

## REFERENCES


BOYKO, E., 1952. The Building of a Desert Garden. *J. Roy. hort. Soc.* **76,** *4.*

BOYKO, H., 1960. The Basic Law of Universal Balance. Presidential Address, Int. Comm. Appl. Ecol. I.U.B.S. Meeting London, July 1960 (see also page 195).

BOYKO, H. & BOYKO, E. 1959. Seawater Irrigation — a New Line of Research on a Biological Plant-Soil Complex. *Int. J. Bioclimatol. Biometeorol.*, **III**, II, B 1, *1—24.*

BOYKO, H. & BOYKO, E., 1964. Principles and Experiments regarding direct Irrigation with Highly Saline and Sea Water without Desalination. *Trans. New York Acad. Sci.*, Suppl. to No. 9, Ser. II, **26,** *1087—1102.*

DESMOND, A., 1962. How Many People Have Ever Lived on Earth? *Population Bull.*, **XVIII,** 1. (also in: World Academy of Art and Science, Vol. II, p. *27—46*).

HEIMANN, H., 1958. Irrigation with Saline Water and the Balance of the Ionic Environment. Int. Potassium Symposium, Madrid, *173—220.*

HELLGREN, G., 1959. Report of Commission VI. Conference on Supplemental Irrigation. Int. Soc. of Soil Science.

HOWE, E. D., 1956. Utilization of Sea Water. UNESCO, Arid Zone Research, Vol. **IV.** Utilization of Saline Water — Review of Research Paris.

JACOBSEN, T. & ADAMS, R. M., 1958, Salt and Silt in Ancient Mesopotamian Agriculture. *Science,* **128,** 3334, *1251—1258.*

NOVOMEYSKY, M. A., 1936. The Dead Sea: A Storehouse of Chemicals. *Trans. Inst. chem. Eng.* **14**, *60—81*.
UNESCO. Arid Zone Research Series, Vol. I-XXI, Paris, 1952 ff.
UNESCO. Tunis Special Fund Project, Report 1964.
VOLCANI, B. ELAZARI, 1940. Studies on the Microflora of the Dead Sea. Hebrew University, Jerusalem.
WHITE, C. LANGDON, 1964. Geography and the World's Population. World Academy of Art and Science, Vol. II, *15—26*.
WHITFIELD, CH. Y. & BROWN, R. L., 1948. Grasses that fix Sand Dunes. U.S. Yearbook of Agriculture, Washington, U.S. Dept. of Agric., *70—74*.
WORLD ACADEMY OF ART AND SCIENCE:
a. Vol. I: "Science and the Future of Mankind" (ed. HUGO BOYKO) 380 p. with 8 figs. Dr. W. Junk, The Hague, 1961. Offset Reprint Ind. Univ. Press, Bloomington, 1964.
b. Vol. II: "The Population Crisis and the Use of World Resources" (ed. STUART MUDD a.o.) 563 p. with 43 figs. Dr. W. Junk, The Hague, 1964. Offset Reprint Ind. Univ. Press, 1964.
c. Vol. III: "Conflict Resolution and World Education" (ed. STUART MUDD) — in Print.
d. Vol. IV: "Irrigation with highly Saline water and Sea water with and without desalination". Proceedings, Int. Symposium in Rome, 5.—9. Sept. 1965 — in Preparation (ed. HUGO BOYKO).

# VEGETATION AND SALINITY

BY

V. J. CHAPMAN

*Auckland University*

In arid regions, water supply is clearly the most important environmental factor. In many areas the next most significant factor is undoubtedly salinity. Sodium chloride is the predominant salt determining salinity followed by sodium carbonate or sodium sulphate and salts of magnesium. Whilst therefore there is no difficulty surrounding the problem of salinity in those cases where there is a clear excess of alkali, be they any one of the above sodium salts, there is considerable difficulty in determining the lower limit of salinization, the point where one passes from halophytic to glycophytic conditions and vice versa. In order to surmount this difficulty numerous investigations, using a great variety of plants, have been carried out in different parts of the world with a view to determining the upper limit of salt tolerance for many species. On the basis of all this work it would appear that so far as sodium chloride is concerned 0.5% in the soil solution can be regarded as representing the critical concentration. With other salts insufficient work has as yet been carried out for any figure to be fixed. At values in excess of 0.5% NaCl one may expect to be faced increasingly with salinity problems, both in respect of soil characters, plant metabolism and vegetation cover, whilst at lesser values such problems should be non-existent.

A major difficulty in trying to reach a satisfactory solution of this problem is that of approach. *The Ecologist and Plant Physiologist regard the issues in terms of tolerance and response of the plant or plants, whereas the soil scientist or agriculturist is more concerned with the soil conditions.* The net result is that salinity and its converse have been defined in a general manner such as that above, or more specifically in a form applicable strictly to the soil. It would seem that the ecologist and plant physiologist tend to think of the plant in terms of the anion associated with sodium or magnesium, whereas the soil scientist thinks more in terms of the cation. Eventually the two points of view should merge, but before that can occur it appears essential for the botanists to carry out many more detailed studies concerning the effect of excess sodium and magnesium on the behaviour of plants. Thus it has recently become evident that in certain littoral marine algae, which have been studied (BERGQUIST, 1959; EPPLEY, 1962) there exists a sodium efflux pump mechanism that comes into operation under certain conditions.

This is clearly one way in which excess salt can be removed and it would be interesting to know whether any comparable mechanism exists in the higher plants. Removal of excess salt by special glands is a feature of some halophytes, e.g. *Limonium* spp., *Spartina* spp., *Frankenia* spp., *Glaux maritima*, though the detailed mechanism of excretion has not been studied. However, from the work of ARISZ et al. (1955) it is evident that energy is necessary for the process of salt removal just as it is in the marine algae studied. It would be very desirable to know whether the mechanism actively concerns only the sodium ion, the anion being associated in a purely passive manner, or whether the mechanism essentially concerns the anion (chloride, sulphate or carbonate), or whether both ions are involved. Not all terrestrial halophytes, however, possess a salt-removing mechanism. In the case of the succulents, e.g. *Salicornia, Batis, Allenrolfea, Halocnemum,* it is believed that compensation for the excess salt is provided for by the uptake of water that is stored in the succulent tissues. The present writer has pointed out (CHAPMAN, 1962) that despite this mechanism the concentration of salt, particularly sodium chloride, may reach a value that is lethal by the end of the growing season, and at that point the excess salt is removed by the sloughing off of some or all of the old succulent material. Other halophytic plants, e.g. *Juncus maritimus*, may also dispose of surplus salt by the loss of the vegetative leaves. Both groups of plants require to be studied in respect of their sodium and chloride efflux mechanisms. Plants growing in soils rich in magnesium salts must be similarly studied.

The removal of salt from the tissues of plants growing in saline habitats is one aspect of the problem: the other is concerned with the absorption of salt by the roots of plants growing in an environment where there is an excess of alkali salts. In certain plants there is clear evidence that optimum growth is only achieved in the presence of sodium salts, especially sodium chloride, in excess of that normally found in soils, e.g. species of *Salicornia* (VAN EIJK, 1939; BAUMEISTER & SCHMIDT, 1962), *Arthrocnemum, Halogeton* (WILLIAMS, 1960) and species of *Atriplex* (BLACK, 1960). Work has been carried out on the uptake of sodium and chloride by halophytes and crop plants and there is clear evidence that the presence of potassium greatly influences the amount of sodium absorbed by plants (BINET, 1962; HEIMANN & RATNER, 1961).

In the meantime, whether plants are enabled to survive in habitats with excess alkali salts by removal of surplus salts or by a capacity to tolerate it, there remains the problem of defining the boundary between the saline and non-saline environment. It has been pointed out many years ago (YAPP, 1922) that the various types of habitat can be recognized, e.g. areal habitat, developmental or successional habitat, habitat of uniform conditions and the habitat of individual plants, and for arid saline areas one may postulate that the areal type of habitat represents the most useful approach. Present day ecological concepts, involving as they do the idea of the continuum or a kaleidoscope of pattern (WHITTAKER, 1956) render it very difficult to set any boundary to a habitat; indeed it could well be argued that if one accepts a continuum, there may be no such thing as a habitat. The concept of habitat, however, represents a valuable working tool and one is still therefore justified in trying to delimit it on an areal basis.

At present it would seem that workers in the salinity field will have to determine whether a habitat is saline or not on the kind of basic determination put forward by STOCKER (1928), who postulated 0.5% NaCl in the soil as the dividing line, or on a soil basis, involving the exchangeable sodium percentage and the conductivity of the soil extract. Ecologists have subdivided halophytic plants into various categories, e.g. VAN EIJK (1939), TSOPA (1939), but in so far as they did not indicate what was or was not a haline soil the categories are not satisfactory. IVERSEN (1936) also published a classification of halophytes but it is unsatisfactory in that the lower limit for halophytic vegetation is taken as 0.01% NaCl in the soil water, a value that is certainly too low. In 1942 the present author modified the IVERSEN scheme in order to incorporate the general thesis that 0.5% sodium chloride or sodium sulphate present in the soil water represents the critical limit between glycophytic and halophytic conditions. This may, in the light of more information, now be modified still further so that we have the following:

A. Miohalophytes

Plants that grow in habitats with a range of 0.01 to 1.0% NaCl, $Na_2SO_4$ or $Na_2CO_3$ in the soil water. Such plants are clearly capable of tolerating more than the limiting value of 0.5% salt, and they are most likely to occur in soils where the soil solution concentration hovers around the critical value. Such plants are, in fact, border halophytes.

B. Euhalophytes

(a) *Mesohalophytes*. Plants that grow in habitats with a range of 0.5 to 1.0% NaCl, $Na_2SO_4$, $Na_2CO_3$ in the soil water.

(b) *Meso-euhalophytes*. Plants that grow in habitats with a range of 0.5 to more than 1.0% NaCl, $Na_2CO_3$ in the soil water.

(c) *Eu-euhalophytes*. Plants that do not occur where the soil water concentration is less than 1.0% NaCl, $Na_2SO_4$, or $Na_2CO_3$. Such habitats are mainly to be found bordering the salt lakes and seas.

At present there is much less information respecting the tolerance of plants to salts of magnesium, and hence it does not seem feasible to propose a similar classification for these plants.

Halophytes do not, of course, always grow in habitats that are saline. Under glycophytic conditions certain halophytes still accumulate salt even if there is very little in the soil environment. This has been clearly demonstrated for *Atriplex vesicaria, A. paludosa* and *A. nummularia* in Australia, and in such cases WALTER (1962) has proposed the term Xerohalophyte. In the Koonamore Reserve of Australia these halophytes can be found growing side by side with non-halophytes (Table I).

**Table I.**

Osmotic pressures ($Cl^-$ as NaCl) of plants from Koonamore Reserve (Australia) (after WALTER, 1962).

| Habitat | Species | O.P. | $Cl^-$ as NaCl | % $Cl^-$ of total O.P. |
|---|---|---|---|---|
| A | *Atriplex vesicaria* | 65.4 ats | 45.2 ats | 69 |
| | *Kochia sedifolia* | 33.7 ats | 13.7 ats | 41 |
| B | *Atriplex vesicaria* | 53.5 ats | 32.0 ats | 60 |
| | *Kochia lanosa* | 22.3 ats | 10.2 ats | 46 |
| C | *Atriplex vesicaria* | 53.5 ats | 32.0 ats | 60 |
| | *Casuarina lepidophloia* | 25.8 ats | 8.5 ats | 33 |
| D | *Atriplex vesicaria* | 55.4 ats | 42.4 ats | 76 |
| | *Myoporum platycarpum* | 33.5 ats | 6.7 ats | 20 |

The determination of salinity upon a soil basis is at present probably best founded upon the following scheme put forward by HAYWARD (1954).

1. Non-saline alkali soil.

Such a soil contains sufficient exchangeable sodium to inhibit the growth of glycophytes but it does not contain appreciable quantities of the salts in a soluble form. The exchangeable sodium percentage

$$\left\{ = \frac{\text{Exch. Na (m.e. per 100 g.m. soil)} \times 100}{\text{cation exch. capacity (m.e. per 100 gr. r. soil}} \right\}$$

is greater than 15. The pH exceeds 8.5 and the conductivity of the saturation extract is less than 4 mhos per cm at 25° C.

2. Saline alkali soil

Such soils contain sufficient exchangeable sodium to inhibit the growth of glycophytes and the soluble salts are present in appreciable quantity. The exchangeable sodium percentage is greater than 15, the pH is usually less than 8.5 and the conductivity of the saturation extract is greater than 4 mhos per cm at 25° C.

3. Saline soil

The inhibition of the growth of glycophytes is here determined by the excess soluble sodium salts. The exchangeable sodium percentage is less than 15, the pH is usually less than 8.5 and the conductivity of the saturation extract is greater than 4 mhos per cm at 25° C.

* * *

Again, it will be noted that there is no provision for soils that are essentially rich in magnesium, but it would clearly be possible to classify a magnesium soil in much the same manner. It will be observed that salinity is based either upon an excess of soluble sodium salts or upon the exchangeable sodium percentage being in excess of 15. The former criterion obviously requires further clarification because one needs to know what represents an excess of a soluble sodium salt. One may suggest that the value of 0.5% in the soil solution could well be incorporated here as the critical value.

The salt of inland saline areas is derived from one of several sources. In some cases it is derived from a marine sedimentary deposit laid down in the Jurassic, Cretaceous or Tertiary periods. In other cases it may be associated with a former arm of the sea which, as a result of land or sea level changes, is no longer inundated. More rarely the salt may be blown inland from the sea over the centuries and then deposited inland (aeolian salt), as for example in South West Africa. A final source of salt occurs in inland lake basins in which the natural drainage outlet ceases to exist, and the subsequent drying up of the former lake brings about an intense concentration of the salt that has accumulated in the waters, e.g. Great Salt Lake in Utah. Because of the varied nature of the origin of the salt, the deposits may be near the surface of the soil or some distance below.

In arid inland areas it is customary to recognize two major soil types that present salinity problems. These are respectively the solonetz and solontchak soils. The solonetz soil generally exhibits a definite structure in the profile and sodium is commonly present as carbonate, with or without sulphate and chloride. The soil water-table is frequently some distance below the surface and the principal accumulation of the salts is also at some depth. The solontchak soils, on the other hand, exhibit very little soil structure and sodium is usually present as the chloride, with or without sulphate and carbonate. The salts tend to accumulate in the surface layers of the soil, and unless there is irrigation or a pronounced rainy season such soils are likely to be more inimical to plant growth than the solonetz. In semi-arid or arid regions it is obvious that either type of soil will be a serious problem so far as plant growth is concerned, though in certain regions, because of local circumstances, there may be no alkali problems (BOUMANS & HUSBOS, 1960).

The primary formation of solonetz soils can take place in one of five ways according to KELLEY (1951), whilst secondary formation as a result of irrigation is also possible by one of three ways (KOVDA, 1947). For details of the processes the reader is referred to Chapter III pp. 43—127.

The salinity of inland areas is dependent upon a number of factors which vary in significance. Some of these also operate in maritime regions (CHAPMAN, 1962). These factors may be listed as follows:

1. Precipitation. The magnitude of this factor will determine the degree of leaching, the depth location of the layer of salt accumulation and the need or otherwise for irrigation. Thus in North Dakota rainfall associated with saline artesian waters brings about a high saline water-table associated with inimical drainage conditions (BENZ et al., 1961).

2. Proximity to drainage channels. In an area as that surrounding Great Salt Lake in Utah the streams flow into the lake where evaporation has resulted in a very high salt concentration in the water, and the salinity decreases with distance from lake and streams. In salt spring areas the salt laden water rises and then flows away to become part of a river system. Again the salt concentration decreases with increasing distance from the spring and the stream bed. WALTER (1962) has also shown how, in the Namib desert, fresh water percolating through a soil comes nearer and nearer to the surface and then, through evaporation at the edge of the stream terrace, salt efflorescence takes place. Under such circumstances salinity increases with distance from the creek bed.

3. Nature of the soil. A soil that initially possesses a high proportion of clay or silt as compared with the sand fraction will be much more unsatisfactory in the presence of excess sodium and magnesium. It is also more difficult to remove the excess alkali from such soils.

4. The vegetation. If the vegetation is dense it will reduce the loss of water from the soil surface but the transpiration demands of the plants may well bring about a continual rise in the soil water, and hence, if the salts are deposited some distance below, it may result in their migration towards the surface. With decrease of vegetation cover surface evaporation will tend to increase, and in the summer very high salt concentrations will be reached in the soil surface layers and frequently the phenomenon of salt efflorescence can be observed.

5. Slope of ground. This will determine the drainage pattern and the soil will be more saline towards the lower levels where the water accumulates.

6. Depth of soil water-table. In general the nearer this is to the surface the more constant will be the soil salinity.

7. The depth of the salt deposit. The nearer this is to the surface, the more saline will be the surface layers unless there are periods of heavy precipitation.

8. Water inflow into region. Generally speaking, water inflow into arid districts comes from surrounding areas of higher altitude and such water is therefore non-saline and its entry will bring about a dilution of the salt in the basin. In many areas, however, the inflow of fresh water still does not exceed the average loss by evaporation so that salinity slowly increases. The effect of slope, depth of soil water-table and water inflow into arid regions is well exemplified in the account by TAGUNOVA (1960) of the relation between soil and plant cover of the north east shore of the Caspian Sea and salinity and moisture conditions.

9. Temperature. This is a factor of profound importance in inland regions. High summer temperatures common in the centres of continents bring about excessive summer salinities which may involve considerable irrigation if crops are to be grown successfully.

* * *

So far as the natural vegetation is concerned the view has repeatedly been expressed that zonations to be observed in saline areas, whether continental or coastal, are essentially related to soil salinity as expressed by the amount of sodium chloride or soluble chloride present in the soil water or in terms of osmotic pressures. It is evident from associated soil analyses which have been made that the vegetation does change as total salt decreases. Thus SHREVE (1942) has described the zonation around Great Salt Lake in Utah in relation to salt content as follows:

| | | |
|---|---|---|
| 2.5% salt: | *Salicornia* spp., *Allenrolfea occidentalis* | |
| 0.5—0.9% salt: | *Distichlis spicata* or *Sporobolus airoides** | |
| 0.8% salt: | *Sarcobatus vermiculatus*<br>*Kochia vestita* | *Atriplex confertifolia* |
| 0.4% salt: | *Artemisia tridentata* | |

* Increasing distance from lake.

In the Moroccan saline steppes the most alkaline areas are occupied by *Salicornia fruticosa* (little accretion) or *Arthrocnemum* spp. (much accretion). With decreasing salt both are replaced by *Suaeda fruticosa* and then by *Juncus* spp. or *Atriplex halimus*. Similar zonations have been described from Northern Egypt (TADROS, 1953) around the Red Sea and in Palestine (Israel) (ZOHARY, 1944, 1952; KASSAS, 1957), and around Tuzgölü in Turkey (BIRAND, 1960). In the Tuzgölü area the zoning and salinities were recorded as follows:

**Table II.**

Mean salinities of soil in vegetation zones at Tuzgölü (after BIRAND).

| Vegetation | 0-2 cm | 2-10 cm | 10-30 cm | 30-60 cm | 60-100 cm |
|---|---|---|---|---|---|
| Zone 1. *Halocnemum strobilaceum* community | (a) (5.25) | (4.75) | 3.1 | | 1.65 |
| | (b) (10) | (10) | (8) | (8) | (8) |
| Zone 2. *Frankenia hirsuta-Statice iconia* community | (a) 0.7 | 2.85 | (3.2) | 2.95 | 1.5 |
| | (b) 0.12 | 0.95 | 1.55 | 1.65 | 1.55 |
| Zone 3. *Artemisia fragrans* steppe | (a) (0.1) | (0.1) | (0.1) | (0.1) | 0.6 |
| | (b) (0.1) | 0.16 | 0.17 | 0.3 | 0.51 |

Values in brackets are approximate owing to the limitations of the instrument used.

Around the Neusiedler See in Austro-Hungary a detailed zonation based upon height above groundwater, distance from lake and frequency of flooding (all three controlling salinity) has been worked out by BOYKO (1931, 1932, 1951) and WENDELBERGER (1950):

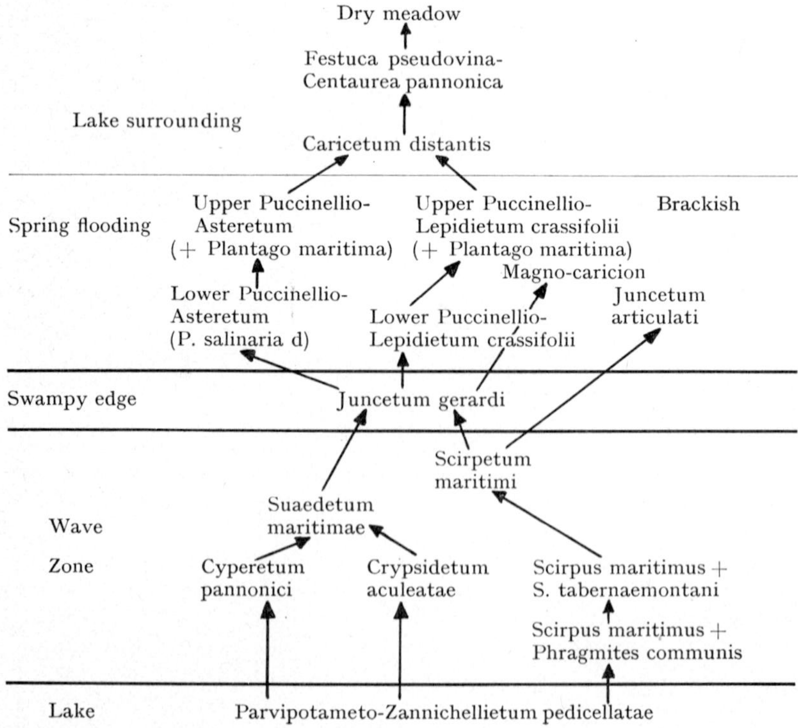

In view of what has been said earlier it would seem that the problem of vegetation zonation in relation to salinity cannot yet be regarded as solved. Before a final solution is reached intensive work will be necessary, using what appear to be key indicator species, to determine whether their apparent response to salinity is based upon response to the alkali metal or to the anion, or whether it is a combination of both. Such work as has been carried out (see p. 29) suggests very strongly that the individual ions are of far greater importance in this particular respect than has been appreciated previously. In particular we need to know very much more about the effect of the sodium and magnesium ions upon plant growth and metabolism. In the case of the grape vine, root development is related to soil salinity, the most active roots occurring in the least saline areas (GOREV, 1957). This is also indicated by the decrease in the whole development of the single plant individuals of grape vine when the vineyard is bordering a salt lake (BOYKO 1934, p. 107).

Those plants that grow in habitats where there is excess salt commonly exhibit certain physical characteristics, some of which give to the vegetation a uniformity of appearance. Such features include the development of succulence, e.g. species of *Salicornia, Halocnemum, Allenrolfea,* a grey, glabrous, glaucous appearance of the foliage, considerable reduction in leaf size, e.g. arid desert shrubs in particular, the development of water storage hairs, e.g. species of *Atriplex,* and the presence of salt excreting glands, e.g. *Statice, Spartina.* It is evident that the presence of the excess salt imposes upon the plants a major physiological problem. SCHIMPER originally expressed this problem in terms of "physiological drought" —namely, abundant water, but of such a high concentration that plants can not readily absorb it; the habitat is therefore physiologically dry. Most physiologists and ecologists would probably agree that this view cannot be upheld because halophytes have the capacity to absorb water of high salt concentration and indeed the concentration of their sap commonly exceeds that of the soil solution. In order that this may happen extra respiratory energy is required and such studies as have been made indicate that the energy is available. However, the concept of physiological drought could properly be applied to plants that are glycophytes and possibly also to some plants that are border line halophytes*.

Not unnaturally, considerable effort has been directed towards determining the tolerance towards salinity of plants that grow in saline conditions. In this connection more work has been carried out on the vegetation of maritime salt-marsh plants than on those of salt deserts, though in the latter case, where utilization is in-

* STROGONOW (1964) points out that this concept can be applied to plants growing on sodium sulphate soils.

volved, much work has been done on potentially economically useful plants. More information on this matter is greatly to be desired because from the published figures it appears that one and the same species may exhibit a different tolerance depending on the locality (Table III).

**Table III.**

Variation in tolerance of coastal halophytic species to salinity.

| Species | HARSH-BERGER | KUNZ & WAGNER | TAYLOR | PENFOUND & HATHAWAY |
|---|---|---|---|---|
| *Spartina patens* | 0.4—4.25% | 0.06—5.38% | 0—3.0 | 0.12—3.9 |
| *Distichlis spicata* | 1.2— | —3.9 | 0—3.2 | 0.45—4.97 |
| *Baccharis halimifolia* | 7. —3.62 | —1.98 | 0—2.8 | 0 — |

(For details of additional species see CHAPMAN, 1962).

In the following table some indication is given of the upper limit of resistance for a variety of cultivated species.

**Table IV.**

Chloride resistance of cultivated plants.

| Species | Parts NaCl/1000 dry soil | Parts Cl/1000 soln. |
|---|---|---|
| Palms | — | — |
| Cotton | 4—8 | 10 |
| Wheat | 4—8 | 4.5 |
| Barley | 4—8 | +2.3 |
| Oats | 4—8 | 4 |
| Cabbage | 8 | 6 —11.5 |
| Beet | 8 | 15 |
| Sweet Clover | 8 | |
| Artichoke | 3—5 | 4 |
| Pomegranate | 3—7 | |
| Tomato | 2—3 | 4 |
| Melons | 2—3 | 1 |
| Marrow | 2—3 | |
| Olive | 2—3 | |
| Peas | | 1.5 |
| Beans | | 1.5—6.0 |
| Vetch | | 1.5—8.0 |
| Almond | | 2.3 |
| Flax | | 3.0 |
| Onion | 1—2 | 2.3 |

Fruit trees and other crops have also been classified by BEEFTINCK (1955) as follows:

Highly resistant: Date palm.

Highly tolerant: Asparagus, beet, spinach, cabbage, cotton.

Resistant: Artichoke, tomato, melilot, lucerne, trefoil, strawberry, clover, fenugreek.

Moderately resistant: Cereals, sunflower, many legumes, species of poplar and eucalyptus, fig, olive, pistachio.

Moderately sensitive: Lettuce, beans, soya bean, sweet potato, tobacco, flax, hemp, citrus, pear, quince, apple, apricot, almond, black currant, American grapes.

Sensitive: French beans, lentil, strawberry, walnut, red currant, raspberry, plum.

Intolerant: Loquat, gooseberry, peaches, cherry.

A comparison of these lists will show that there is some divergence in the findings. This may in part be due to the different methods of expressing the tolerance, but more probably it is concerned with the actual horticultural variety that has been tested. In many cases the age of the plants tested has not been stated and seedlings may often be very much more intolerant of high salinity than are the adult plants. This aspect of tolerance appears to have been largely neglected except in the case of maritime salt marsh plants. As the present author has pointed out (1962) many reports do not distinguish between the different kinds of saline soil, and data expressed in terms of the air-dry soil are meaningless without information on the soil moisture. This can be overcome by the use of the electrical conductivity of a saturated soil extract. If this does not exceed 8 mhos per cm crops are likely to grow well. Even when agreement is reached on the method of expressing results they are still not meaningful unless we know what is meant by sensitivity. Is it expressed in terms of survival, power to grow, capacity to reproduce or productivity in terms of crop yield? It may be suggested that the last named, expressed as a relative value in relation to a standard*) non-saline soil, is probably the best.

In so far as halophytic plants grow in a medium where there is an excess of certain ions in the soil solution, physiologists have interested themselves in determining the amount of such ions that are absorbed by the plants. Frequently these results are expressed in terms of osmotic pressure, but these figures may be meaningless because Arnold (1955) has pointed out that only about 62% (mean value) of the osmotic pressure is caused by the chloride ion, with a range from 42% in *Atriplex hastata* to over 90% in some species of *Salicornia*. Analyses of whole plants are likewise misleading because there is clear evidence that parts of plants can differ greatly in the amount of salt that they contain. It appears that in most plants maximum values are likely to occur in the leaves.

* Standard only in respect of the alkali and halide ions.

A problem that seems to have been little studied, though clearly one of great significance, is the degree of permeability of the protoplasm to sodium, magnesium, chloride and carbonate ions. It has become increasingly evident over the last few years that an intensive attack upon the uptake, absorption and excretion of the ions mentioned above is essential if we are to reach an adequate understanding of the problem of salinity.

Whilst the sodium ion is of paramount importance in soils that are rich in sodium chloride or sodium sulphate because of its effect upon the dispersion of the soil colloids, it appears to be the chloride ion that is of great importance so far as cellular physiology is concerned, though more work is needed upon the influence of the sodium ion. There are several reports (ARNOLD, 1955) of the effect of increasing the chloride ion upon the pH of the cell sap. In some examples, increasing the chloride content in the external solution results in increasing the sap acidity. Even very high values of sodium chloride do not result in a great change of pH to the alkaline region through sodium ions being absorbed. If there is a sodium efflux pump in halophytes, as some work suggests, it appears to be extremely effective, even with high concentrations of sodium in the external medium.

**Table V.**

Osmotic pressure and pH of cell sap in relation to chloride content of external medium (After HARRIS, 1934).

| Species | O.P. (ats) | pH | gCl/l |
|---|---|---|---|
| *Allenrolfea occidentalis* | 44.1 | 5.94 | 23.34 |
| | 39.7 | 6.6 | 17.61 |
| *Artemisia tridentata* | 21.9 | 4.87 | 2.76 |
| | 25.5 | 4.03 | 1.93 |
| | 13.9 | 5.3 | 1.53 |
| *Lepidium integrifolium* | 17.0 | 4.52 | 5.83 |
| | 21.3 | 4.9 | 5.27 |

In most halophytes the chloride ion seems to have little effect upon the sugar content of the cell sap. An exception to this is found in the succulents where there is evidence that increasing the chloride content results in a decrease of the sugar content (ARNOLD, 1955).

Succulence in halophytic plants has been related to the salinity of the soil and to the amount of salt present in the plant tissues. Whilst ultimately this contains a large measure of truth, VAN EIJK (1939) has shown that succulence is induced by the action of certain specific ions. Thus $NO_3^-$, $Cl^-$ and $Na^+$ ions all increase the degree of succulence whereas $SO_4^{=}$ and $Ca^+$ reduce it. Doubt has been cast by ARNOLD (1955) on the interpretation of these results and he believes that it is the ratio of absorbed/free ions in the cells of the plants that controls the phenomenon, an increase in the free ions promoting succulence. Support for this view is provided by his work on succulence in tomato plants where he showed that an increase in succulence was accompanied not only by an increase of $Cl^-$ in the ash but also by an increase in the ash expressed as a percentage of the total water content. Whichever proves to be correct, whether it is a specific ion or the total of all free ions or the absorbed/free ion ratio, we still do not know how the ions act. In some way or another they must increase the capacity of the cell to absorb water and to enlarge, thus producing the water-storage tissue. Not only is our knowledge of the amounts of sodium, magnesium, chloride and carbonate present in halophytes very inadequate but also we have little or no information about the seasonal fluctuation of such ions. This information is urgently required and must be considered in conjunction with the amounts of the specific ions and the total amount of free ions in the cells.

In some species, e.g. *Allenrolfea* and *Suaeda divaricata*, the amount of potassium exceeds the amount of sodium, a phenomenon also recorded by BEADLE et al. (1957) for some species of *Atriplex* in Australia. This is a problem that invites attention, especially since we now know that certain of the marine algae possess a sodium efflux pump and a potassium influx pump. It may well be that some of the halophytic plants possess similar mechanisms. From the view point of potential soil reclamation, it is also important to note that in the presence of potassium, sodium uptake is materially reduced (HEIMANN & RATNER, 1961).

Salinity may not only exert an influence upon the morphology of species that grow in such habitats but it can also affect their rate of growth and the success of their seed germination. In the case of crop plants, the growth rate of beet, spinach, turnip and cabbage is stimulated by moderate levels of NaCl (up to 4 ats) (NIEMAN, 1962). Up to the present the results of the various studies that have been carried out show that the great majority of halophytes germinate best under fresh-water conditions or at least under conditions of salinity that are less than those under which the adult plant grows. It is quite evident that high salinity is in most cases only an inhibitor, because if ungerminated seeds from conditions of high salinity are subsequently placed in lower salinities they germinate quite freely (POMA, 1922; MACKAY & CHAPMAN, 1954). In a number of cases, principally succulents such as *Salicornia*, germination will take place in concentrations up to 2% NaCl but generally the amount of salt needs to be less. STOCKER (1928) and SCHRATZ (1936) have both indicated that the pre-history of the seed parent can be of great significance in its effect upon seed germination. This is a matter that justifies further attention.

Insofar as low salinity favours germination, it is clear that the ecologist is greatly interested in factors that can bring about a reduction in the salinity of the surface layers of the soil. In the Mediterranean minimal values are reached in Winter (MOLINIER, 1947), but from most other arid regions the necessary information is not yet available. It is evident that the incidence of rainfall is likely to be of paramount importance in determining the time and percentage of seed germination. Whilst information on germination is available for a number of maritime salt marsh species, very little is available for species of alkali deserts. Salinity *per se* is not likely to be the sole factor controlling germination. TSOPA (1939) has reported that temperature may be important, low values around 15° C favouring germination. The same author also lists the following species in which germination is affected by light, *Triglochin maritima, Beckmannia ericiformis, Spergularia marginata, Glaux maritima, Gratiola officinalis.* In view of current work on light, especially red light, in relation to germination, it is not unlikely that many more species are affected by this factor.

Apart from variations of salinity upon seed germination there is also the problem of subsequent growth of the seedlings. Here again very little attention has been given to the various species, but from the meagre amount of information available it appears that with increasing age there is an increasing tolerance towards salinity. It would be of some interest to know whether there is any ontogenetic change in protoplasmic resistance to uptake of ions or whether it is merely slow acclimatization of the cells to increasing salinity.

Apart from changes that may occur in the resistance of plants to salinity with increasing age there is also a general effect of salinity upon growth. This commonly manifests itself in the size to which the plants attain. In general, increasing the salinity reduces the size of the plant and organs may be affected proportionately. Thus, HARSHBERGER (1911) gives figures for the length and width of the spikes of *Typha angustifolia* under different salinities.

Whilst the majority of plants undoubtedly grow better under slightly brackish or freshwater conditions there are certain genera,

**Table VI.**

| Site | Salinity (sp. gravity) | Mean ht. of plant | Mean length of spike | Mean dia. spike |
|---|---|---|---|---|
| Sea edge | 1.0145 | 1.066 m | 8.9 cm | 1.9 cm |
| Middle of marsh | 1.014 | 1.427 m | 10.8 cm | 2.3 cm |
| Head of lake | 1.005 | 1.855 m | 15.8 cm | 2.2 cm |

e.g. *Salicornia, Allenrolfea, Halocnemum,* and some species of *Suaeda* which are obligate halophytes and grow better when more than 0.5% salt is present.

Whilst most workers have ascribed the general reduction in size of plants under natural conditions of salinity to the great osmotic pressures induced within the plant and hence the difficulty of water absorption, this cannot by any means be regarded as the sole factor involved, and in some cases it may not be operative. Acceptance of this factor as a major one would also involve acceptance of physiological drought as a factor (see p. 31 and below), and this is by no means proven. Within the soil itself the presence of sodium ions affects the dispersal of soil colloids, and as a result such soils can readily become waterlogged and reducing conditions then ensue that inhibit plant growth. The absorption of certain ions against a gradient may reduce the amount of energy available for normal metabolic processes, or the excess of sodium or magnesium, chloride or carbonate may inhibit the metabolic activities of the cell protoplasm. VAN EIJK (1939) has also shown that in *Salicornia* dry wt. is related to certain anions independently of sodium (Table VII). Growth data of plants growing under saline conditions need therefore to be very carefully scrutinised in relation to other ions present, whether in excess or not.

**Table VII.**

Effect of ions on mean dry weight of *Salicornia* plants in culture solutions with $Na^+$ concentration constant at $^4/_{12}$ mols/litre. (After VAN EIJK).

| $SO_4^=$ : $Cl^-$ in solution | Mean dry weight | $NO_3^-$ : $Cl^-$ in solution | Mean dry weight |
|---|---|---|---|
| 0:4 | 0.12 g | 0:4 | 0.12 g |
| 1:3 | 0.13 | 1:3 | 0.12 |
| 2:2 | 0.10 | 2:2 | 0.12 |
| 3:1 | 0.08 | 3:1 | 0.07 |
| 4:0 | 0.04 | 4:0 | 0.03 |

The presence of such ions in excess also raises problems of ion antagonism. MAGISTAD (1945) has reported that under certain conditions unfavourable ratios, especially of calcium and magnesium, may be produced and poor growth results. The calcium in the plant is of great importance because of its relationship to cell wall formation and hence permeability. Any ion that may adversely affect the calcium metabolism of cell wall formation will obviously affect cell growth. Alkali soils invariably have a high pH and there is good reason to believe that under such conditions the availability of iron, phosphorus and manganese may be greatly reduced. The elements are admittedly only required in small quantity, but if the soil is already deficient and physically unfavourable then the small amount available may not be absorbed by the plants. It can be seen from the above outline that a variety of possible mechanisms can be invoked to account for reduction of growth under conditions of high salinity*. At the present time it can only be said that we have very inadequate information about any of these mechanisms under conditions of high salinity and it is a problem that requires urgent attention.

Whilst it is perhaps not unnatural that attention is generally directed to the effect of excess ions upon growth, the possible effect of other factors must not be overlooked. Though most plants growing in areas of excess salinity might, because of their low stature and general spatial separation on the ground, be regarded as sun plants, this may not be true of all species. In the case of saline desert plants we have no information on this matter but at least one salt marsh plant, *Suaeda novae-zelandiae,* has been shown to be a specific sun plant because growth is greatly reduced in 50% sunlight and ceases at lower light levels (TURNER in CHAPMAN, 1962).

In view of SCHIMPER's hypothesis of physiological drought in relation to salinity it is pertinent to consider the effect of salinity upon transpiration. If physiological drought exists, then increasing salinity should steadily lower the transpiration rate, and in any case the transpiration of plants growing in soils with more than 0.5% NaCl in the soil solution, should show a transpiration rate that is generally less than that of glycophytes. Of those halophytes that have been investigated it seems that the leaves generally have a smaller number of stomata per sq. mm than do glycophytes. An exception to this would appear to be provided by *Plantago maritima*

**Table VIII.**

| Plant species | Habitat | Stomata per sq.mm (av.) | |
|---|---|---|---|
| *Suaeda maritima* | damp | 63 | Saline |
| *Salicornia stricta* | marsh | 104 | |
| *Plantago maritima* | marsh | 187 | |
| *Aster tripolium* | marsh | 90 | |
| *Limonium vulgare* | moist | 126 | |
| *Populus alba* | damp | 315 | Non-saline |
| *Populus nigra* | dry | 135 | |
| *Veronica chamaedrys* | dry | 175 | |
| *Veronica beccabunga* | damp | 248 | |

* STROGONOV (1964) considers the $Cl^-$ion has a definite toxic effect.

(see Table VIII), though even in this case the total leaf area is small as compared with, say, *Populus nigra*.

One of the principal difficulties in comparing transpiration rates is that the result depends on the basis used for expressing the data. This was first appreciated by AUBERT in 1892 and was confirmed by SCHRATZ later (1934—37). SCHRATZ, using refined methods of measurement, showed that the transpiration rate was less for his halophytes than for glycophytes when it was calculated on a fresh weight basis. If, however, leaf area was used then the halophytes had a higher transpiration rate. STOCKER (1928) reported a similar variation in transpiration rate for certain Egyptian salt desert plants. In Table IX this can be observed by comparing the rates for *Arthrocnemum glaucum* with those for *Typha* or for *Salsola* and *Suaeda* with those for *Phragmites*.

**Table IX.**

Mean transpiration rates.

| Species | g/hr/ $dm^2$ | g/f. wt. | g leaf dry-wt. | Midday T.R. (Total $H_2O$ content) | Plant above ground ($H_2O$ content) | Succu-lence grade |
|---|---|---|---|---|---|---|
| *Typha latifolia* (very wet area) | 0.26 | 0.057 | 0.33 | 6.9 | 83 | 3.8 |
| *Eragrostis bipinnata* (grass swamp) | 0.26 | 0.083 | 0.42 | 11.6 | 72 | 2.2 |
| *Phragmites communis* var. *stenophylla* (drier) | 0.66 | 0.122 | 0.96 | 25.1 | 65 | 2.7 |
| *Arthrocnemum glaucum* (lagoon) | 0.61 | 0.056 | 0.4 | 6.9 | 80 | 8.7 |
| *Salsola Kali* / *Suaeda maritima* } * | 0.52 | 0.10 | 1.49± 0.35 | 11.0± 0.8 | 90± 1 | 4.8± 1.2 |
| * German salt marsh. | | | | | | |

It may also be noted that in the case of the first two species the roots of the *Typha* were in a soil solution with an osmotic pressure of 8 ats., and the roots of the *Eragrostis* in a solution with an O.P. of 27 ats.

It is evident that a new basis for expressing transpiration results is desirable. SCHRATZ in 1937 suggested the "Wasserumsatz" or the time taken to transpire the total weight of water contained by the plant. Unfortunately few workers have so far adopted this basis. More recently ARNOLD (1955) has proposed a productivity* basis, and from his experiments it seems that the chloride ion independently of sodium brings about an increase of productivity.

* mg dry substance produced per g transpired water.

It appears, therefore, as if salinity does not necessarily involve a reduction in transpiration nor is there any evidence that it affects water uptake. Certainly excess of the chloride ion does not affect water uptake (ARNOLD 1955). Salinity would not, however, be the only soil factor affecting transpiration rate. The rate could well be more closely related to the moisture content of the soil than to the total quantity of absorbable free ions. Although his data came from three separate species BRAUN-BLANQUET (1931) considered that they demonstrated this fact. Plants of *Salicornia radicans* from the wettest habitat had the highest transpiration rate, *S. fruticosa* from a drier habitat a somewhat lower rate and plants of *S. macrostachya* from the driest had the lowest rate. However, no attempt has really been made as yet to distinguish between the effects of water content of the soil and salt content of the soil solution on transpiration of halophytes. The actual transpiration rate is probably a result of the interaction of available soil moisture with the total quantity of ions, in which the free ions play a more important part than the absorbed ones. Thus work by VAN EIJK (1939) has shown that for *Salicornia* and *Aster* increasing the sodium, calcium and chloride tended to lower the transpiration rate, whereas raising the sulphate brought about an increase. It is to be regretted that no recent work of this nature has been carried out on plants from arid regions.

An interesting approach to the problem was made by ADRIANI in 1945, though it was applied solely to plants of the salt marsh. He proposed a classification of halophytes based upon the factors which he believed to control the transpiration rate. As the present author has pointed out (1962) the classification requires further study before final validation but it certainly provides a basis for future investigators. ADRIANI suggested that relative humidity, light intensity, changes in water content and wind velocity were the major factors controlling transpiration, the first-named acting alone or in combination with one or more of the remainder. At the present time insufficient work is available, particularly on the transpiration rate of arid zone halophytes, to justify such a basis of classification.

From what has been said in the foregoing pages it is evident that whilst much data has been accumulated on the effect of salinity on vegetation and individual plant species, and the nature of the saline habitat, there are many questions left unsolved. A recent contribution by STROGONOV (1964) emphasises the importance of the kind of salinity. He shows that for certain species the chloride and sulphate ions of sodium exert quite different effects, the former promoting halo-succulence and the latter halo-xeromorphism. This work re-emphasises the great importance of individual ions and shows how little we still know.

## REFERENCES


ADRIANI, M. J., 1945. Sur la Phytosociologie, la Synécologie et le bilan d'eau de Halophytes de la Région néerlandaise méridionale, ainsi que de la Méditerranée française. *S.I.G.M.A.* **88,** *1—217.*

ARISZ, W. H., CAMPHUIS, I. J., HEIKENS, H. & TOOREN, A. J. VAN, 1955. The Secretion of the salt glands of *Limonium latifolium* Klze. *Acta bot. neerl.* **4** (3), *322—338.*

ARNOLD A., 1955. Die Bedeutung der Chlorionen für die Pflanze. *Bot. Stud.* 2. Jena.

BAUMEISTER, W. & SCHMIDT, L., 1962. Über die Rolle des Natriums im pflanzlichen Stoffwechsel. *Flora* **152** (1), *24—56.*

BEADLE, N. C. W., WHALLEY, R. D. B. & GIBSON, S. B., 1957. Studies in Halophytes. II. *Ecology* **38** (2), *340—344.*

BEEFTINK, W. G., 1955. Examination of soils and crops after the inundations of 1st February 1953. III: Sensitivity to salt of inundated fruit crops. *Netherl. J. Agric. Sci.* **3** (1), *15—34.*

BERGQUIST, P. L., 1958. Evidence for separate mechanisms of sodium and potassium regulation in *Hormosira banksii. Physiol. Plant.* **II**: *760—770.*

BIRAND, H., 1960. Erste Ergebnisse der Vegetationsuntersuchungen in der Zentral-anatolischen Steppe. I. Halophytengesellschaften des Tuzgölü. *Bot. Jb.* **79** (3), *255—296.*

BLACK, R. F., 1960. Effects of NaCl on the ion uptake and growth of *Atriplex vesicaria* Heward. *Austr. J. biol. Sci.* **3** (3), *249—266.*

BOJKO (BOYKO), HUGO, 1931. Ein Beitrag zur Oekologie von *Cynodon dactylon* Pers. und *Astragalus exscapus* L. *S.B. Akad. Wiss. Wien* (Math.-naturwiss. Kl. Abt. 1, **140,** 9. u. 10. Heft).

BOJKO (BOYKO), HUGO, 1932. Über die Pflanzengesellschaften im burgenländischen Gebiete östlich vom Neusiedler See. *Burgenländ. Heimatbl.,* **1**: *43—54,* Eisenstadt.

BOJKO (BOYKO), HUGO, 1934. Die Vegetationsverhältnisse im Seewinkel. *Beiheft z. Bot. Cb.* **51,** II: *600—747,* Prag-Dresden.

BOJKO (BOYKO), HUGO, 1951. On regeneration problems of the vegetation in arid zones. In: Les bases écologiques de la régénération de la végétation des zones arides (Ed. H. BOYKO), Stockholm juillet 1950 — Union Int. des Sci. Biologiques, Série B (Colloques) **9**: *62—80,* Paris, 1951.

BRAUN-BLANQUET, J., 1931. Zur Frage der "Physiologischen Trockenheit" der Salzboden. *Ber. schweiz. bot. Ges.* **40,** (2).

CHAPMAN, V. J., 1962. Salt Marshes and Salt Deserts of the World. Leonard Hill, London.

EIJK, M. VAN, 1939. Analyse der Wirkung des NaCl auf die Entwicklung, Sukkulenz und Transpiration bei *Salicornia herbacea,* sowie Untersuchungen über den Einfluss der Salzaufnahme auf die Wurzelatmung bei *Aster tripolium. Rec. Trav. Bot. Neerl.* **36,** *559—657.*

EPPLEY, R. W., 1962. Major Cations in: Physiology and Biochemistry of Algae, Ed. R. A. LEWIN, Academic Press, New York.

GOREV, L. N., 1957. *Dokl. Akad. Nauk. UZSSR.* **1957** (3), *55—57.*

HARSHBERGER, J. W., 1911. An hydrometric investigation of the influence of sea water on the distribution of salt marsh and estuarine plants. *Proc. Amer. phil. Soc.* **50,** *457—496.*

HAYWARD, H. E., 1954. Plant growth under saline conditions. Reviews of research on problems of utilization of saline water. U.N.E.S.C.O. Paris, *37—72.*

HEIMANN, H. & RATNER, R., 1961. The Influence of potassium on the uptake of sodium by plants under saline conditions. *Bull. Res. Coun. Israel.* Sect. A. **10** No. 2, *55—62.*

IVERSEN, J., 1936. Biologische Pflanzentypen als Hilfsmittel in der Vegetationsforschung. Copenhagen.

KASSAS, M., 1957. On the Ecology of the Red Sea Coastal Land. *J. Ecol.* **45** (1), *187—203.*

KELLEY, W. P., 1951. Alkali soils, their formation, properties and reclamation. New York.

KOVDA, W. A., 1946—47. Proiskhozdenie i rezhim. Zasolenykh poch. vols. 1,2. Moscow.

MACKAY, J. B. & CHAPMAN, V. J., 1954. Some notes on *Suaeda australis* Moq. var. *nova zelandica*, var. nov. and *Mesembryanthemum australe* Sol. ex. Forst. *Trans. Roy. Soc. N.Z.* **82,** (1), *41—47.*

MAGISTAD, O. C., 1945. Plant growth relations on saline and alkali soils. *Bot. Rev.* **12,** *181—230.*

MOLINIER, R., 1947. La végétation des Rives de l'Etang de Berre. *S.I.G.M.A.* **103.**

NIEMAN, R. H., 1962. Some effects of Sodium chloride on growth, photosynthesis and respiration of twelve crop plants. *Bot. Gaz.* **123** (4), *279—284.*

POMA, G., 1922. L'influence de la salinité de l'eau sur la germination et la croissance des plantes halophytes. *Acad. Roy. Belg. Bull. Class. Sci.* 5th ser. **8** (2), *81.*

SCHRATZ, E., 1934. Beiträge zur Biologie der Halophyten II. *Jb. wiss. Bot.* **81,** *59—93.*

SCHRATZ, E., 1936. Beiträge zur Biologie der Halophyten III. *Jb. wiss. Bot.* **83,** *133—189.*

SCHRATZ, E., 1937. Beiträge zur Biologie der Halophyten IV. *Jb. wiss. Bot.* **84,** *593—638.*

SHREVE, F., 1942. Desert vegetation of North America. *Bot. Rev.* **8** (4), *195—246.*

STOCKER, O., 1928a. Das Halophyten-Problem. *Erg. der Biol.* **3,** *265—354,* Berlin.

STOCKER, O., 1928b. Der Wasserhaushalt ägyptischer Wüsten- und Salzpflanzen vom Standpunkt einer experimentellen und vergleichenden Pflanzengeographie aus. *Bot. Abtr.* **13,** *200.*

STROGONOV, B. P., 1964. Physiological basis of salt-tolerance of plants. Israel Prog. Sci. Transl., Jerusalem.

TADROS, T. M., 1953. A phytosociological study of halophilous communities from Mareotis (Egypt). *Vegetatio* **4,** *102—124.*

TAGUNOVA, L. N., 1960. O sryazyakh pochvenno-restitel'nogo pokrova severovostochnogo poverezh'ya Kaspiiskogo morya s usloviyami zasoleniya i uvlazhneniya. *Biull. Moskov. Obsh. Ispyt.* Prirody Otdel. Biol. **65** (1), *61—76.*

TSOPA, E., 1939. La Végétation des Halophytes du Nord de la Roumanie en connexion avec celle du reste du pays. *S.I.G.M.A.* **70,** *1—22.*

WALTER, H., 1962. Die Vegetation der Erde. vol. 1. Jena.

WENDELBERGER, G., 1950. Zur Soziologie der Kontinentalen Halophyten-Vegetation Mitteleuropas. *Oest. Akad. Wiss.* Mat.-Natur. Kl. **108** (5), *1—180.*

WHITTAKER, R. H., 1956. Vegetation of the Great Smoky Mountains. *Ecol. Mon.* **26** (1), *1—80.*

WILLIAMS, M. C., 1960. Effect of sodium and potassium salts on growth and oxalate content of *Halogeton. Plant Physiol.* **35** (4), *500.*

YAPP, R. H., 1922. The Concept of Habitat. *J. Ecol.* **10** (1), *1—17.*

ZOHARY, M., 1944. Vegetational transects through the desert of Sinai, *Israel Exp. J.* **2** (4), *201—215.*

ZOHARY, M., 1952. Ecological studies in the vegetation of the near eastern Deserts. *Palest. J. Bot.* **3,** *57—78.*

# ARIDITY AND SALINITY
(A survey of soils and land use)

BY

P. C. RAHEJA

*Director*

*Central Arid Zone Research Institute, Jodhpur, India.*

## Introduction

The arid zones of the world are characterized by the minimum of annual precipitation and the maximum of heat and aridity. The precipitation in arid parts varies from 5 to 15 inches and in semi-arid regions from 15 to 30 inches. According to SHANTZ (1956), based upon climatic maps of MEIG's (1952), the approximate areas of extreme arid, arid and semi-arid regions of the world are 2.24, 8.42 and 8.20 million square miles respectively, i.e. about 36% of the gross land area of the world. The development of salinity is far more extensive in the extreme arid and arid lands than in semi-arid and humid lands. Based upon the characteristics of vegetation, SHANTZ classifies the areas as follows:

**Table I.**

Area of arid lands based on vegetation.

| | |
|---|---|
| *Semi-arid:* | |
| Sclerophyllous brushland | 1,180,000 square miles |
| Thorn Forest | 340,000 square miles |
| Short grass | 1,200,000 square miles |
| *Arid:* | |
| Desert grass savanna | 2,300,000 square miles |
| Desert grass, desert shrub | 10,600,000 square miles |
| *Extreme arid:* | |
| Desert | 2,430,000 square miles |

According to this classification, the area under arid vegetation is far greater than that under extreme arid or semi-arid vegetation. One of the characteristics of these lands is that the moisture seldom percolates to the sub-soil. The soluble carbonates once percolated with the soil moisture, are not returned to the surface and gradually accumulate at the bottom of the moist layer. This layer in due course is stratified. Above it, sodium salts tend to accumulate, rise to the surface by evaporation and form a saline zone, particularly in view of the very sparse plant cover. Most of the arid area belongs to the pedocals group of soils.

The most acute problem of the arid areas is the lack of water. The precipitation received during the winter months is less evaporated from the soil but is not sufficiently utilized by the vegetation as the temperatures of the surrounding atmosphere are low, and in consequence the rate of photosynthesis is low too. In summer rainfall areas evapo-transpiration is highest during the rainy season, although the effective utilization of water by the plants is also high. Besides the insufficiency of precipitation, the wide variability in rainfall is a characteristic feature of arid regions: In several of the arid areas rain falls mainly in thunderstorms; in others, the showers are either meagre in amount, or the precipitation is heavy in a brief period. This leads either to rapid evaporation of moisture or to heavy storm run-off. Consequently, the precipitation received is not fully utilized by the vegetation. In other areas the yearly precipitation is partly received in the winter season and partly in summer. This is neither efficiently utilized by the vegetation in the winter nor in the summer. In such regions, annual species with a short life cycle predominate. In extreme arid regions the lack of moisture in certain years may be absolute or be caused by insufficient precipitation. Under such conditions the ground remains bare of vegetation or covered over by psammophile shrubs in the runnels and rivulets.

In arid habitats, vegetation is normally in balance with precipitation and the other macro-climatic factors. There is seldom an accumulation of salts in any section of the profile, unless water from the surrounding areas accumulates in depressions or the drainage of the large area is centripetal. But when the environment is artificially changed and irrigation is brought to the arid areas, a fluvial environment is established and an agricultural-irrigation complex develops. The rapid surface evaporation by heat and movement of the soil's moisture up and down the profile profoundly alter the balance of salt in the soil particularly if it contains a considerable part of clay particles. Salt tends to accumulate in certain horizons and steadily appears in the top horizon of the profile, which is the medium from which growing crops draw their water and nutrients. Conditions of waterlogging accentuate the process of salinization. There are, however, certain regions which have developed from geological materials free from sodium and magnesium elements. But, by and large, most of the arid soils possess these elements derived from the minerals from which such soils have developed.

The soils of extreme arid regions are somewhat different from the soils of arid regions. The former are raw mineral soils which have weathered to a small extent, although the physical breakdown is dominant. The three major types are the skeletal soils, the several wind-blown soils and the deposited soils. These are often extremely shallow. The skeletal and several wind-blown soils have coarser material than the deposited soils, the latter having received the fine and medium sand blown by wind.

In contrast to these, the arid zone soils are better developed, especially the sierozems. Besides these, the desert steppe soils are to be found. The soils of the sierozem group are mature and have a clearly developed profile down to 1.5—2 m. This group is frequently associated with saline soils and gypsum crust soils, but in the sierozems the accumulation of soluble salts and gypsum is more marked at depth (sometimes as much as 1.5 m) than at the surface. The desert steppe soils generally have a sandy texture. The surface crust is cemented to varying degrees by lime or gypsum. They are redder than the "Grey and Red Desert steppe soils". The former have developed in a slightly damper climate than the latter. In the arid zone soils, the content of soluble salts is generally low due to leaching. In silty and clayey soils salts may accumulate in certain periods. The hydromorphic Brown soils, which have a clayey texture, very often become saline or alkali soils. Except in Grey Desert Steppe soils which have a crust of gypsum, this mineral as well as lime is very frequently washed down into deeper horizons, the latter to a lesser extent, particularly in the Brown soils (AUBERT, 1961).

## Historical Development of Arid Zones

### a) General Remarks

UNESCO has recently published a monograph on "A History of Land Use in Arid Regions". The global arid lands have been discussed in sequence for the different regions as follows:

1. South Western Asia
2. Indo-Pakistan
3. Eurasia
4. Nile Valley in Egypt
5. Maghrib
6. Sahara—Sahel
7. South Africa
8. South West Africa
9. North-America
10. Meso-America
11. South-America.

These regions not only contain arid lands but are also contiguous to extreme arid and sub-arid lands. The history of land use has been traced on historical evidence produced by extensive archaeological investigations of the prehistoric settlements, the published records of the historical periods, and the photogrammetric analysis of the aerial photos which have shown the developments in irrigation, the reclamation works and the drainage pattern in ancient and current times.

## b) Pattern of Land Use

### *1. South-West Asia*

This region consists of Anatolia, Syria, Lebanon, Israel, Jordan, Iraq, South Arabia, Yemen, Aden, Iran and Afghanistan. The history of land use in this region has been traced by WHYTE (1961). In ancient times the land use pattern has varied substantially in different tracts of this region. In the Tigris-Euphrates valley of ancient Mesopotamia, in the initial stages non-sedentary groups hunted for food. In the second stage, semi-sedentary communities acquired basic agricultural techniques for settled life. This was succeeded by upland subsistence and village subsistence patterns, when cultivation of wheat, barley and domestication of sheep, goats and probably cattle started. The plough-irrigation agriculture began with the fourth stage, and storage and trading in grain in the fifth stage. The same sequence, it appears, has been repeated in the Jordan Valley and other riverine tracts. The change from food-gathering to food-producing occurred 6,606 ± 380 years to 11,240 ± 300 years ago. Thus, the present day land use began about 5,500 B.C. In Iran, in the tract lying between the Elbruz Mountain and the Southern shore of the Caspian Sea, the reaping of grain started about 5330 B.C. ± 260, which period is identical with that in the development of agriculture in the Tigris-Euphrates Valley. In the Negev desert of Israel, the Nabataeans developed a flourishing agriculture based upon conservation of run-off water in the period from about 200 B.C. to A.D. 600. At one time in the various valleys of Elbruz, Fars, Zagros, Khorassan and the steppes of Mughan and Atrek, agriculture was practised, the antiquity of which has not been fully traced. These areas were subsequently overrun by nomadic tribes and agriculture was substituted by nomadic cattle rearing.

The land use in more fertile valleys consisted originally of monoculture. It became more complex with the introduction of legumes and other crops. From the Bronze age until medieval times, cereals, mainly wheat and barley, were the mainstay of the people. Market crops such as fruits and flax were produced under irrigation. A strong centralized control of irrigation in the entire Tigris valley and part of the Euphrates existed in 1760 B.C. (DROWER, 1954). The archaeological investigations conducted by JACOBSON (1955) have shown that salt on the surface appeared in 2400 B.C., being noticed first in scattered patches. Then, extension continued up to 2100 B.C. in Southern Babylonia. The references to "white" salt are not noticed from 2100 B.C. to 1200 B.C. From the latter period to 600 B.C. "wet" salt references are found in old records from northern Babylonia. The two critical factors which arrested agricultural development were siltation in the delta region and the rise of the water-table. The consequence of the latter was the actualization of the salinity problem which simultaneously affected the prosperity of the region.

The valleys of Iran, which were flooded and marshy in the pluvial period began to dry up between 15,000 and 10,000 B.C. In these fertile areas grasses began to flourish. This was followed by peasant agriculture. The draught animals were in use by 4000 B.C. for cultivating land for dry farming. This was followed by clearing scrub and irrigation. In the Khuzistan area of Iran, irrigation was developed by setting up temporary diversion dams in rivers and streams. From the canals the water was artificially lifted to the fields, thus preventing to a considerable extent over-irrigation and salinization of land. This type of land use was practiced in the period from 900 B.C. to 400 B.C. Further irrigation, especially in the fertile areas between Kermanesh, Hamadan and Burujad, was developed subsequently. A great weir with masonry structure was built across the Dujayl(Karun) River in the 7th century A.D. A dam over the river Kur irrigated 300 villages in Fars. However, by present day standards these irrigation works were small.

*2. Indo-Pakistan Region*

BHARDWAJ (1961) and WHYTE (1961) have traced the history of land use in this region. Throughout the Palaeolithic period the chief means of subsistence were hunting and food gathering. In the second Glacial Phases, people lived in small and isolated groups along river terraces. In the fourth and fifth millenium B.C., agriculture was introduced, according to PIGGOT (1950), from the Middle East through Baluchistan into Sindh and further north throughout the Indus Valley. It appears that then the climate of that region was more humid than at present.

Agriculture in Baluchistan was largely confined to valleys. Carl SAUER (1952), however, is of the opinion that the knowledge of agriculture spread from the Bay of Bengal region to the Indus Valley and migrated from there further to the Middle East. In the Chalceolithic period in the riverine areas, which were either irrigated or inundated, irrigated agriculture was practised in large settlements. In dry areas or "bars", the agriculture subsisted around water points. Here agriculture was subsidiary to nomadic herding. The Harrapa culture which extended from the Makran Coast to Kathiawar in the South and from the Gulf of Cambay to the Himalayan (Siwalik) foot-hills and eastward to Jamna Basis goes back to the third millenium B.C. and flourished on the banks of the Indus and the Ravi (PIGGOT, 1950). During the Vedic period (c.2000—600 B.C.)

the peasants dug wells and constructed canals to supply water to crops (MAJUMAR et al., 1955). Till this period there is no record of the rise of salinity in these irrigated tracts. In Sindh, agriculture continued to flourish between 600 B.C. and 325 B.C. In the subsequent period up to 700 A.D., which corresponds to the Hindu period of Indian history, irrigation systems were expanded and equitable distribution of irrigation was ensured (DUNBAR, 1949). In this period, irrigation in Kashmir (WALTER, 1904) and periodic irrigation in Afghan and Mughal were further extended. During the reign of Ferozshah Tughlak, extensive irrigation works were planned and executed to stabilize agricultural economy. Five great canals were dug to distribute the water of Jhelum and Sutlej for irrigation. Sher Shah and after him Akbar further gave impetus to irrigated agriculture in the 16th century. During the British rule, the linking canal system consisting of Upper Jhelum—Lower Chenab and Upper Chenab—Lower Bari Doab and connected irrigation works were set up in the Western Punjab. In 1935, Lloyd Barrage was opened to cover 3.2 million acres in Sindh. In Eastern Punjab, the Western Jamna canal and Sirhind canal, set up already in the Mughal period, were remodelled and extended. In the thirties and forties of the present century, the Haveli and Gang canal systems were established to irrigate the areas in the west Punjab and Rajastan, and already before this in the North West Frontier Province several canals were dug, diverting the waters from the rivers Swat and Kabul in order to irrigate large tracts in the Kabul river valley.

Simultaneously with the rapid extension of irrigation in the Indus valley and Jamna basin, a rise in salinity occurred and the problem of waterlogging assumed serious proportions. It was first noticed in 1915 and by 1942—43 the waterlogged area covered more than 700,000 acres.

**Table II.**

Area damaged by salinity and waterlogging in some of the main canal irrigation tracts.

| Canal system | Acreage under salinity | Acreage under waterlogging | Commanded area Acres |
|---|---|---|---|
| Lower Jhelum canal | 33,948 | 5,214 | 1,449,674 |
| Upper Jhelum canal | 9,916 | 4,293 | 540,755 |
| Lower Bari Doab canal (except Lahore District) | 43,567 | — | 1,460,659 |
| Lower Chenab canal | 228,039 | 5,100 | 2,934,324 |
| Upper Chenab canal | 208,339 | 2,840 | 1,444,922 |
| Haveli canal | 24,263 | — | 1,007,650 |

This rapid salinization has been brought about by the rise of the water-table. There is a sub-alluvial ridge running from Delhi to Shaphur. Since the general flow of groundwater goes from north-east to south-west, this ridge impedes free drainage and checks the flow of underground streams with the consequent rise in the water table. In fact, by 1960 in the 16 canal districts of the erstwhile Punjab more than 3 million acres, out of 12.5 million acres irrigated, have been seriously affected by salinity, of which 1.3 million acres have completely gone out of cultivation. Salinization is taking place at the threatening rate of 100,000 acres per year (ASGHAR, 1960).

*3. Central Eurasia*

The pattern of land use development of the Eurasian arid zone has been traced by KOVDA (1961). This region consists of a very large and extensive area alluvial in nature. This alluvium has been deposited in the Black Sea area by the rivers Danube, Dniester, Dnieper, Don, Donetz and Kuban; in the Caspian area by Volga, Ural, Kura, Terek and Emba; in the Aral-Caspian depression by Amu Darya, Sir Darya, Murghab and Zeravshan; in western Siberia by Ob and Irtysh; in western China by Turufan, Tarim, Ili and Chu, and in eastern China by Hwang Ho and Hwai Ho. This entire region is far away or cut off by the mountain ranges of the Alps, the Balkans, the Carpathians, the Caucasian Mountains, the Urals, the Himalayas, the Tien-shan Mountains, the Kunlun, the Stanovoi Peaks and the Yablonai Mountains which act as barriers against moisture-bearing winds from the oceans.

In the second half of the Quaternary period, the level of Eurasian arid zone plains was uplifted and these alluvial plains were desiccated. The climate of the central parts of the Eurasian continent, though perhaps becoming colder, stayed arid. In the interglacial periods, the glaciers melted, river flow was increased and led to strong erosion in the upper reaches of the rivers and to deposition of huge amounts of alluvial material in the middle reaches, particularly in the deltas. Primitive Man migrated to fresh fertile areas along river banks and into the deltas near the seas where the climate was more congenial. The earliest settlements occur in valleys of the Murghab and the Zeravshan; the Kura and the Arass; the Amu Darya, and the Sir Darya; the Hwang Ho, the Hwai Ho and the Yangtse. After the second half of the Quaternary period, when xeric conditions had gradually set in in the region of the alluvial plains, the ancient sandy areas provided more favourable habitats, inducing ancient man to shift there. Food gathering and stock breeding were now adopted as the main means of his subsistence. Soon, the scrub vegetation was denuded for fuel, building material and grazing by livestock, and in a relatively short time the vegetation-covered sandy areas gave place to shifting sands.

In the early Neolithic period (in the third and second Millennium B.C.), a primitive system was adopted in southern Siberia, the southern parts of the Dnieper and Dniester basins, the Doab between Volga and Kama rivers and the Kuban plains (GREKOV, 1946 and KISELEV, 1949). The implements in use in the middle Dnieper basin, from the lower Danube to the middle reaches of the Dnieper, were spade and hoe. Agriculture was combined with hunting and fishing. At that time domestic animals were not yet in use for farming. This pattern lasted from the Neolithic times to the early Bronze age. The crops cultivated were wheat, barley, rye and millet. Farming in the Black Sea area began in the second Millennium B.C. On the steppes, millets were cultivated then (PASSEK, 1949). Stock breeding developed slowly during the hoe-farming period, predominating in the dry eastern parts of the Russian Plain, while cultivation was practised in the more favourably situated tract in the north—the western Black Sea region (LIBEROV, 1952).

The change-over from hoe-farming to field-crop farming in eastern Europe in the non-irrigated sections, occurred through the invention of bronze and iron tools in the middle of the first Millennium B.C. By then, ancient man had already domesticated strong draught animals such as horses and oxen. Iron replaced bronze on a large scale in the seventh to second centuries B.C. The long-term fallow system was adopted, steppe grass being burnt to clear the land. The major crops were hard wheat and millet, the subsidiary crops cultivated were barley, hemp and flax, onions and garlic, viticulture and bee keeping. Agriculture flourished more in the Black Sea area than in the other regions of eastern Europe. The Scythians were the Antes Tribe, an important sedentary farming people (RYBAKOV, 1945).

The iron plough was very much improved in the second half of the first Millennium A.D. It was capable of cutting the soil horizontally and thus destroying the grass and weeds. In the Kiev area, the large mould board plough was in use in the eighth-ninth centuries, making possible a highly developed system of farming, stock breeding and vegetable culture. The three-field system of farming, consisting of fallow, winter crops and spring crops, predominated. Deep cultivation of fallow resulted in storage of soil moisture and destruction of weeds. This economy was upset by the Mongolian invasion of Genghis Khan. Several farming settlements were abandoned and nomadic life was introduced into this region, lasting for about 200 years.

The rivers Volga, Ural, Emba, Kura, Terek, Sir Darya, Amu Darya, Ili and Tarim are wholly confined within the Eurasian continent. Some of these giant rivers deposited alluvium in lakes and deltaic regions and were conducive to the formation of saliferous rocks and saline deposits. Primitive Man settled along these rivers at favourable spots and primitive agriculture began first in the flood plains and river deltas. This kind of farming was started in the ancient delta of the Amu River, between approximately 3000 B.C. to 2000 B.C. The land was either inundated or artificially flooded by a network of canals, from which water distribution was controlled by primitive check dams. An extensive irrigation network was developed in the second quarter of the first Millennium B.C., being improved and extended as population increased. The main irrigation canals on the Amu Darya consisted of Heri, Kurder and Baghdad, which were set up in the first century A.D. Besides the uncontrolled high floods which partly washed them away, the invasions of Arabs, Mongols and other tribes contributed to their final destruction (TOLSTOV, 1945).

Irrigated farming in the dry deltas of Murghab and Tedzhan began in the third and second Millennium B.C., when primitive canals were built. In the fifth and fourth century B.C., there existed extensive reservoirs equipped with sluices. Due to uneven deposits of silt in some parts of the deltas of Amu Darya, Sir Darya, Murghab and Kara the relief changed and, in consequence, some parts were flooded whereas others suffered from lack of water and salinization (ANDRIANOV, 1951).

Irrigation from shallow streams and rivers flowing down the hills was also practised in Central Asia by Uzbeks, Tadjiks, Turkmens and Kirighies. Their primitive irrigation works were destroyed by Mongol invaders in the 13th century and in subsequent invasions by Tamarlane. They introduced nomadism, the whole hydrology of the region was changed and the lands were salinized due to desiccation. Large patches of solonchaks developed in the ancient irrigated areas of Ferghana, Bukhara and Khorzem largely due to over-irrigation. The drainage canals had been built to leach out the salts.

In Siberia and Kazakhstan, irrigated farming was started in arid steppes in ancient times. These too were destroyed by Mongol conquerors, but were reconstructed in the eighteenth and nineteenth centuries A.D.

In Transcaucasia, reservoirs and canals were set up for irrigation in the ninth and eighth century B.C. from the rivers Kura, Arake and small mountain rivers. Flood water farming was actually started in ancient times.

At the end of the nineteenth century when the Central Asiatic territories were joined to Russia, about 3 million hectares were irrigated. In Caucasia, the irrigated area was 300,000 hectares. This was increased by another 2.7 million hectares by the end of 1917. Since then, new irrigation systems have been developed in Uzbekistan, Kazakhstan, Tadjikistan, Azerbaijan, Armenia, the Volga River basin, the Terek area, Southern Siberia, Southern Ukraine, Crimea and northern Caucasus. The irrigated areas increased from

4.08 million hectares in 1917 to 12 million hectares in 1958. The programme to augment irrigation of about 30 million hectares is in progress in the desert regions of the Volga, Don, Southern Ukraine, northern Crimea, Caspian, Turkmenia and Kara-kalpal.

Although irrigated farming in China was taken up in more ancient times, controlled irrigation was started by Wei 4,000 years ago from the river Hwang Ho. The first large scale irrigation system was constructed about 2,200 years ago on the river Ming Tsiang, a tributary of the Yangtse. Thereafter, similar irrigation projects were set up on the Tsintsvi and Khangtsai. These large scale irrigation works, measuring over 700 hectares each, made irrigation farming possible on about 2.2 million hectares (VORONIN, 1955; BUDARIN, 1957).

Since the water-table is fairly close to the surface in the northern Chinese Plain and in the deltas of the Yangtse, the Chchiang and other rivers, and since these ground waters are non-saline, irrigation by wells has been practised from time immemorial. The construction of shallow reservoirs for irrigating crops goes back four to five thousand years.

The arid lands of China lying in the Sinkiang and Inner Mongolia region are, however, mostly saline. In this area canal systems were started in the Han epoch in the Shansi province. Even today, they irrigate an area of 20,000 hectares in Sinkiang. According to an estimate, the average life of this system is between 100 and 150 years, so that new lines have to be and are in fact dug out at such intervals. Altogether in 1958 the irrigated area of China was 53.37 million hectares constituting 48% of the country's total of 112 million hectares of cropped land.

*4. North Africa (Morocco, Algeria and Tunisia)*

Four-fifths of Maghrib lies in the "desert" arid zone. The remaining one-fifth covers the north of Morocco, the Atlantic Coast with the slopes of the Middle Atlas facing the Atlantic, and the northern side of the Algerian and Tunisian Tell Atlas mountains from Algeria to Bizerta. The mean rainfall is more than 600 mm in a temperate climate. The rainy season coincides with the winter months. In the arid zone, the rainfall ranges from 150—200 mm to 300—400 mm. The soil is mainly chalky. It is largely composed of continental sediments often encrusted, and of recent alluvial deposits which, as they become saline, result in the formation of shotts and selekhas (solonchak). In the greater part of this region dry farming is practised. The irrigated area is meagre. The sources of irrigation are wells, springs, run-off from wadis and underground galleries (foggara). Small dams or rather diversion dikes are provided in the main mountain massifs and in the foot-hills to irrigate gardens. These are maintained on a communal basis. Canalised flooding is most widely distributed in the dry plains at the foot of mountains which receive adequate rain—i.e. the foot of the Atlas mountains in Morocco, the Nodna plains in Algeria, the plains and the mountains of the Sahara and the Kairouan plains in Tunisia.

Depois (1961) traced the history of development of land use of this region.

In ancient times agriculture was started on a dry farming basis. Irrigated agriculture was initiated in the second Millennium B.C. by Phoenicians and expanded by Carthagians in the north-eastern third of Tunisia (Camps, 1960; Gsell, 1913—28). But it was largely in the Roman period from the second century B.C. to the fifth century A.D. that agriculture and irrigation expanded very fast (Baradez, 1949). Wells were constructed, springs were tapped and flood water or run-off were diverted for irrigation. This was extended to almost all the wadis. The Foggara system was in use in Roman times. The influence of Arabs and Jews, Syrians and Egyptians on irrigated agriculture in Maghrib was less important than in Spain. There, between the eighth and eleventh centuries, the orientals constructed pipes for carrying water for irrigation over long distances.

The invasions of eastern nomads into Maghrib had disastrous consequences for land under cultivation. Their effect was most serious in Cyrenaica which became almost entirely a Bedouin region. The small dams were neglected and terraces were gullied. In modern times the position of agriculture began to be retrieved during the Turkish rule. But there was not much revival or expansion of irrigation except in very small areas adjacent to towns.

The whole of Maghrib came under the influence of France in the period from 1811 to 1962, when the economy was built up fast. In Algeria, the waters were diverted from the wadis in the semi-arid region by constructing stone dikes. In the arid zone 7 dams were built between 1860 and 1883 and another seven after 1920 and completed in 1939. These dams store surplus water from winter rains for use in the summer. After 1939, two more dams have been completed. The total irrigated area from these reservoirs is 170,000 hectares. Thousands of wells have been sunk for market gardening. The gross area which can be ultimately developed under irrigation is in the neighbourhood of 350,000 hectares. In Tunisia, after the Medjerda works are completed, the irrigated area will rise to 100,000 hectares. In Morocco, irrigation can be extended to over one million hectares.

*5. Sahara—Sahel Region*

Under this name, the intermediate zone between the extreme arid desert and the Sudan is understood. In this region, a series of alternating comparatively wet (pluvial) and dry (inter-pluvial) periods have occurred. The humid Neolithic phase, from about 5000 to 2,400 B.C., constituted the last Saharan biological and human population optimum, when vegetation, people and livestock were in equilibrium (Butzer, 1958). Thereafter, there has been little change for the past 4,000 years. In the humid phase, settled farming communities were established in the north and stock rearing was in vogue in the south. After this neolithic humid period, the climate deteriorated and vegetation was seriously exploited both by man and beasts for fuel, fodder and building materials. Cultivation was confined to small isolated places around water holes, where people adopted sedentary life.

Owing to paucity of archaeological and historical evidence, the account of the Sahara-Sahelian agriculture remains conjectural. The dearth of resources and a deteriorating environment forced the people to migrate to the Sahelian wadis, to develop a linear pattern of farming and a punctiform one in the Saharan oases (Malaurie, 1953). Some cultivation is practised in inter-dune areas where water is available for crops to grow and to complete their life cycle, otherwise most of the dune areas are barren. In the oases, sedentary farmers have been using wells for irrigation with counterpoised sweeps. Flood water farming is in vogue in beds of wadis, which are clayey or sandy loam. In the upper reaches of wadis, dams have been built to retain water and extend cultivation on the flooded area. Thus, the five types of sources of irrigation for farming are (a) Bour type of cultivation; (b) irrigation from wadis by run-off water or spring water; (c) percolation wells; (d) foggaras; and (e) artesian wells (Capot-Rey, 1953). These means of irrigation have been developed as early as the twelfth century A.D. In some of the oases the salt problem has become serious. In recent times, under French rule, new water resources have been developed in southern Algeria. A large number of artesian wells have been drilled with great success and foggaras have been re-aligned. In flood areas, unstable traditional embankments have been replaced by concrete dykes equipped with draining outlets and over-falls. The persistence of obsolete methods of irrigation has, however, led to development of salinity in several areas.

### 6. *Egyptian Valley*

In the Middle Palaeolithic period, man moved down into the Nile Valley some 10,000—20,000 years before the advent of the Neolithic agriculture. Thus, for 10,000 years man was a food gatherer and hunter. According to Butzer (1959), agriculture began to be practised in Egypt about 5,000 years ago. The valley was inundated annually into natural basins, the natural levee and the desert margins. Habitations existed on elevated spots. This was the period when man turned from food-gathering to food-producing, which is a crucial one in the history of land use in the Nile Valley. HAMDAN (1961) has termed it the "Cotechnic" period.

In the next period, Menes (3400 B.C.) took steps to control the river and probably started basin agriculture on its left bank which had a nine times larger cultivable area than the right one (WILLCOCKS & CRAIG, 1913). During the twelfth dynasty, the Fayum project as a regulator on the right bank was undertaken, wherefrom, in the years of deficient floods, water could be led back to the valley. These basins ranged in size from 1,000 to 40,000 feddans (1 feddan = 1.04 acres) and filled up successively from south to north. This kind of land use was essentially a winter crop system, as the floods arrived in Egypt in July and reached their peak in September, subsiding quickly until January. This type of cropping was essentially an extensive type of land use and the rise of salinity was not very serious. The limits of basin agriculture extended up to the northern littoral of the delta during the reign of the Pharaos. According to one estimate, 6 million acres were under cultivation in ancient Egypt. But in the extreme north, there was always a hard core of marshland and saline coastal lagoons, inhabited by fisherfolk (AUDEBEAU, 1924—25; TOUSSOUN, 1924). In the vicinity of Beraris, in the late and early Arab periods, about 1.5 million acres were converted into alkaline lands (HUME, 1924). HAMDAN (1961) states that there is strong evidence to indicate that neglect and apathy of the cultivating classes to maintain canal networks hindered regular drainage to the sea. The excess water was left over the land until a saline crust developed by evaporation. At the time of the French (Napoleon I) expedition, the actual cultivated area did not exceed 3.217 million acres. By and large, in the area where basin agriculture was practised, the problem of salinization never assumed any serious proportion in spite of the very arid climate, because filling up of the basins drained away the dangerous salts from the surface and subsurface (MOSSÉRI, 1922—23).

In 1820, steps were initiated to supply perennial irrigation and these were completed by 1890. The first summer canal, Ibrahimiya, was set up in 1873 and supplied irrigation to sugar-cane plantations of Ismail and a summer supply to Fayum. The Aswan dam was erected in 1902 to store the flood water for summer supply. Subsequently, barrages were constructed at Assyut (1902), Zifta (1903), Isna (1909) and Nago Hamadi (1930) and with the increasing water requirements the Aswan Dam had to be heightened twice: 1912 (2,500 million $m^3$) and 1933 (5,700 million $m^3$). The opening decades of this century may well be designated the era of dams and barrages (HAMDAN, 1961). Now four-fifths of the cultivated area has perennial irrigation. Basin irrigation is practised in Upper Egypt. In 1955—56, the perennially irrigated area was 5.698 million feddans and the gross irrigated area over 10 million feddans.

In the wake of perennial irrigation, the intensity of cropping adopted is high and fallowing of land has been given up. Thus there is a constant downward movement of water in the soil. This constant accumulation of water in the subsoil has persistently increased the water-table, leading to progressive salinization of the land. The Nile water contains some soluble salts and according to HUME (1924) every acre receives an estimated amount of 96 kg salts, mostly chlorides, under perennial irrigation. Thus Egyptian agriculture has become more than ever a constant struggle with salt. The salinization has been more in the sandy southern delta area than in the more clayey north and along canals rather than away from them because of heavy seepage from the canals themselves (WILLCOCKS & CRAIG, 1913). The problem of a raising water-table has been solved largely by the development of a network of drains and drainage canals after 1918, draining the endangered area into the northern lakes or the Mediterranean Sea. "Good land has an average salt content of about 0.3%, medium land 0.5%, while land rated poor has 0.8% and barren land may contain anything up to 25%. The gradual transition from good land in the south through medium to bad and barren land in the north is a fundamental feature" (HAMDAN, 1961).

### 7. *Trans-American Region*

#### a) North America

The history of land use has been traced by ARMILLAS (1961). Although man entered the New World some 25,000 years ago, settled life started about 13,000 years thereafter. Even 12,000 years before the current times, man was still in the predatory-foraging stage of economy. Game hunting was the principle means of deriving food, mainly east of the Rocky Mountains from the borders of present-day Canada to Central Mexico before the desiccation of the area. As game became scarce, this changed to diversified food gathering in stages, which remained in vogue till 6000 B.C. from Oregon to Mexico and from the foothills of the Rockies to the Pacific Coast. In the arid centre, several adjustments in agricultural economy occurred in relation to riverine, lacustrine or littoral habitats (WEDEL, 1956).

Xerothermic conditions became prevalent in the Great Basin, the South-West and Southern California about 5,000 B.C. and lasted up to 2000 B.C., when the climate turned cooler than hitherto, and moist. It was within this period, when "desert culture" developed there. In the N.W. Great Plains, hunting and food gathering persisted until the horse was introduced and re-domesticated in about 1500 A.D. In the Anasazi country or Colorado Plateau, which includes the San Juan river drainage-basin, north-eastern Arizona, south-eastern Utah, south-western Colorado and north-western New Mexico, there is no evidence of agriculture until the beginning of the Christian era. Cultivation of crops was taken up in successive stages from the first century to 900 A.D. In that period, crops taken up for cultivation were maize, *Cucurbita moschata*, *Phaseolus vulgaris* and cotton. In the subsequent period, crops taken up for cultivation were *P. lunatus*, *P. aconitifolius* (WEDEL, 1953).

In the regions of the Mesas, cultivation was restricted to dry farming. In the Jeddito valley, ancient sand dunes have been identified by stone lines which date back to the 13th century. This type of farming is still practised in the Hopi country. The moisture stored in the dunes is utilized for planting and maturing the crop. Maize and beans are the chief crops grown on sand dunes, particularly where the sub-soil is less pervious. The sand-dune fields situated against the escarpments of the mesas utilize the water seeping from the porous sand-stone cap rocks. Such sand-dune agriculture was in vogue on the eastern slopes of the San Francisco Peaks and in north central Arizona at about 875 A.D. This country was, however, converted into a desert by the eruption of the Sunset Crater, the lava of which covered a large tract between the peaks and the small Colorado river with a mantle of basaltic cinder. The stone lines indicate the practice of setting up wind-breaks in this region (COLTON, 1932).

In the Hopi region, flood water farming was practiced in the beginning of 1000 A.D. (Pueblo II). Possibly it was in vogue in the whole of the Anasazi country. The Hopi country provided a steep gradient which flattens at the foot of the slopes. The shallow fans had come into existence due to constant flooding from the upper reaches. In this period, the high terrain flood water was diverted through channels. In southern Arizona, such places have been called ak-chin (arroyo mouth) by the Papago Indians (HACK, 1942). The farmers spread this water by earthen spreaders (BRYAN, 1929). The Navajos, who learnt their agriculture from Pueblo Indians, built low flood terraces of large arroyos, or in the dry beds of intermittent streams where water spreads evenly without affecting the crops adversely. Sometimes they constructed temporary earthen diversion dams. Such check dams set up in prehistoric times are found in Mesa Verde (south-western Colorado) (GREGORY, 1916; BREW, 1946).

Canal irrigation was practiced in the Mesa Verde region during 1100 A.D. to 1300 A.D. (Pueblo III). The use of reservoirs and ditches dates back to that period in the San Juan area, although the irrigated area itself was small (BREW, 1946). The agriculture of the Pima Indians developed upon canal irrigation at the close of the seventeenth century in the upper reaches of Sonora and the San Miguel river and down the San Ignicio and Altas. Rich irrigated land was found by early Christian missionaries as far west as Cabora. Temporary wooden barrages were built as diversion dams and from these water was diverted into a network of conveyance and distribution canals. The Pima people, though largely dependent upon irrigated agriculture in abnormal and even in normal years, gathered beans of *Prosopis* species and cactus fruits and seeds *(Carnegica gigantea, Lemaireocereus thurbei, Opuntia fulgida, Opuntia echinocarpa)*. In abnormal years, the relative importance of farming, food-gathering and hunting as the basis of subsistence economy was reversed. This reversion was much more in evidence in the early fifteenth century, when prehistoric irrigation works dwindled (CASTETTER & BELL, 1942).

Sedentary farming in southern Arizona dates back to the early Christian era when the Hohokem culture, which evolved out of the food-gathering Cochise Desert culture, established itself. Then these people took to flood water farming and in successive generations by 800 A.D. had developed a local canal system from the Gile river. This was extended and by the fourteenth century it flourished 10 km south of the Salt river (HAURY, 1945). During the span of 500 years from 800 A.D. to 1300 A.D., these canal systems were the cause of high salinity. The canals were silted and frequently reexcavated (HAURY, 1936). A similar type of canal irrigation was established also in prehistoric times in the Salt River valley (HAURY, 1956).

The agriculture of the Lower Pima people also was developed upon canal irrigation. These canals had as their sources of supply the middle and upper sections of the Yaqui river and the mid-course of the Sonora River. The Sonora desert's Papago section of the Pima people, due to very arid conditions (rainfall below 150 mm) west of the Santa Cruz River, subsisted more on gathering wild plants and on hunting than on agriculture (CASTETTER & BELL, 1942). The Yuman tribes of the lower Colorado river valley depended on flood-farming. This system of farming was likewise the source of subsistence of the pre-Spanish tribes of Southern Sonora and the Sinola Coastal Plains in North-Western Mexico. There, however, dependable summer floods and less dependable winter floods of light intensity rendered possible the taking of 3 crops a year in succession. The drainage of this region originated from rivers in the high western Sierra Mandre (CASTETTER & BELL, 1951).

The history of land use in the arid regions of the United States during the post-Columbian period has been reviewed by LOGAN (1961a). The white man began to move into the vast arid and semi-arid regions between the Rocky Mountains and the ranges bordering the Pacific Ocean in the middle of the nineteenth century. Most of them trekked in search of beaver pelts. Potential settlers moved in late in the 1860's. Most of the land was occupied for its good conditions for cattle rearing and this second half of the nineteenth century has been termed as the era of the "open ranges". Several million acres were set aside for forestry, but by and large ranchers had their way. In 1916, the stock-raising Homestead Act was passed but it had only small effect. Indiscriminate grazing brought about a deterioration of the vegetation, diminishing the value of the area as range, which was always overstocked and lacking any effective grassland management. In 1934, however, the Taylor Grazing Act stabilized the situation and the potential of cattle and sheep production rapidly increased.

Although irrigation was practised by various American Indian tribes, the better Spanish techniques were introduced by the first missions in southern Arizona. Anglo-Saxons, who moved into coastal California, with its Mediterranean climate, learnt the art from the Spanish missionaries. This came about in the 1840's. The Mormons brought additional techniques, when they moved into the Salt Lake area in Utah, and because of their theocratic culture they established a communal irrigation system, diverting the water from the Wasatch Range. This community spread into Idaho and southern Nevada. "Later expansion carried the Mormon communities across the plateaux of south-eastern Utah and Arizona into parts of New Mexico and the adjacent state of Chihuahua in Mexico". Several companies began to operate in order to control the streams and to dig canals over long distances. They were either mutual water associations or water companies for profit by selling water to the farmers for their irrigable land.

Large-scale development of irrigation started at the turn of the nineteenth century, and it was mainly for this reason that these decades have been termed the "Reclamation Era". Several multipurpose dams were set up on the major streams of the West. The chief ones are the Grand Coulee Dam in Washington, the Boulder or Hoover Dam on the lower Colorado River, and the Shasta and Friant Dams of the California Central Valley Projects. As a consequence, the whole agriculture was stabilized in this region. Now cropping could be and has been oriented to production of irrigated crops, fruit, sugar-beet, vegetables—potato, tomato, peas and lettuce, cotton, grapes and dates. The irrigated areas in the Imperial Valley and a part of Coachella Valley carry crops of lucerne which are primarily raised for hay, and irrigated pasture. The animals are fattened here and shipped to Los Angeles market. The hay is exported to dairy farming areas for breeding stocks. In the Nevada region, where the climate is cooler, not only alfalfa but also temperate fruit are grown. Here too sheep and cattle are raised on irrigated pastures and are herded into high mountain areas during the summer season when good grazing is available there.

Large reservations have been provided for the Indians of the American deserts. The Papago tribe which has lived by hunting and food-gathering, is taking to irrigated agriculture and some of them work as labourers in the adjacent irrigated lands along the Salt and Cila Rivers. Apache tribesmen have taken to cattle industry.

They let the cattle graze in the mountains during the summer and winter them in the lowlands. Navajos have not changed their subsistence farming. Recently, some of them have settled along the River Colorado in areas which were formerly subject to flooding, and have taken to mechanised agriculture. The Pueblo Indians of the Rio Grande Valley continue to practice agriculture. Irrigation has been provided in some areas in their reservations, and besides maize, beans, squash and melon, they now raise deciduous fruit introduced by the Europeans. Hoe-farming is still prevalent, although modernisation of agriculture is proceeding apace.

b) Meso-America

In Mexico the highlands facing the Pacific, the interior of valleys, and the coastal areas, i.e. great parts of central, western and southwestern Mexico, suffer from seasonal droughts. Since these unfavourable conditions coincide partly with the summer season, the evapo-transpiration and run-off losses are high. Here, the rain-farming-fallowing system was in vogue prior to the Spanish conquest, but thereafter, the interest in irrigated farming was awakened. In the mid-fifteenth century, during the reign of Nezahualcoyotl, a reclamation project was undertaken in Central Mexico to irrigate the gardens of the king as well as several communities. It is likely that other works were also carried out in this period in the Mexican Valley (Wolf & Palerm, 1955).Prior to that, a minor irrigation project was set up in a small valley between the Maravilla and Altatonga hills in the period A.D. 800—1200 (Armillas, Palerm & Wolf, 1956). Another area where irrigation existed in medieval times was the Tetzcocan piedment. In the Nexapa valley of the Coatlalpan and the adjacent province of Amilpas, irrigated farming was in vogue at the time of the Spanish conquest. However, the techniques of reclamation and conservation, principally hydraulic works, were technologically less developed and were limited to small-scale enterprises that could be accomplished with the resources of a single community or a small cluster of local communities, politically integrated in a small principality. In the Nexapa River valley in southern Pueblo, central control of water use was exercised. The largest and most centralized area of intensive agriculture was the Valley of Mexico, because of the canal irrigation, the Chinampas in shallow water ("floating gardens") and of the terracing of land on the slopes. This type of land use developed during the period 200 B.C. and 700 A.D. (Sauer, 1948).

(c) South America

Intensive agriculture by flood farming in the coastal region of Peru dates back to the first Millennium B.C. There tapping of rivers started already before the middle of this Millennium. In the latter half, large-scale irrigation works were constructed. By the late

Gallinazo period, which corresponds to the early centuries of the Christian era, the Viru Valley canal irrigated 9,000 hectares which formed 40% of the area. Crops of cotton and sugar-cane were cultivated. The area continued to flourish from 500 B.C. to 500 A.D. In the subsequent period, the lay-out of the canal network changed somewhat but the acreage remained the same. A similar network of canals existed in the valley oasis of Moche and Chicama to the north (WILLEY, 1953).

From about 800 A.D. to the Spanish conquest, the trend was to move away from the previously irrigated lands to the coastal dune belt in order to utilize the groundwater sources there by means of the so-called Pukio basins. These basins measured 100 × 50 m to 30 × 30 m sunk about 1 m below ground and separated by ridges 2 to 4 m high. From 1000 A.D. habitations based on basin irrigation clustered in the upper narrow valley and along the dune zone. The shifting of agricultural settlements from irrigated sections to the coastal dune belts was necessitated by soil salinization and excessive rising of the water table due to prolonged use of irrigation, since there was no provision for drainage facilities. Thus, the water resources were indiscriminately managed, a fact resulting in salinization in the districts of Piora and Pisco (REPARAZ, 1958).

The Inca empire extended from northern Ecuador to south central Chile and north western Argentine in the period from 1438 A.D. to 1527 A.D. The Quechua Indians preferred to settle at an altitude intermediate between the puna pastures and the cultivated land of the valleys and canyons. In these areas they practised agriculture by means of altitudinal zoning of land use, land reclamation and soil conservation by terracing and irrigation. Maize was the main crop on which these people subsisted and potato was the subsidiary crop grown in an altitude between 3,000—3,500 m. Micro-thermic frost tolerating crops such as quinoa *(Chenopodium quinoa)*, little potato, and oca *(Oxalis cremetu)* were cultivated at heights ranging from 3,500 to 4,250 m. The higher punas were left for pasturing. For the production of these crops irrigation was necessary because of the long dry season and the low efficiency of rain water in consequence of the very high run-off losses. The valleys are deep and narrow, and, therefore, irrigated areas were small. The Quechua Indians built stone faced irrigated terraces. The irrigation canals often ran for miles along the side of a valley to irrigate a comparatively small terraced area (ROWE, 1946). They used this method in about the first Millennium of the Christian era (BENNETT, 1946a).

In the Chilean desert the early inhabitants depended upon marine subsistence. The agricultural expansion was limited by the small amount of water available for irrigation and the small size of valley bottoms. There is no evidence to indicate that climatic conditions have changed during the past 4000 years (WILLEY, 1953). Along the coast from Arica and Coquimbo marine fishing and plant cultivation subsisted side by side. In the northern tracts, cultivation started in the early centuries of the Christian era. The crops domesticated in this period were maize, cotton and gourd as a result of the expansion of the Tiahuanaco culture from the central Andes to this tract (BIRD, 1943).

In the latter half of the first Millennium of the Christian era, agriculture began to develop along the river Loa in the Calama oasis in the Atacama hinterland and the water was simultaneously conveyed by aqueducts. Livestock industry flourished, meat and various raw materials being obtained from llamas and alpacas. These animals were grazed in the Punas, as the high bleak plateaus in the Andes are called (BENNET, 1946 (a); 1946 (b) and BENNET & BIRD, 1949).

At the same time intensive agriculture and grazing was in vogue in north western Argentine. The inhabitants adopted a pattern of agriculture which had evolved in the Central Andes, and in part from regions where the Tiahuanco culture developed (WILLEY, 1946). At the time of the Spanish exploration, river flood plain farming was already practised in Tierra Conade along the Dulce and Salado rivers of Santiago del Estero. Beyond, the Pampas and Patagonia were roamed by nomadic hunting bands (APARICIO, 1946).

### *8. Arid Africa South of the Equator*

#### a) South Africa

The arid regions of Africa south of the equator consist mainly of two parts, namely, South Africa and South-West Africa. The history of land use in the former has been described in detail by TALBOT (1961). In two-thirds of the basin of South Africa the rainfall is less than 20 inches, most of which, in 85% of the area, comes during the summer season when its effectiveness is limited due to high evapo-transpiration. The arid region, south of the Orange River consists of the Little Karoo (between the Langebergen—Outeniqua Range and Swartbergen), the Great Karoo (between the Swartbergen and the Nieuweveld—Sneeuwbergen escarpment) and the Upper Karoo (north of the escarpment), and the Down Karoo (between Cedarbergen Range and the Bokkeveld Range on the west and the Roggenveld escarpment on the east). The vegetation of the Karoo is desert-scrub.

The Bushmen, survivors of the Mesolithic hunting people, still inhabit the Karoo and the Kalahari. They are excellent hunters, in spite of their primitive tools, and collect also a variety of edible roots, fruits, herbs, insects, reptiles, eggs, honey and other veldkos (wild food) (TOBIAS, 1964 and STORY, 1964). The Hottentots moved across the lower Orange River with their cattle and sheep in the fourteenth century (ROMYN, 1924). Although mainly pastoral nomads, they supplemented their living by hunting, fishing and food gathering.

Europeans occupied the first settlements in the seventeenth century, but did not move into the hinterland till 1730. The natives brought the slaughter stock to the coast, where the settlers bartered it with the sea-farers at the re-victualling station. These Dutch outposts were already developed in 1652, but as the population increased they set up farms on well-water piedmont sites within 40—50 miles of the Table Bay.

Slaughter stock was in great demand particularly during the Anglo-French war period in 1744, when several warships and troopships touched Table Bay, and this opportunity induced pioneer colonists to take to stock rearing in the arid inland areas, as the demand exceeded the supply available from trade with the Hottentots. Butter and fresh meat industry flourished for over 200 years until the Suez Canal was opened and the steamship changed its sea route between the Atlantic and the East (NEUMARK, 1957). Tender mutton was available from the hardy fat-tailed sheep which grazed in 10,000 sq. miles of the Karoo (THUNBERG, 1795). The graziers steadily spread to the north, north-west and eastern sectors of the arid lands along the streams, and set up habitations close to water points where their advance was stopped by Bushmen and Bantus. Recurrent droughts and locusts sometimes destroyed the vegetation so much that the graziers had to seek fresh pastures. Mobility was the key to survival. Until the nineteenth century, the products in demand were Cape sheep wool, mutton, slaughter stock, the tail-fat, soap, candles and ostrich eggs. Soap was manufactured from the ash of Karoo bushes, particularly *Salsola aphylla*. A sudden change came in 1867, when diamonds were first discovered, and thus started "a revolution that would transform the basis of the country's economy within a generation".

The pattern of land use in the nineteenth century and early twentieth century is intimately connected with the domestication of Merino and Karakul sheep, Angora goat and the domestication of the wild ostrich for its feathers. The demand for these primary products continued to increase till the onset of World War I. Then, the ostrich feather industry declined, but thereafter it steadily picked up. The Angora goat population, rapidly increased after its introduction in 1836, was, however, severely reduced by the draughts of 1915—1916 and 1918—1919. Besides, demand for mohair decreased due to world competition, and since then Merinos occupied the place formerly held by Angoras.

Adequate feeding with protein is essential in order to obtain quality feathers from ostrichs. Lucerne proved to supply an excellent feed for the delicate young birds (NOBLE, 1875 and EVANS,

1911). "Consequently the Karoo and particularly the Little Karoo—which could most readily provide adequate water and extensive areas of irrigable land for lucerne—were the areas that benefited most from a form of land use, that, in its hey-day, was incredibly remunerative." Lucerne in the Little Karoo is now grazed by Frisian cattle and also in the Great Karoo lucerne land provides fodder and hay.

With the discovery of diamonds, and later on of gold, means of communications were rapidly developed by the colonial Government. This opened up the hinterland and also brought in its wake a large number of immigrants who engaged in various associated activities. Pedigree cattle were introduced to meet the growing demand for milk, milk products and meat. Numerous ranches under irrigation were developed in the Hardveld. In the Sandveld of the Kalahari the ground water resources began to be tapped in 1903 and from then on these formerly almost waterless areas were developed for livestock industry. After the Second World War, the Union Government tried to set up settlements in the Kalahari, particularly along the dry courses of the Molopo and Kuruman Rivers and in the Bushveld of the northern and eastern Transvaal. This Bushveld provided grasses of high nutritive value such as various species of *Digitaria, Eragrostis, Panicum* and *Aristida* (Acocks, 1964 and Davidson, 1964). Also fodder trees play an important part and consist of *Acacia heteracantha, A. nigrescens, Boscia albitrunca, Grewia flava, Combretum apiculatum,* and *Copaifera mopane.* Grazing in these areas was, however, restricted by Tsetse flies, the transmitters of nagana or trypanosomiasis, whereas the remaining areas of Bushveld did not attract many white settlers.

"Irrigation exerted only a very limited and localized influence on the development of the arid regions of the Union; its application was stringently curtailed by relief, by soil and by climate." The two main rivers Orange and Vaal have a steep gradient and traverse narrow valleys. The surrounding area is very much higher than the river surface. The sites suitable for storage dams and weirs are few and the irrigable land is extremely limited. Also, the groundwater availability is limited, and these waters have a high mineral content and their salinity renders them unsuitable even for stock water (Levinkind, 1940; Stander, 1954). Springs and streams provide some irrigation, particularly the Oliphants River in the Little Karoo and some tributaries of the Great Fish River (Von Reenen, 1925). In the Upper Karoo and in the Great Karoo, water storage works were started in 1864, particularly the one storing the waters of the Sak River. In spite of this, the amount of water available for irrigation was limited. Sometimes flood basin irrigation was possible, and thus cultivation of wheat on a few thousand acres, until the salinization of the land forced it out again (Union of South Africa

Report, 1912). At Olivenhouts Drift, a weir was established on the Orange river and the canal therefrom irrigated a large area of alluvial land along the river's north bank (Union of South Africa Report, 1914). During the period from 1918 to 1925 permanent reservoirs were set up in the former ostrich-farming and flood irrigation areas in the Little Karoo and also on the Sunday and Great Fish Rivers. These dams rapidly silted due to overgrazing in the catchment areas. The Hartebeestpoort Dam was completed in 1924 and so were also several other reservoirs in the period 1929—1943. The Vaalbank Dam provides irrigation to 80,000 acres in the Hart Valley and wheat, groundnuts and lucerne hay are produced there for sale. Another but smaller dam on Riet River provides irrigation for 15,000 acres. This general trend towards irrigated farming indicates intensification of land use everywhere.

An interesting enterprise in the direction of industrialization of a large rather arid area in the Limpopo region has been started by a mining concern with a view to benefitting the native population. There, the cultivation of various arid plant raw materials has been started in the late 1950's as advised by the Israeli experts H. and E. Boyko, and some of the cultivated species are now already in the state of industrial processing. New aspects for the future of arid land use may stem from these beginnings.

b) South-West Africa

The arid land of South West Africa "extends from the Bechuanaland border westwards along the twenty-third parallel to the intersection with the meridian of 17° East and from that point northwestwards parallel to and about 100 miles distant from the coastline." The history of land use has been traced by Logan (1961b). Nomadic bushmen populated this region in ancient times and were, in general, hunters and food gatherers. Recurrent droughts placed a limit on the growth of population and settled mode of life. White immigrants started settling there in 1890, when the German Government initiated a large-scale colonisation programme, which proceeded apace for the following 25 years.

The Germans had to take to stock rearing because of insufficient water for irrigation or inadequate rainfall. "Agriculture was limited to narrow patches of the flood plain along the lower course of the Swakop River, where the raising of a variety of crops on a small scale was carried out under irrigation." The European breeds of cattle could not adapt themselves to the dry environment as they were acclimatized mainly to moist and cool regions. Therefore the white settlers took to sheep and goats. Karakul sheep introduced in 1907 proved very successful. After World War II rapid immigration has increased the demand for more farms and quite a large proportion of the arable and grazing areas is now under cultivation in pastoral industry.

*9. Australia*

The history of land use in Australia has been traced by WADHAM (1961). According to this author the population of the whole of Australia numbered not more than 300,000 before the arrival of the colonizing British fleet in 1788. These people inhabited areas which had a more moist and equable climate along the northern, eastern and south-eastern coasts. They hunted, fished in streams, extracted roots with wooden sticks and gathered all types of other food. They lived a nomadic life, but did not in any way impair the natural vegetation or cause degradation of soils, for the simple reason that their number was exceedingly small in relation to the vast areas and their tools were very primitive (TINDALE, 1959).

The first immigrants settled near Sydney. The inward march into the dry interior did not begin till about 1820, when production of fine wool was taken up in the southern coastal areas. The pastoralists advanced along the streams and river courses into the interior. "Gradually the occupied areas were extended. Apart from the waterless regions, the only sheepless parts were those where the vegetation was too thick, or the terrain too rough, or the places where sheep failed to thrive owing to maladies which were then obscure." This extension went hand in hand with the development of artesian wells which made feasible the exploitation of many areas of arid zones, especially in Queensland. About 80,000 bores were drilled there for this purpose (HILLS 1953; Rural Reconstruction Commission Report, 1945).

The rainfall occurrence is very undetermined and, in addition, seasonal in character, which makes cropping uncertain and risky. Therefore, there is very little of cropped area in the extreme arid and arid regions which cover by far the greater part of this continent. In the semi-arid regions a system of mixed farming has been evolved. It consists mainly of mixed pastures (e.g. *Phalaris tuberosa* and various strains of subterranean clover both introduced from mediterranean countries) followed by wheat. The ley is given top dressing of phosphate in order to increase its production and to augment the nitrogen status of the soil. The area of wheat growing is limited by the 10 inch Isohyet. In Western Australia, Woomera rye grass substitutes for the used ecotypes of *Phalaris tuberosa* endemic to mediterranean climate. The average yield of wheat hardly exceeded 10 bushels in the period 1890 to 1905 (CROOKES, 1917). With the introduction of better adapted wheat varieties and application of phosphate, to cover phosphorus deficiency in the soil, and the introduction of tractor ploughing in New South Wales and Queensland the average yield increased by 10—25% and still more by the ley-farming system. Small irrigation schemes were first taken up in Victoria and then in other states. Victoria has already harnessed all its water resources which can be tapped for irrigation. In New South Wales, most of the rivers flowing westwards have small catchment areas and the flow from year to year is very variable. In Queensland, small irrigation schemes have been executed. The Burdekin project is still in its early stages. In Western Australia the water of most of the streams south of the Tropic have been impounded. South Australia has a small irrigated acreage, but there is the possibility of utilizing the Murray River water near its mouth, when salinity will have decreased. The Snowy Mountain scheme will further increase the irrigated area. The total irrigated area in 1958 was 1.75 million acres of which two-thirds are pasture lands and the rest is for orchards, vineyards, market gardening, sugarcane, linseed and tobacco. Due to mismanagement of irrigation, however, salinity has developed in some of the areas, and to overcome this menace drainage is being provided now in such areas in order to leach out the salts which have accumulated in the top horizon due to waterlogging, particularly in the Murrumbidgee area.

### c) Geology as Causal Factor of Salt-affected Arid Lands (with Indo-Pakistan as an Example)

Among the many geophysical factors responsible for the accumulation of salts in the soil, the geological and orographical conditions in space and time belong to the foremost ones. Their complexity is, of course, specific for each catchment area and to describe them in this survey for all regions would transgress by far the size of one single volume.

In the following, a short sketch of the geology of one of the smaller areas of salt-affected arid lands may, therefore, serve as an example for the complexity of these factors and their close correlation with the salinity problem as a whole.

The areas selected for this purpose are those of the arid lands in the subcontinent of Indo-Pakistan.

The four main parts of the Indian Peninsula, dealt with in this connection are:

1) The Indo-Gangetic Plains
2) The coastal regions
3) The Table lands
4) Pakistan.

#### *1. The Indo-Gangetic Plains*

Prior to the rise of the Himalayan ranges, the present Indo-Gangetic plains were submerged under the sea. The major part of Western Rajasthan up to the Aravallis and the south-eastern Punjab was part of the Tethys sea. The climate of this region was temperate and the vegetation was adapted to marine conditions. With the uplift of the Himalayas, the foredeep began to fill up with the detritus from the Himalayas. Some of the sea-salts were precipitated in the process of deposition of the alluvial and other materials and by and by absorbed by their clay particles. The salts were precipitated in the order of their charge. Sodium salts were the last to be precipitated or absorbed by the soil fractions. The process of deposition of the detritus occurred in the earlier estuaries in that geological time and is still in progress in the estuaries of the Bay of Bengal and the deltas of the Indus river. The alluvium deposits in some areas are over 4000 feet deep and in others as shallow as a few hundred feet. The exchangeable sodium is not present uniformly throughout the depth of the alluvium, but its presence is associated with the habitat factors and the climate of the recent past. Where conditions of waterlogging have prevailed, due to stagnation of water in various periods, there, as a matter of course, the salts have accumulated more than in periods when drainage was free and unhampered. In the north-western sections of these plains the conditions have been more arid or less humid respectively than in the central or eastern sections. The rate of evaporation of the sea brine being much higher in the arid parts than in the humid sections, the salts entrapped in the alluvium, lying mainly in the more arid parts, have a high concentration. On the contrary, the streams were more free of salt at their sources than at the mouth of the deltas formed by them. All these factors have combined to bring about a specific regional distribution of salts. In the central Uttar Pradesh, the salt accumulation in the profile is much higher than either in the Terai, eastern or western regions.

The rock material composing the Himalayan ranges belongs principally to the groups of shales, phyllites, calcareous dolomite and limestone. All these contain a high content of lime, and this lime, together with the other disintegrated and decomposed rock materials, has accumulated and interacted with the brine of the sea. In most areas, therefore, the ratio of Ca to Na is high. For instance, in the Eastern Uttar Pradesh and North Bihar, some of the soils have over 30% lime in their profile. These are highly calcareous soils. In spite of the low content of exchangeable sodium their pH is, therefore, high and even a slight increase in the exchangeable sodium value shows effects of salinity.

The foredeep of the Tethys Sea filled up comparatively slower than that of the Ganges basin. The conditions of aridity evolved in the former region were more severe than in the regions now comprising Uttar Pradesh, Bihar and West Bengal. The out-flow of the flood from the Himalayan ranges through streams has been relatively small. Therefore, lagoons were formed and in them magnesium and calcium salts were the first to precipitate. That accounts for the gypsum deposits in this region and the high salinity in such areas as Pachbhadra, Sambhar, Lunaksar, Didwana, etc. In these shallow lakes, the gypsiferous material appears at a depth of 100 to 250 feet, whereas the profile above it is largely saturated with sodium salt. These lagoons and low lying areas contain up to 6% sodium chloride and the brine is at present used to manufacture sodium salts by evaporation (KRISHNAN, 1952).

In the Gangetic plains, the underground flow of water is much more unrestricted and there is free drainage through the underground streams which are formed by annual recharge from precipitation. The deeper profiles are, therefore, relatively free of salts. On the other hand, in Western Rajasthan, the amount of underground flow and rate of recharge of underground water is limited for climatic reasons. Besides, there is an underground ridge, which separates the northern from the southern section, stopping the flow of underground water from the north. Consequently, the salts accumulated in the entire profile up to the parent rock. This concentration of sodium salts is in some of the profiles very high and the underground waters are usually brackish. In Jodhpur, Bikaner and Jaisalmer, the chloride and sulphate content is more than 1000 p.p.m. The variation in chloride content ranges from 200 to 1500 p.p.m. The ratio of Cl: $SO_4$ is 4 : 1. This high salinity is sometimes indicated by geographical names. Thus the river Luni, which has its source in the Aravallis, becomes saline, after running a course of about 250 miles, and has been termed Luni meaning the saltish stream.

## 2. *The Coastal Regions*

In the coastal regions, where tidal action is predominant, salt-affected areas develop as the sea-salt is continuously deposited on the beaches. Thus, in the Rann of Cutch, there are extensive marshes which have a high level of salinity. They have developed because of the tidal action and fanning out of the flood water from seasonal rivers and streams that discharge their effluent into the sea. The Khar or Kshar lands on the Western Coast in the Gujarat and Maharashtra State are good examples of such a development because of tidal action. The "A" horizon in the typical profile contains sodium carbonate and is of brown colour. The "B" horizon is more silty and sandy than clayey in texture and has a black colour. The total salt content varies from 3.04 to 3.60% with a pH between 8.6 and 8.76 (BASU & TAGARE, 1954). In the coastal Konkan region of Maharashtra and the coastal tract of Kerala, the soils alongside the estuaries of the streams have a high salinity. A similar process is in progress in West Bengal, Orissa, Andhra Pradesh and Madras states, which form the coastal eastern sea-board of India. "The coastal plain of Coromandel and the Circars is the typical uplifted plain of marine erosion with the usual inland-facing questas, isolated granite or gneiss hills, old offshore islands and coastal lagoons. At the mouths of great rivers these features are marked by deltaic formations" (SPATE, 1960).

The wind-blown salts from the sea do not travel very far. Over short distances, they accumulate in depressions, but in the arid region, the coastal fine grain particles carry some of the salt inland. These salts stem from the wind-blown sea spray which deposit on the alluvial fans of the deltas. HOLLAND & CHRISTE (1909) have advanced the theory of wind-blown replenishment of salt in the desert lakes. This is based upon the fact that the analysis of first rains showed amounts of chloride greatly in excess of the quantities indicated by the U.S. Isochlor lines at similar distances from the coast. Later studies have not substantiated this theory. It is principally the very poor drainage, low precipitation and high rate of evaporation which account for the salinity in the desert tracts of Rajasthan.

*3. The Tablelands of the Indian Peninsula*

In these regions, extensive volcanic activity with series of eruptions threw out great amounts of lava through wide fissures at geological intervals. Thus, a basaltic rock formation was built, ranging from 2000 to 10,000 feet in thickness in almost horizontal layers. This formation is known as the Deccan trap, because of the usual step-like aspect of weathered hills of basaltic rocks. Basalt contains feldspar, augite and magnetite minerals and the weathering of this material led to the development of the Black Cotton soils.

The southern plateau section consists of the Dharwar and the Cuddapah systems. The Cuddapah rocks are within the big belt on the east of Deccan, between the Krishna and Pennar rivers and in the valley of the upper Mahanadi. The Dharwar system consists of "series of narrow belts, the trough of tight pack synclines in the Mysore—Dharwar—Bellary area; flanking the Chota Nagpur Plateau on the N. and S. and in patches westwards as far as Nagpur city and in the Aravallis". There are very different kinds of rocks to be found here, consisting of clastic sediments, chemically precipitated rocks, volcanic and plutonic rocks. They all show a high degree of metamorphosis (SPATE, 1960). The crystalline granite and gneiss rocks vary in texture from crystalline to homogenous fine grained ones. The primary minerals found in these rocks are feldspar, quartz, biotite, and hornblende, while the secondary ones are chlorides, tourmaline and kaolin. The soils derived from these rocks are the extensive Red soils of India.

The Vindhyan system lies over the Cuddapahs in the lowest part of the Krishna—Pennar trough. The chief belt extends along the northern flanks of the Peninsula from Chambal to the Son rivers.

Patches of lava of Lower Vindhyan age are found around Jodhpur. The Vindhyan system consists of shales, slates, quartz and limestones. In the Lower Vindhyan marine shales, limestone and sandstones are the main components of the geological profiles (KANITKAR, 1944).

Wherever the rocks contain orthoclase feldspar, in peninsular India, the soils derived from them contain sodium salts, principally sodium chloride. Most of the Red soils of Peninsular India have neutral to slightly acidic reaction except in small pockets where the reaction is alkaline. There salinity is caused by irrigation which brings about a rise of the water table and, consequently, an accumulation of salts on the soil surface. As mentioned above, the basalt of the Deccan Trap contains feldspar, and in the Black soils, we find, therefore, sodium salts and all these factors together lead to the fact that even a slight accumulation of salt in the soil profile causes already characteristics of salinity, and affects the vegetation adversely.

*4. Pakistan*

From the Makran coast of Baluchistan through the mountainous frontier tracts of Sind and the North West Frontier Province, to Kashmir and further along the Himalayan Foot-hills to the Brahmaputra gorge, we are confronted with a great development of tertiary rocks in Cutch and north Gujarat. In the Sind-Punjab region, which at present forms the Indus basin, the geological strata change from South to North with regard to their origin from a marine character to brackish water and fresh water. In the Siwalik region the water is already almost sweet. Thus the Indus Plain of west Pakistan is partly of marine origin and partly of a brackish water origin. Most of the canal-irrigated area is of this marine-brackish water origin (KRISHNAN, 1960), and naturally the solid profiles contain, therefore, larger or smaller amounts of sodium and magnesium salts. The alluvial material brought down by the Indus and its tributaries brings less salt with it, but the minerals included are shales and phyllites, and these contributions have enriched the soils to a considerable degree. The groundwater table in this region ranges from 1 m to 50 m but with the supply of canal water and mismanagement of land, it has risen fast in the irrigated areas, and brought up the salts to the surface. In consequence, extensive salinity has developed in these tracts. In others, however, where irrigation is practised from the wells, the water table remains stagnant and the upward movement of salt is much restricted.

## Nomenclature and Classification

It is difficult to obtain a uniform nomenclature, the main reason being that conditions differ widely from region to region and in addition, different points of view are used for the various classifications. Some of these may be reviewed here as a basis for comparison.

1. U.S.A.

HILGRAD (1906) recognised two classes of salty lands and named them "white" and "black" alkali. The former denotes soils which contain neutral salts of sodium chloride and sodium sulphate since these salts form a white efflorescence on the surface when the soil is dry and uncultivated. In "black" alkali soil, sodium carbonate is the predominant component. This salt reacts with organic matter and from this reaction stems the black appearance of the soil. Such soils are also characterized by their high proportion of exchangeable sodium. The U.S.A. Soil Salinity Laboratory in Riverside (1954) subdivided alkali soils into three classes, namely, (i) Non-saline alkali soils; (ii) Saline-alkali soils; and (iii) Saline soils. In non-saline alkali soils, the exchangeable sodium percentage is higher than 15, and the conductivity of the saturation extract is less than 4 millimhos per centimeter at 25 °C. The pH of the saturated soil usually exceeds 8.5. In the saline-alkali soil, the exchangeable sodium percentage exceeds 15, the pH of saturated soil extract is less than 8.5 but conductivity exceeds 4 millimhos/cm at 25 °C. In saline soils, the conductivity of the soil extract is higher than 4 millimhos/cm at 25 °C, and the exchangeable sodium percentage is less than 15. Its pH is usually less than 8.5 (HAYWARD, 1956). In 1958 the term alkali was replaced by "sodic". "According to the morphological classification of soils, the saline soils are solonchaks and the sodic soils are generally solonetz". In sodic soils, the exchangeable sodium content is high enough to affect plant growth adversely (BERNSTEIN, 1961).

According to more recent investigations and experiments, this general statement has, however, to be modified. As far as the influence of Na in soil and water on plant growing is concerned, it has been shown that the right chemical balance of the ionic environment is of major importance and to a lesser degree only the absolute amount of Na (HEIMANN, 1959), whereas the experiments of BOYKO and BOYKO (1959, 1962) have proved that for these same reasons even irrigation with sea-water of oceanic concentration with its very high NaCl-content is possible for many economic plants under certain physical soil conditions i.e. soils of great permeability with a negligible percentage of clay particles (e.g. sand).

2. Russia

GEDROIZ (1917) classified the saline alkali soils into three classes, namely, (i) Solonchak, (ii) Solonetz; and (iii) Solod. He based this classification on the degree of salinity and the leaching of sodium from the soil. Solonchaks are saline soils. In the solonetz type, a high level of exchangeable sodium modifies the soil profile appreciably in its physico-chemical properties. They have a columnar structure in the B Horizon and a platyform structure in the A Horizon. The solod soils are degraded solonetz. This terminology is largely based on certain profile characters rather than on chemical ones which form the basis of the U.S.A. Riverside Laboratory classification. Solodization depends upon the amount of silica in the A Horizon (or $A_2$ Horizon for solods). The accumulation of sesquioxides in the $B_1$ Horizon is highly correlated with both solonchakic and solonetzic characters (POPAZOV, 1956). In solonetz, exchangeable sodium is not the chief distinguishing feature, but mainly a morphological one. In a solonetz profile, the layers where particles less than 1 $\mu$ have been transported below the upper humus horizon, are solonetzized. It is the degree of this transport which serves as a measure of solonetzicity (PERSHINA, 1956).

The takyr soils of the U.S.S.R. have a non-saline layer of 10 to 30 cm, but underlying to it is a highly salinized layer containing over 2% salt. Since gypsum is present in this layer, the ratio of sodium to calcium is about equal and their chloride and sulphate are in equal proportions (BAZILEVICH et al., 1956). The formation of these takyr soils is preceded by the formation of meadow-solonchak soils.

Classification of saline and alkali soils in U.S.S.R. is generally based on the amount of total soluble salt and the quantity of chloride, sulphate and carbonate and their relative predominance in the soil. Recently KOVDA (1961b) has given a detailed classification based on these characters. The various types of solonchak soils recognised are: (a) soda-solonchak; (b) sulphate solonchak; (c) chloride solonchak; (d) nitrate solonchak. In the soda-solonchak the predominant salts accumulated are $Na_2CO_3$, $NaHCO_3$ and $MgCO_3$. The sulphate solonchak soils contain predominantly $Na_2SO_4$, $MgSO_4$ and $CaSO_4$. Solonchak soils of this type have the lowest toxicity for plant growing. The chloride-solonchaks contain NaCl, $MgCl_2$ and sometimes $CaCl_2$. In such soils the total salt contents are high. Normally they do not contain gypsum. If, however, a large amount of gypsum is present, these soils are easy to reclaim. The nitrate-solonchaks contain predominating $NaNO_3$ and $KNO_3$. Their toxicity is identical to that of the chloride solonchaks. In nature, pure accumulation of any such salts is rare, but the predominance of one or the other group of anions is rather frequently encountered. Besides solonchaks in which the total soluble salts usually exceed 1—2%, saline soils such as solonchak-like soils (with 0.3—0.8% total soluble salts) and slightly saline meadow soils are also frequently recognized.

The sub-division of solonetz soils is based on their hydrological regime. Broadly speaking they embrace two groups: (a) Meadow solonchak-like solontsy; and (b) steppe solontsy. In the former group, the groundwater table occurs at a depth of 3—5—8 m, and these waters are slightly or moderately mineralized. "The lower parts of Horizon B and Horizon C in the meadow alkaline-solonchak-like solonetz are always characterized by the more or less marked presence of easily soluble salts and sometimes of gypsum, and contain always a significant quantity of calcium carbonate". They may be crusty or deep, depending upon the thickness of the A Horizon. "The fertility of the second group of solonetz soils is much higher than that of the crusty ones". The steppe solontsy soils have developed independent of the rise in water-table in ancient watersheds or old flood plains due to the residual alkalinization (KOVDA, 1961b).

### 3. Hungary

In Hungary, DE SIGMOND (1927) classified alkali soils from the utilitarian aspect and sub-divided them into three main types, namely, (i) soils rich in alkali and $CaCO_3$, (ii) soils having less alkali but considerable amounts of $CaCO_3$ and (iii) soils rich in alkali but free of $CaCO_3$. He revised his classification in 1932 and enlarged this sub-division to 5 types:

1. Saline soils.
2. Salty alkali soils.
3. Leached alkali soils.
4. Degraded alkali soils.
5. Regraded alkali soils.

A short description of these five types is given by the following picture:

The saline soils have sulphate and chloride as dominant anions. The soluble salts exceed 0.1%, sodium less 12% of the exchangeable bases, the bulk of the cations being sodium. The soluble salts also include magnesium salts to a high proportion.

Salty alkali soils have a considerable amount of sodium carbonate and the proportion of exchangeable sodium is high. Soluble salts exceed 0.1% and the pH of these soils is about 8.5.

The third type, leached alkali soils, has a very low percentage of chlorides and sulphates, whereas the proportion of sodium carbonate is high. The pH value ranges from 8.5 to 10. A further important characteristic is that the B Horizon is solonetzized and has a columnar structure. Usually a hard pan exists in the sub-soil.

The degraded alkali soils have a pH less than 6 and the $A_2$ Horizon is highly leached. This soil type is differentiated from acid soil by the presence of exchangeable sodium.

The regraded alkali soils are those which have again developed alkalinity after having been leached previously. These soils are characterized by the general poor physical properties of leached soils and by a very high salt content. DE SIGMOND's classification was revised by ARANY (1956c) on the pattern of that recognized in U.S.A. to alkali, saline, saline-alkali, leached-acid, and regraded types.

More recently the alkali (szik) soils of Hungary have been classified on the Russian pattern into six types: solonchak, solonchak-solonetz, meadow-solonetz, steppe-like meadow solonetz, solod and solonetzic-meadow soils. This classification depends upon the nature and amount of salts present, the thickness of the A Horizon and the depth of the calcium carbonate layer (SZABOLCZ & JASSO, 1959).

4. Regional Nomenclature and Classification with India as Example.

The very different types of salty soils have been recognized as such by peoples connected with agriculture since times immemorial. The many specific popular names for these different types are the best proof of it. India may serve as an excellent example for this fact and is dealt with here in more detail for this purpose.

At the same time, the mention of these names may also be of some value to the non-Indian soil scientists, giving these data from a subcontinent which is generally much less known from this point of view than the analogous areas in Eastern Europe, Central Asia, North Africa and North America. This applies also to the parts dealing with the extent of salty lands, in which India may serve as a more detailed example for all similar regions, showing the close interrelation with the general aridity and salinity problem of the various geographical, geological, climatic and soil factors as well as of the anthropogenic influence of mismanagement in land use.

The nomenclature of salty soils varies a great deal from state to state in India. In the Punjab these soils are variously called Reh, Thur, Bari, Bara and Rakkar. In Utar Pradesh, Rajasthan, Bihar and Madhya Pradesh, the terms current are Usar, Reh and Kallar. In Gujarat these are called Khar and Lona soils, whereas the terms prevalent in Majarashtra are Kshar and Chopan. In Andhra Pradesh these are usually named Chaudu and Ippu. In Madras state, the Tamil term for alkali soil is Kalar.

The following definitions show that not only the old words are different but that the actual concepts to be understood by them vary to a high degree. The term Reh is equivalent to saline soils which have developed both in irrigated areas and in unirrigated low-lying areas. The major portion of salinity is due to the high content of sodium sulphate and to a small extent of sodium chloride. Frequently an efflorescence is to be seen as accumulation on the surface. Thur soils are equivalent of saline alkali soils which have a high pH. The soluble sodium salts have been converted from anhydrous to crystalline ones under humid soil conditions because of waterlogging. Rapid formation occurs where the water-table is high. Here too, the salts appear as a white efflorescence on the soil's surface (MEHTA, 1940). Bari soils constitute an intermediate stage of alkali development between normal and Bara soils. When a normal soil has been partially alkalized, it forms this type. The predominance is that of sodium bicarbonate and sodium carbonate. It is usually a poor but not altogether a barren soil. This type corresponds with the Russian concept of solonchak-solonetz. Bara (Solonetz) soils differ from Bari soils in that they are already completely alkalized. The laminated superficial horizon overlaying a deep layer exhibits a columnar or prismatic structure. The rounded tops of the columnar elements and their vertical sides frequently carry a white deposit of salts. The top horizon has a high pH and forms a compact impervious layer. The soil horizon below this layer has a medium pH and contains calcium carbonate nodules (Kankar). The soil profile underneath this horizon has a higher permeability and a low pH. Further down in the profile, it is succeeded by a sandy layer and with the depth of the profile the exchangeable calcium increases (MCKENZIE-TAYLOR, PURI & ASGHAR, 1935; PURI, MCKENZIE-TAYLOR & ASGHAR, 1937; MCKENZIE-TAYLOR 1940; and MCKENZIE-TAYLOR & MEHTA, 1941).

The terms Rakkar and Kallar soil are synonymous. This soil has a very high pH (9.4 to 10.4) and a high percentage of exchangeable sodium. Its salt is principally sodium carbonate. Rakkar and Kallar have a high clay content and are therefore very compact soils. The compact and hard crust is of a dirty white to dark-brown colour. These soils are completely devoid of vegetation unless water stagnates for some time over them. The Usar soils do not have high encrustation of salts on the surface, but the sub-soil is highly alkaline. Usually a compact layer of sodium clay exists, preventing root penetration or at least making it very difficult. These soil types correspond with those termed saline-sodic soils. The encrustation in the Reh soils of Upper Pradesh consists of sodium chloride and sodium sulphate, the latter being of a crystalline nature. At some places the efflorescence is so thick that only few highly salt tolerant plant species can withstand it. The water-table is fairly high and in spite of the high precipitation there the salts continue to accumulate on the surface.

The Khar (Kshar) lands or tidal type of saline soils occur on the coastal areas of the Gujarat and Maharashtra states. In the Little Rann of Cutch, both saline sodic and saline non-sodic types are met with. There, the saline-sodic soils contain large quantities of gypsum. In the saline non-sodic soils the calcium content varies from 10 to 15% and the amount of exchangeable calcium is low. In the other coastal areas these soils are saline sodic in nature.

The typical profile there has a high content of sodium carbonate in the brown "A" Horizon and also in the black "B" Horizon. The latter Horizon is more silty and sandy than clayey in texture. The Lona soils have the same characteristics as the Reh and Usar soils. In Kutch, the "A" Horizon is dark brown in colour and has a predominance of sodium carbonate, whereas the lower horizons contain gypsum and lime and these horizons show, therefore, a predominance of sulphate. The Lona soils in the Goraru tract of north Gujarat and in the Saurashtra areas accumulate an efflorescence of sodium sulphate and sodium chloride. These salts have developed here by irrigation or in low-lying areas by similar but natural causes.

The Lona soils of south Gujarat and the Chopan soils of Deccan-Maharashtra occur in the deep Black Cotton soils, where irrigation facilities are in common use. Two types are recognized there: In the first type, both the "A" and the "B" Horizons are hard and compact, while in the second type the "B" Horizon is loose and friable. These soils have a good depth and a high reserve of calcium carbonate. A further characteristic of them is that a zone of accumulation of soluble salts is to be found in the sub-soil depending on the depth of the water table. Normally this depth ranges from 0.75 to 1.50 m. The degree of sodium saturation is high in the surface soil, whereas the content of calcium is proportionately low. The

**Table III.**

Classification of salty lands.

| Chemical classification of soil | Characteristics | Indian nomenclature |
|---|---|---|
| Saline (Solonchak) | Encrustation of white efflorescence; sodium less than 15% of exchangeable cations; the chief anions are chloride and sulphate, rarely nitrates; soluble carbonates in low proportion only; conductivity of saturation water extract not less than 4 millimhos/cm at 25 °C; pH less than 8.5. | *Reh; Appiu* |
| Saline-sodic (Solonetz-Solonchak) | White encrustation on highland and mixture of white and black in depressions; predominant anions chloride, sulphate and carbonate; high proportion of sodium and low proportion of calcium cations; pH around 8.5. | *Bari; Usar; Khar; Kshar; Lona; Chopan Thur.* |
| Non-saline-sodic (Solonetz or Alkali soils) | "A" Horizon saturated with sodium in the adsorbed state; "B" Horizon generally of a columnar or prismatic structure in heavy soils; exchangeable sodium greater than 15%; predominant anions are carbonate and bicarbonate; pH up to 10. | *Bara; Kallar; Rakkar.* |

Telagu term Choudu corresponds to sodic soils. It has a high proportion of sodium carbonate and sodium bicarbonate. Another soil type is called Appiu soils, characterized by a predominance of sodium chloride. They are found close to the coastal area in the delta regions.

## Mode of Development of Alkalinity under Arid Environment

Four successive phases were distinguished by DE SIGMOND (1932) in the formation of salt affected soils. These phases are 1.) accumulation of salt, 2.) the formation of alkali soils, 3.) natural or artificial washing out of the excess of alkali salts, and 4.) the degradation of alkali soils. He termed these phases salinization, alkalinization, desalinization and degradation. The fifth phase recognized was regradation. A slight modification of this terminology has been suggested by MAGISTAD (1945) namely: salinization, alkalinization, desalinization, solodization and regrading. The main sources of the salts are shales, sandstones, volcanic ash, etc., and they are released in the process of weathering of minerals under arid conditions. When these salts tend to accumulate in the soil, the process is known as salinization. Such accumulation occurs when the balance of leaching into the sub-soil is upset or the process of movement of salts is reversed. Waterlogged conditions under irrigation accelerate this process. Under humid conditions, the soluble salts of potassium, magnesium, sodium and calcium are leached out rapidly from the soil down to a great depth. But under arid conditions, where the rainfall is usually inadequate and evaporation frequently surpasses by far the amount of effective precipitation, the leaching process is restricted to the top horizons. In consequence, the process of salinization is very fast.

All the cations and anions are in an adsorbed state on the soil particles, particularly the clay fractions. Sodium tends to displace the other cations as its concentration in the soil increases and as a parallel function, the replacing power of sodium very rapidly increases by dilution. Consequently, under waterlogged conditions Ca and Mg ions are displaced, the cationic exchange occurring as follows:

$$\text{Ca-clay} + 2\,\text{Na} \rightleftarrows \text{Na}_2\text{-clay} + \text{Ca}^{++}$$
$$\text{Mg-clay} + 2\,\text{Na} \leftrightarrows \text{Na}_2\text{-clay} + \text{Mg}^{++}$$

In due course, the calcium and magnesium salts are precipitated and the clay becomes highly sodiumized (KELLEY & CUMMINS, 1921 & KELLEY, 1951). REITEMEIER (1946) reported for several soils that the relative replacing power of $Ca^{++}$ and $Mg^{++}$ is 1.6 as against 5.0 of $K^+$ and $Na^+$. The excess of $CO_2$ given out by the roots reacts with sodium and forms sodium bicarbonate and carbonate and thus alkalinization is brought about. The anions associated with the sodium salts have also their influence in the formation of alkali soils, as the sulphate and chloride anions associated with sodium help displace the calcium by sodium. ARANY (1956b) found that the displacement of calcium by sodium is less with $CO_2$ and $HCO_3$ ions.

When the excess of alkali salts is leached out of the soil either artificially or naturally, the process is known as desalinization. It is, however, a very slow process in normal agricultural soils and may be reversed at any stage. In the leaching process, the exchangeable sodium is replaced by divalent cations of $Ca^{++}$ and $Mg^{++}$, which may be present in the soil and the ratio of exchangeable cations is thereby altered. In the salinized state, the colloidal clay particles tend to disperse and leach down to the B Horizon of the soil and to form there a dense sub-soil horizon. Due to this cementing action, columnar or stratified layers are formed. Such conditions are evident in solonetz soils. In the "A" Horizon calcium replaces the sodium and then the soil remains no longer alkali but becomes chemically normal.

When the H ions replace the Na and Mg ions, the degradation process of alkali soils takes place. In this process, sodium hydroxide is formed accompanied by partial hydrolysis of the adsorbed sodium on the clay particles. The sodium hydroxide reacts with carbon dioxide in the following manner:

$$\begin{matrix} & Na & & & & & Na \\ Ca\ clay & Na & + H_2O = & Na\ OH & + & H\ clay & Ca \\ & Na & & & & & Na \end{matrix}$$

$$CO_2 + H_2O\ (H_2CO_3) + 2NaOH = Na_2CO_3 + 2H_2O.$$

Gradually, the sodium carbonate is leached out, leaving hydrogen clay in the soil (CUMMINS & KELLEY, 1923). It is by this process that the pH value is lowered to 6.0 (MAGISTAD, 1945). This reaction occurs particularly in soils which contain very little calcium carbonate. The resulting soil is called solod, and this process of degradation has been termed solodization (GEDROIZ, 1926). After such a degradation, sometimes a reaccumulation of salt occurs. This is called the process of regradation, and the sodium ions start again to accumulate in the soil (DE SIGMOND, 1932).

It is a well known fact that under irrigation accumulation of salts in the soil occurs at a much faster rate than under arid conditions. This process of salinization is particularly rapid when the soil's crust has sodium in its chemical composition and the quality of irrigation water is saline in character. When the soil is impermeable, even a good quality water brings about its early salinization. According to ARANY (1956b), "A water is alkalizing if it gives up sodium to the soil; it is reclamatory if it takes sodium from the soil" (BERNSTEIN, 1961). In the meadow soils in Hungary alkalization has occurred because they lie lower than the surrounding chernozems. In Russia, SCHAVRYGIN (1956a; 1956b) has observed that meadow and meadow-solonchakic stages precede takyr formation. Takyr soils have a non-saline layer of 10 to 30 cm, the horizon below it is, however, strongly saline and contains over 2% salt (BAZILEVICH et al., 1956).

### Extent of Salty Land

Also in this respect, India may serve as an example for the significance of this important problem, particularly for the reason that the concept of India is outside of it mostly associated with that of a humid tropical country where salinity plays only a minor role.

Some other countries, where the salinity and aridity problems are of particular importance, may be mentioned below in order to round up the global picture in this respect.

1. India

According to an estimate, 20 million acres of saline sodic soils in different stages of degradation exist in India. The extent of salinity of these lands varies from State to State. In States where extensive irrigation facilities have been developed, the extent of such lands is surprisingly large. This is, of course, particularly the case where aridity is also high. In the following, these problem areas of India's arid zone may be described in more detail:

*a) Punjab*

The alluvial soils of the Punjab contain 10 to 15% clay and sodium is present almost everywhere in the soil crust. But most of the soils have a high content of calcium carbonate which sometimes occurs in the form of Kankar. The rainfall in the plains varies from 10 to 25 inches, and the climate is, in accordance with these latitudes and with the low altitude, arid to semi-arid. These areas together with similar ones unmapped up to now constitute about 5 million acres. Their soils show a wide diversity in physico-chemical characteristics. The Doab, for instance, between the Ravi and Beas rivers covers less extensive saline areas than the Doab between the Sutlej and Ghaghar rivers in the northern zone. In the southern zone, extensive areas irrigated by the Western Jamna Canal have already developed a dangerous salinity. In some tracts, over 25% of the area have become salty. In the Karnal district, which has more extensive areas suffering from salinity than other districts, the typical salt scraping showed the following composition:

**Table IV.**

Nature of soluble sodium salts in soil scraping.

| Nature of salt | percentage |
|---|---|
| Total salts | 22.8 |
| Sodium carbonate | 48.3 |
| Sodium bicarbonate | 33.4 |
| Sodium sulphate | 13.3 |
| Sodium chloride | 3.9 |
| Sodium nitrate | Traces |

In these soils, either the degree of alkalization decreases with the depth or there is practically no change in its value. The decrease occurs due to the formation of calcium sulphate or calcium chloride in the reaction between sodium sulphate, sodium chloride and calcium carbonate by double decomposition which percolates down during the rainy season. In other areas of the Punjab where salinity occurs, the predominant salt in the profile is sodium sulphate. In these areas there is, however, also a high content of calcium carbonate leading to a certain balance in the soil, whereas sodium carbonate and sodium chloride occur in fractional quantities only (MEHTA, 1940; UPPAL, 1963).

*b) Uttar Pradesh*

In the saline areas of Uttar Pradesh, in the alluvium of the Western and Central districts, the main salts appear to be sodium carbonate and sodium chloride and in the eastern districts sodium sulphate. Highly sodiumized or sodic soils occur in the western districts of the State, particularly in Bulandshahar and Aligarh. The main characteristics of these soils are a relatively very high percentage of sodium carbonate and that their hard and compact crust shows a dirty white to dark brown colour. Their pH varies from 9.4 to 10.4. Saline sodic soils are in evidence through western and central Uttar Pradesh. Their pH varies from 8.7 to 11.3, and they show a high percentage of sodium. As a remarkable fact it has to be mentioned that there is complete absence of gypsum. In several localities they have a compact layer of sodiumized clay at 0.75 to 1 m depth which prevents penetration of roots into deeper layers and is conducive to conditions of waterlogging. The arid climate and poor internal drainage are the responsible factors for the alkalization of these soils. The main source of salt seems to be the weathering of minerals. These soils generally occur in moderately low-lying areas where the water-table is high and drainage is poor due to indurated clay. Sometimes a cemented calcium carbonate nodule (Kankar) pan with or without a highunder ground water-table is to be found. In Eastern Uttar Pradesh saline soils occur in pockets. Similar soils occur also in other parts of the state, but there much better conditions prevail. The soils usually have free drainage and a high amount of free lime. A slight amount of salinity is indicated by a white efflorescence formed on the surface during the hot weather. In addition to the more favourable physico-chemical conditions of the soils the rainfall is much higher and varies in these areas from 25 to 40 inches (Anonymous, 1939; AGARWAL & YADAV, 1954). The extensively flooded regions of Ganga, Jamna and their tributaries, called Khadar lands, remain submerged in water over long periods. During the hot summer months, when the surface dries out and the groundwater table is still high, a white efflorescence characterizes these areas already from a wide distance. Such naturally highly saline soils are usually shallow and occur where there is no natural flow of ground water in lateral or vertical direction.

Man-made saline or saline-sodic soils have been developed in almost all the irrigated tracts of Uttar Pradesh. These are the areas commanded by the Agra Canal, the Eastern Jamna Canal, the Ganga Canals and the Sarda Canals; but it seems necessary to remark here that the source of salt is not the canal water. The reason is, that the inflow of canal water into the soil has considerably raised the groundwater level, and due to the impeded underground drainage otherwise very fertile soils have under these circumstances either become salinized or are in the process of salinization and alkalinization.

**Table V.**

Average pH, total solids and E.Ce. of water of Aligarh and Kanpur districts.

| Source of water | pH | | Total solids p.p.m. | | E.Ce. $10^6$ | |
|---|---|---|---|---|---|---|
| | Aligarh | Kanpur | Aligarh | Kanpur | Aligarh | Kanpur |
| Canal | 8.0 | 8.52 | 165 | 325 | 235.7 | 444 |
| Well | 8.1 | 8.64 | 501 | 943 | 714.8 | 1400 |
| Drainage | 8.8 | 8.75 | 1821 | 943 | 2601.6 | 2882 |

In Kanpur although the canal waters have a higher pH, more total solids and a higher electrical conductivity than the Aligarh canal water, the Na—K exchangeable values are extremely low and, therefore, these waters present very low salinity and alkali hazards. The drainage waters, however, contain an excessive quantity of salts and, obviously, it is the salt in the soil's uppermost layer which is responsible for developing salinity there.

*c) Maharashtra*

In Deccan, all mature deep black soils have a profile zone of salt accumulation. The concentration of sodium increases with the depth with a simultaneous decrease in free lime. The salts present are generally sulphates and chlorides. Leached from the "A" Horizon, they have built a "B" Horizon. Very wide variations in the Ca/Na ratio occur from one soil to the other. The sub-soil water-table is found between 2 and 3.5 meters below the surface and depending upon the water-table, the zone of salt accumulation is to be found at 1 to 2 meters depth. The surface soil develops a high degree of sodium saturation, the calcium being proportionately low. The former varies from 1 to 40 and the latter from 55 to 85% of the base exchange complex. Two such categories of soils have been observed. In the first category of soils, which is comparable to solonetz, both the "A" and the "B" Horizon are hard and compact. In the second category, the "A" Horizon is hard and compact, but the "B" Horizon loose and friable. Of these two categories, the former requires a more prolonged treatment than the latter (BASU & TAGARE, 1943a).

In 1929 the salty lands in the irrigated areas of the Right and Left Bank Godavari canal tracts comprised 42% of the commanded area. The survey in 1930 of the Nira Left Bank-Canal, indicated 34% of the area in different stages of deterioration. Under high aridity during the summer months, with excessive irrigation the zone of salt accumulation moves up and down the profile with consequent salt encrustation on the surface. The worst types of sodic soils have developed where alkalization has proceeded to the extreme, but it should be emphasized that the irrigation water and the sub-soil water have not contributed to this development of alkalinity. It is the residual salts, accumulated in the soil's crust, which have caused this alkalization of the soil profile by alternate wetting and drying (BASU & TAGARE, 1943b).

*d) Gujarat*

Saline sodic and saline non-sodic soils in the Little Rann of Cutch occupy an area of 800 sq. miles. The saline sodic soils contain, however, as a favourable factor an appreciable quantity of gypsum and also the lime content is relatively high and varies from 10 to 15%.

Khar lands or the tidal type of saline lands occur in Saurashtra or the Western coast of Southern Gujarat. The typical profile has a brown "A" Horizon containing sodium carbonate. The "B" Horizon is black in colour and of a more silty than clayey layer. The variation in total soluble salts ranges from 3.04 to 3.60%, pH ranges between 8.60 and 8.76 and $CaCO_3$ between 3.2 and 6.6%. The exchangeable calcium is low (BASU & TAGARE, 1954).

*e) Andhra Pradesh*

In this State, both non-saline sodic and saline sodic soils occur in the Black soil region particularly where irrigation facilities have been developed, which is the case in relatively small pockets only. In the delta regions of Godavari and Krishna saline soils occur but, as paddy is grown extensively there, the effect of salinity is not much in evidence.

2. Australia

In the rainfall regions of less than 17″ in the states of Western Australia, South Australia, Victoria and N.S. Wales salinized brown soils cover large areas. These soils have been named "mallee" type because of the typical *Eucalyptus* species inhabiting this area particularly *Eucalyptus oleosa* and *E. dumosa*, building together with a number of sclerophilous scrubs the so-called "Mallee" Formation (STEWART, 1959). The soils are highly alkaline with a pH exceeding 8. Calcium averages 15%. Na and Mg occur in the sub-soil in equal proportion of about 40%. Some soils of this type cover approximately 55% of the total areas (TEAKLE, 1950; Anonymous, 1949; PRESCOTT, 1944). Solonetz soils which show a strong prismatic structure in the "B" Horizon occur largely in Western Australia, in upper south-east South Australia and in the sub-coastal valleys of Queensland. They cover only 0.6% of the total area (PRESCOTT, 1944).

In the irrigation districts of N.S. Wales, Victoria and South Australia, because of the rising water-table, large areas have turned saline, and extensive drainage works have been set up to reclaim these areas in the Murray and Murrumbidgee river basins. These great reclamation works have successfully prevented further loss to vineyards and orchards (PENMAN, TAYLOR, HOOPER et al., 1939; TEAKLE, 1937).

3. Egypt

With the introduction of permanent irrigation in the Nile Valley, salinization set in in different areas. A recent survey has shown that about 300,000 acres have turned saline and some of the most valuable irrigated areas have already gone out of cultivation. The Reclamation Bureau has surveyed these areas and has classified them according to the pattern of the U.S. Salinity Laboratory. A programme of reclaiming these areas has been laid down at the rate of 20,000 acres per annum. In these areas, a drainage system is being provided for by the state and by the peasants on the basis of the recommendations of the Reclamation Bureau of the Department of Agriculture.

4. Hungary

The salt affected lands in Hungary are predominantly solonetz, but in certain localities solonchak and solod also occur. The extent of these lands is about half a million hectares. Almost all of these areas lie in the Great Hungarian Plain. This basin is surrounded by mountains, and the seasonally fluctuating water-table favours the process of alkalization (SZABOLCZ, 1956b). Most of these soils are of a meadow type and have therefore been termed meadowsols (SZABOLCZ & JASSO, 1959).

5. U.S.S.R.

In the U.S.S.R., the saline and alkali soils comprise nearly 3.4% of the land area or 75 million hectares (BERNSTEIN, 1961). In the irrigated areas of the Central Asian Soviet Republics of Uzbekistan, Tadjikistan, Kazakhstan, Azerbaijan, Georgia and Turkmenia, solonchak soils are the most common. Solonetz soils are broadly distributed in the southern regions of recently ploughed virgin lands in western Siberia and Kazakhstan. Solod soils have a relatively less wide distribution and occur in the western parts of the steppe region in the European part of the U.S.S.R. and in western Siberia. Meadow solonchaks are of less significance and occur mainly in the valleys of all rivers in the southern region of the U.S.S.R.

In the arid zone areas of the Central Asian Republics of Uzbekistan, Tadjikistan, and Kazakhstan, there are large reclamation projects on saline soils in progress in the Golodkaya steppe. About 230,000 hectares of land were reclaimed in this area until 1955 and another 175 thousand hectares are proposed to be reclaimed by 1965. The total area of this Golodkaya steppe is nearly one million hectares of which about 800,000 hectares are proposed to be reclaimed for irrigation and cultivation of cotton, by laying down a collossal net of underground drainage works in the near future. The groundwaters in this area are to be found at a depth of 0.5 to 20 m and all of them are invariably saline waters. Large canal systems, the Kiroo Main canal and the Southern Golodnostepsky canal, have been constructed in the Syr Daria region for the purpose of soil leaching, reclamation and irrigation.

In Kazakhstan, spots of solonetz soils occur in depressions in the chernozem and chestnut soil area. They are generally divided into three types: Meadow solonetz with a groundwater table at 3—5 m; Meadow steppe solonetz with a water-table at 5—7 m and steppe solonetz with a water-table at more than 7 m deep. The meadow steppe solonetz soils are of a solonchak type, containing a considerable amount of soluble salts in the upper horizon.

6. North America

The alkali soils in the U.S.A. exist mainly in the South Western region where, due to non-provision of drainage, about 300,000 acres have turned saline. This development can best be observed in the

San Joaquin, Sacramento, Coachella and Imperial Valleys of California where irrigation has been provided on an extensive scale. In the arid tracts of Utah and Nevada in the Great Basin, such soils occur in the depressions where the salts tend to accumulate as a natural consequence of the inadequate rainfall there. Weathering and intensive evaporation cause the salts to accumulate in the top horizon. Similar soils are met in the drainage basin of the Colorado River in parts of Colorado, Arizona and California, and also in the Rio Grande drainage area in the states of New Mexico and Texas. Solonchaks occur even as far North as North Dakota in the surrounding humid gley or Chernozem soils, and also in south-western Idaho so called "slick spots" cover rather extensive areas (Harris, 1920; Hayward, 1956; and Bernstein, 1961).

7. Meso-America

In Mexico we are confronted with saline soils in the whole arid region along the west coast, where they are spread out over large areas particularly in the States of Sonora, Sinoloa, and Durango. Also the dried out Texcoco lake beds have a high percentage of salts and a high pH value. Saline soils also occur in the irrigated areas of the Mexicali district, in the Chihuahua State along the Rio Grande south of El Paso, and in the State of Tamaulipas along the lower Rio Grande. Almost all soils of the Rio Grande flood plain area are highly alkaline (pH 8.9) and have about 89% exchangeable sodium in the base exchange complex. In this flood plain we encounter several almost completely barren spots caused by these unfavourable soil conditions (James, 1942, and Hayward, 1956).

8. South America

In accordance with the high aridity along the Pacific regions of Northern Chile, Peru and the Southern parts of Ecuador we find large areas of saline soils. This desert region, on account of its extremely low rainfall, particularly in Northern Chile, is partly without any vegetation and extends over about 25 latitudinal degrees i.e. from 6° South to 31° South.

In the Atacama province of northern Chile, saline soils prevail in the alluvial deposits along the streams, but also the extensive sand dunes near the coast include many interdune depressions with soils of a solonchak type. Saline soils occur also in the vicinity of Iquique (Chile, 20° 5.) but there they have only a moderate salt concentration and a high lime content (James, 1942; Storie & Mathews, 1945).

The arid zone of Ecuador and Peru shows somewhat higher rainfall figures, namely about 10 inches, but here we encounter salinization over large areas caused by man. Most of the irrigated lands in the pacific parts of Peru are in close proximity of rivers in the 40 or more oases there. Since the drainage is centripetal in these oases, extensive parts of them have turned into saline, saline alkali and non-saline alkali lands (ANDERSON, 1950; JAMES, 1942).

The Atlantic parts of South America are climatically much more favourable for good soil conditions, but even there, in north-east Brazil, for instance, where irrigation has been provided from reservoirs, we find salty lands in patches which account for about 25,000 hectares. Some of these reservoirs have brackish water and the use of these waters has, of course, accentuated the salinity problems in these areas (HAYWARD, 1956).

## Vegetation

### General Remarks

Since the natural vegetation is dealt with in this book in more detail by V. J. CHAPMAN, we shall restrict ourselves here to a few general remarks in this respect.

The vegetation of arid regions is by and large a xerophytic one, and the anatomical and physiological characters of this xerophytic vegetation are further subdivided into (a) ephemerals; (b) succulents; (c) non-succulent perennials; and (d) evergreen xeromorphics. Allied to xerophytes are the halophytes, i.e. plants specifically adapted to habitats of high salt concentration. They may be succulent or non-succulent, and occur along sea coasts, salty lakes, former lagoon beds and similar localities. They are also widely distributed over all dry regions where salt appears. There the soil itself is impregnated with salts (DRAR, 1955), which may also be visible in many cases on the soil surface as precipitation or efflorescence.

Halophytes are truly salt resistant plants. These can be further sub-grouped into euhalophytes, crinohalophytes and glycohalophytes. The euhalophytes have a low rate of respiration and low activities of oxidative enzymes, peroxidases, polyphenoloxidases and catalase, which remain stable in the presence of added salts. Species belonging to this group are, for instance, *Salicornia herbacea, Suaeda prostrata, Petrosimonia Litwinowii, P. triandra, Salsola soda* and *Echinopsilon hyssopifolium*. This group has been further subdivided into halo-succulent and haloxeritic species. To the former sub-group belong *Suaeda prostrata* and *Salicornia herbacea* and to the latter *P. triandra, P. Litwinowii, Echinopsilon hyssopifolium* and *Salsola soda*.

The crinohalophytes have a higher rate of respiratory activity than other halophytes and do not accumulate salt in their body. Good examples of this type are *Atriplex salina* and *A. tartarica*. (HENCKEL, 1954; HENCKEL & ANTIPOV, 1956; POKROVSKAIA, 1957.)

In the distribution of vegetation in salty areas in the arid zone, BOYKO's Geo-ecological Law of Distribution is of great significance.

This law reads in its shortened form: "Microdistribution (topographical distribution) is a parallel function of macrodistribution (geographical distribution) since both are dependent on the same ecological amplitudes" (BOYKO, 1947). Climatic influences, morphology, anatomy and physiology of individual plants and superimposed on that, the salty habitat further influence these variegated manifestations.

This physiological adaptability restricted only by the general limits of their ecological amplitudes, is dealt with by AKOPIAN (1957), who shows that in some species like *Goebelia alopecuroides*, under conditions of high salinity in their habitat, the concentration of disaccharides is appreciably increased with a simultaneous disappearance of monosaccharides. This enables the plant to flourish under conditions of high osmotic concentration of soil solution as their photosynthetic apparatus continues to function normally. This may be the factor operating in species such as *Butea monosperma*, *Prosopis spicigera* and so on. Possibly *Acacia arabica* also falls into this category.

Russian agricultural scientists in the Central Asian Republic of Uzbekistan have defined the ecological conditions of naturally occurring plant communities on different saline soils with respect to the degree of soil salinity and to the groundwater conditions. Some details of these five plant communities may be quoted here: (see Table VI)

Generally it can be said that the effect of salinity or alkalinity consists in increasing the osmotic concentration of the soil solution which in return reduces the physiological availability of water for the plants, and leads to an accumulation of toxic amounts of ions

**Table VI.**

Plant communities occurring on soils of different degrees of salinity.

| Bog soil Groundwater table 0—1.0 m. | Meadow soil Groundwater table 1—2 m. | Meadow Seirozem soil Groundwater table 2—3 m. | Seirozem soil. Groundwater table above 3 m. |
|---|---|---|---|
| 1. | 2. | 3. | 4. |
| Group I: | Plants characterizing low degree of soil salinity. (Total soluble salt 0.4—0.8%, chloride 0.01—0.4% and sulphate 0.05—0.3%). | | |
| | (a) *Plants with shallow root system.* | | |
| Panicum crusgalli; Plantago lanceolata; Setaria glauca; Setaria viridis. | Andropogon halepensis; Portulaca oleracea; Cynodon dactylon. | Bromus tectorum; Centaurea pieris; Triticum aegilops. | Carex stenophylla; Poa bulbosa; Ceratocarpus arenarius. |

| 1. | 2. | 3. | 4. |
|---|---|---|---|
| | (b) *Plants with deep root system.* | | |
| Phragmites communis; Trifolium fragiferum. | Melilotus alba; Alhagi camelorum; Mentha arvensis; Xanthium strumarium; Imperata cylindrica. | Erigeron canadensis; Amaranthus graecizans; Erianthus ravennae. | Salsola turkestania Artemisia scapaeformis Cichorium intybus; Convolvulus arvensis; Cyperus rotundus; Chenopodium album; Diarthron vesiculosum Grigensohnia oppositiflora. |
| Group II: | Plants characterizing medium degree of soil salinity. (Total soluble salts 0.8—1.2%, chloride 0.04—0.10% and $SO_4$ 0.3—0.45%). | | |
| | (a) *Plant with shallow root system* | | |
| | | Hordeum murinum; Agropyrum prostratum. | |
| | (b) *Plants with deep root system.* | | |
| | Karelinia caspia; Lepidium perfoliatum; Lepidium latifolium; Inula caspica; Lactuca tartarica; Lactuca scariola. | Polygonum argirodolum; Agmiofesa pubescens. | Atriplex tartarica. |
| Group III: | Plants characterizing high degree of soil salinity. (Total soluble salts 1.2—1.6% chloride 0.10—0.2% and sulphate 0.4—0.6%). | | |
| | (a) *Plants with shallow root system.* | | |
| | (b) *Plants with deep root system.* | | |
| Aster tripolium. | Cynanchum acutum; Suaeda paradoxa. | Halocharis hispida; Tamarix gallica; Populus pruinosa; Zygophyllum fabago; Kochia hyssopifolia. | |
| Group IV: | Plants characterizing higher degree of soil salinity. (Total soluble salts 1.6—2.0%, chloride 0.2—0.3% and sulphate 0.6—0.8%). | | |
| | (a) *Plants with shallow root system.* | | |
| | — — — — — | Bromus sewerzowi | — — — — — |
| | (b) *Plants with deep root system.* | | |
| | Suaeda heterophylla; | Suaeda arcuata; Petrosimonia brachiata. | |

| 1. | 2. | 3. | 4. |
|---|---|---|---|
| Group V: | Plants characterizing highest degree of soil salinity. (Total soluble salts 2.0—2.5%, chloride 0.3—0.4% and sulphate 0.8—1.0%). | | |
| | (a) *Plants with shallow root system.* | | |
| Salicornia herbacea. | Cressa cretica | Salsola crassa; Salsola lanata. | |
| | (b) *Plants with deep root system.* | | |
| — — — — — | Aeluropus litoralis | — — — — | — — — — — |

within them. Magnesium salts are more toxic than sodium salts (MAGISTAD et al., 1943; GAUCH & WADLEIGH, 1942). It is also well known that different plant species show also a different critical concentration of substrate at which water absorption ceases. Thus, for instance, for *Allium cepa* it is 6.5 atmospheres, and for bean seedlings 2.4 atmospheres (ROSENE 1941; TAGAWA 1934). But the critical limit is not associated with any salts such as NaCl, $CaCl_2$, $Na_2SO_4$, sucrose or mannitol (HAYWARD & SPURR, 1944). WADLEIGHT & AYERS (1945) have therefore proposed the concept of "total soil moisture stress" which is a summation of the osmotic pressure of the soil solution and the soil moisture tension expressed in atmospheres.

In addition to the "total soil moisture stress" the injurious action of salts varies and each has its specific effects. Some plants accumulate sodium while others do not absorb it. The sodium content of native species of New Jersey in the U.S.A. varied from 0 to 2.4% and of cultivated varieties from 0 to 3% in dry matter (WALLACE et al., 1948). Besides alkali soil conditions as distinguished from salinity may cause toxicity too (LILLELAND et al., 1945).

The toxicity of $NaHCO_3$ is about half of that of $Na_2CO_3$ expressed in terms of Na. Normally $NaHCO_3$ seldom occurs in toxic concentrations. The salt of $NaHCO_3$ is more toxic than NaCl and it appreciably reduces the intake of Ca by the plants (BREAZEALE, 1927; HELTER et al., 1940). Similarly, chloride salts are more toxic than sulphates at the isosmatic concentration (HAYWARD & SPURR, 1944; EATON, 1942 and WADLEIGH et al., 1946). It was early recognised that salt tolerance of plants is quite different in the presence of other salts from that which is indicated for each salt separately (HILGRAD, 1906). Other factors of influence are the type of soil, the climatic conditions and the available soil moisture, and of course, the specific properties of the species and varieties of plants as such.

## Relative Salt Tolerance in Economic Plants

### 1. *General Remarks*

The U.S.D.A. Salinity Laboratory has found that electrical conductivity of the soil-water extract is a better measure of the salinity tolerance of various plants than other criteria. The influence on plant growth at various levels of electrical conductivity of soil-water extract is shown in the following table VII:

**Table VII.**

Interrelation of E.C. and Plant performance.

| Influence on plant growth | Electrical conductivity Millimhos per cm at 25 °C |
|---|---|
| 1. Salinity effect most negligible | 0—2 |
| 2. Growth of very sensitive plant species restricted | 2—4 |
| 3. Growth of many plant species restricted | 4—8 |
| 4. Only salt tolerant plant species grow satisfactorily | 8—16 |
| 5. Few very salt tolerant plant species grow satisfactorily | above 16 |

In some species the salinity effects are more marked at the germination and seedling stage than in others. In general, the degree of delayed germination and injury to seeds and seedlings is in direct proportion to the osmotic pressure of the saline solution (Hayward, 1956). Those seeds which do not develop high imbibition pressures are incapable of absorbing a sufficiently large quantity of water to initiate the process involved in germination. In general, cereals are more salt tolerant than legumes. The descending order of tolerance in cereals is: barley, rye, wheat, oats, corn; and that in legumes: peas, red clover, alfalfa, white clover. Another important economic crop which has a higher salt tolerance than legumes other than alfalfa is sugar-beet (Stewart 1898; Harris, 1915).

Ayers & Hayward (1948) have standardized the technique of determining germination of various crop seeds in soils which involves moistening and salinizing a non-saline soil. Tests on alfalfa, barley, red kidney beans, corn, onions and sugar-beet revealed that no seeds germinated at a salinity level of 0.4% NaCl on a dry soil basis, but an 80% germination rate could be obtained with barley at the 0.3% level (20 atmospheres osmotic pressure), and 80% germination with alfalfa at the 0.1% level (7.3 atm. O.P.). (Ayers, 1952).

Chloride salts are the most toxic for germination and seedling development; $CO_3^{=}$ intermediate and $SO_4^{=}$ least toxic besides the toxicity due to concentration of salt and soil moisture stress. The

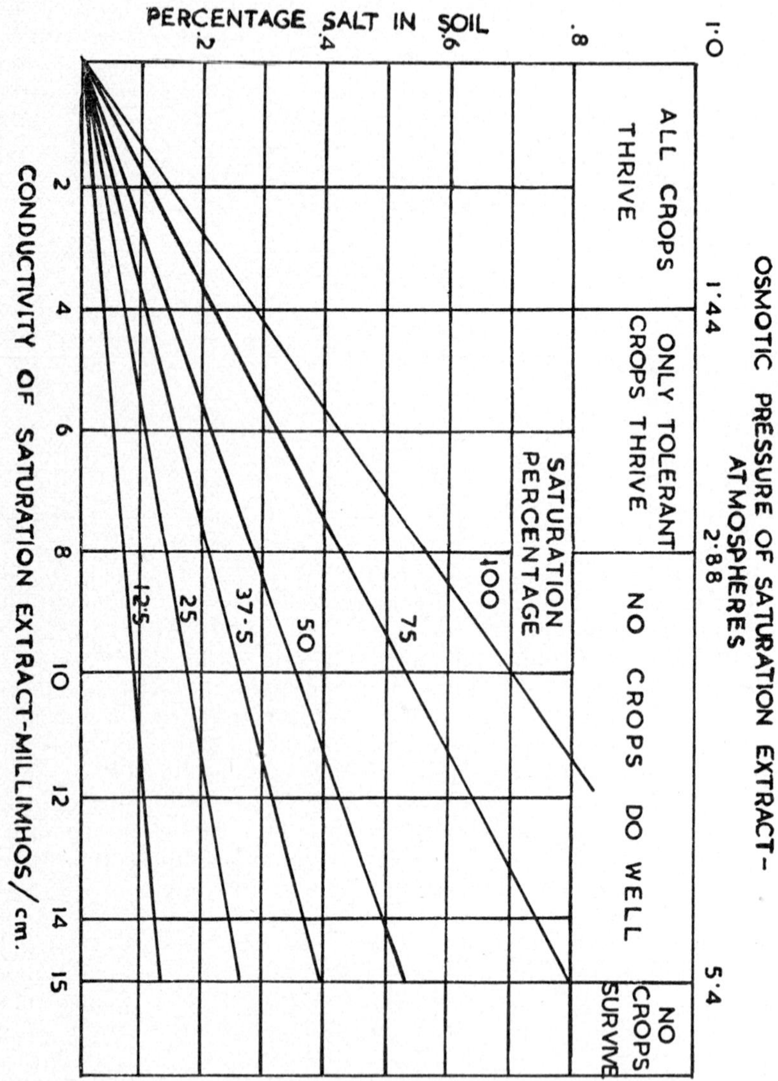


Fig. 1. Graphical Representation of the Relation of Crop Growth to the Conductivity and Osmotic Pressure of the Saturation Extract and to the per cent Salt in Saline Soil. After U.S. Regional Salinity Laboratory (1954).

toxicity of $Na_2CO_3$ is mitigated when organic matter is present in the soil (HARRIS & PITTMAN, 1919). The descending order of toxicity of various soils is as follows: NaCl, $CaCl_2$, KCl, $MgCl_2$, $KNO_3$, $Mg(NO_3)_2$, $Na_2CO_3$, $Na_2SO_4$ and $MgSO_4$. But it is of high practical importance that the effect of combined salts is less due to their antagonism in the soil (HARRIS, 1915).

## 2. *Economic Plants*

The U.S. Soil Salinity Laboratory (1954) has classified the crops on the basis of relative salt tolerance into three categories, namely, good, moderate and poor.

Species of native halophytes may continue to maintain their level of growth when the ECe values are higher than 15 millimhos. This, however, is not to be expected of crops and this relationship has been graphically represented in Fig. 1 taken from HAYWARD (U.S. Salinity Laboratory Staff, 1954). The salt tolerance is also condi-

**Table VIII.**

Relative salinity tolerance of various crops.

| Crops | Degree of relative salt tolerance | | |
|---|---|---|---|
| | Good | Moderate | Poor |
| Cereals | Barley, rye<br>Italian millet | Wheat, oats<br>Sorghum | Corn, finger millet |
| Legumes | Alfalfa,<br>Ladino clover | Strawberry clover,<br>berseem, shaftal<br>hubam clover, white<br>and yellow sweet<br>clovers, broad bean | Bean *(Ph. vulgaris)*<br>peas, gram<br>pigeon pea<br>*Ph. aconitifolius*<br>*Ph. aureus*<br>*Ph. mungo*, cowpeas |
| Oil seeds | *Brassica campestris*<br>var. *taramira* | *Brassica campestris*<br>var. *sarson.* | *B. campestris*<br>var. *toria*<br>castor, linseed<br>sesame niger |
| Sugar crops | Sugar-cane | Sugar-beet | |
| Fibre crops | | Sann hemp | Cotton, jute |
| Green manure crops | *Sesbania aculeata* | Sann hemp | Cluster bean,<br>*Phaseolus trilobus*<br>*Sesbania speciosa.* |
| Grasses and fodder crops. | *Cynodon dactylon*<br>*Sporobolus* spp.<br>*Distichlis spicata*<br>*Puccinellia cristallina* | *Chloris guayana*<br>*Bracharia mutica*<br>fodder sorghum<br>*Agropyron smithii*<br>*Paspalum vaginatum* | Giant star grass,<br>Napier grass,<br>Sudan grass<br>*Teosinte sindicus*<br>*Cenchrus* species |

tioned by the pH of the soil i.e. (a) whether the soil is high in exchangeable sodium, but has a moderate pH; (b) it is high in exchangeable sodium but with a pH of 8.5 and above; and (c) it is high in exchangeable sodium but with a considerable accumulation of trituratable carbonate. AGARWAL & YADAV (1956) observed that the higher the pH, the narrower is the range of salinity tolerance in paddy, sorghum, wheat and barley. Of these paddy tolerated a somewhat higher pH and barley a higher salinity than the other crops. The relative salinity tolerance of various crops under varied conditions of soil and water as reported in literature and observed in the field is to be seen in Table VIII.

This grouping is based on the observations of several authors (GAUCH & WADLEIGH, 1951; HARRIS, 1920; KEARNEY & SCOFIELD, 1936; MAGISTAD & CHRISTIANSEN, 1944; MILLIGAN et al., 1951; U.S. Salinity Laboratory, 1954; AYERS et al., 1948; WADLEIGH et al., 1947; EATON, 1942; BERNSTEIN, 1961).

Frequently some varieties of the same crop are more tolerant to salinity than others. Several studies in the U.S.A. have been conducted on alfalfa and sugar-beet; on rice in Egypt; rice, barley, wheat, sorghum and pearl millet in India. As a result of such studies in India the varieties of paddy which have shown a high relative tolerance to salinity are presented in Table IX (GHOSH et al., 1956) and similar examples of other crops are shown in Tables X and XI.

**Table IX.**

Salinity tolerant rice varieties recommended in different states of India.

| Particulars | States of India | | | |
|---|---|---|---|---|
| | Orissa | West Bengal | Madras | Punjab |
| Names of varieties | SR.26B; SR; SR14 | Chin. 13 Chin.19 | SR.26B T.892. | Jhona 34 N.P. 130 |

These varieties are often raised in the nursery and transplanted into saline soils in order to reclaim them. Generally, late varieties are more salinity tolerant than medium or early varieties. In Egypt the varieties Agami and Nahada have shown a higher salinity tolerance than other varieties under cultivation. OTA & YASUE (1958) tested the salinity tolerance of 58 varieties of rice and observed significant differences in their germination in culture medium containing NaCl up to 6% concentration. The more resistant varieties tolerated 2 to 3 times more salinity than the sensitive ones. This shows that in such experiments the stress has to be laid not only on the species but to a no lesser degree also on the variety.

Kharchi wheat, for instance, commonly grown with saline water irrigation in sandy soils of Rajasthan, India, is the most salt tolerant wheat. In Egypt the varieties found salt tolerant are Giza 144, Giza 145, and Balladi 116. Ota & Yasue (1957) observed significant differences in germination of wheat varieties ranging up to 1% NaCl concentration. Ayers et al. (1952) have reported data on salinity tolerance of four varieties of barley and two varieties of wheat in artificially salinized plots and found that Atlas barley had the highest salt tolerance. Marshall (1942) observed good stand and satisfactory maturation of Regal barley. In Egypt the salinity tolerant varieties of barley recommended are Belladi 16, Giza 117, and Hybrid 11.

Salt tolerance of several varieties of sorghum and pearl millet has been determined at the germination stage by Abichandani & Bhatt (unpublished). The data are summarized as follows:

**Table X.**

Relative salt tolerance of pearl millet and sorghum varieties.

| Crop | ECe in millimhos/cm at 25°C | | | | | |
|---|---|---|---|---|---|---|
| | 0.7 | 2.0 | 4.0 | 8.0 | 12 | 16 |
| Pearl millet | | | | R.S.K.; | Bikaner (local) Sardarshahr (local); P6.T55; Jetsar (local); Gadra Road (local); Agra (local). | Manihari (local); Pal (local); Khatwni (local) |
| Sorghum | M. 35; Sweet Sudan 1.S.704 | Tl | Nebraska j5809; Delhi 1.S.541 J.718; R.J.4R. | Texas hyhr. 610; Texas milo 25245 | M.47-3; Gadra Road (local); Pokran (local) | |

Obviously, millet varieties are relatively more tolerant to salinity than sorghum varieties which show a very wide pattern of salinity tolerance.

Observations on salt tolerance of sugar-cane varieties have been recorded at different Sugar-cane Research Stations in India. The data are summarized as follows:

**Table XI.**

Relative salinity tolerance of sugar-cane varieties in the field.

| Particulars | Relative salinity tolerance | | |
|---|---|---|---|
| | Good | Medium | Poor |
| Varieties of cane | CO.205; CO.285 | CO.290; CO.210; CO.312; CO.331; CO.421; CO.453; CO.513; CO.L.9; CO.S.109; CO.S.321; B.O.3; B.O.4; B.O.10; B.O.24. | CO.213; CO.214; CO.281; CO.313; CO.419; CO.499; CO.527; CO.622; CO.S.245. |

Besides restricting the growth of crop, salinity also influences the quality of cane i.e. its polarization and mineral content. When varieties such as CO. 214 and CO. 281 accumulate excess of sodium in the soil, the sugar recovery is adversely affected. But in general, sugar-cane from the aspect of germination and growth, can withstand a high degree of salinity in the soil.

It is notable that in cotton the short staple varieties are more tolerant to soil salinity than the medium and long staple strains. In case the salinity is in the sub-soil, the late sowing of medium staple (2.25 cm) cotton has proved distinctly advantageous in the Punjab (Dastur & Singh, 1942). Wahhab et al. (1957) found less striking differences in cotton varieties during the germination stage than at later phases of its life cycle. Shchipanova (1955) observed that when $SO_4$ constituted 50–60% of the solid residue, the maximum limit permitting growth of cotton was 0.3% of total solid residue. Similar observations were recorded by Pensokoi (1956) at the shoot emergence stage. However, at the maturation stage 0.6% was the tolerable salinity level.

Saline grazing land:

With regard to halophytic plant communities used for grazing it may be added here that for more than 30 years numerous papers on plant sociology have been dealing with this subject. In many of these papers, particular attention is paid to the striking zonational pattern and the distinct plant associations according to the salinity and the fluctuating groundwater depth or the open water level of shallow salt lakes (see page 133 ff).

One of the first detailed plantsociological studies describes such zonations near the Neusiedler Lake on the Eastern border of Austria, investigated in the years 1928—1935 (H. Boyko, 1931, 1947). Most of the numerous salt lakes there dry out completely in late summer, leaving a white salt cover looking like a snowfield in the centre, with differently coloured vegetational zonations around it, until a Cynodontetum belt marks the transition zone to the semi-arid steppe plant communities higher upwards. The greater part of these areas are good grazing grounds. The highly salt resistant Suaedetum zone near the centre, with *Suaeda maritima* and *S. salsa* as the main components, constitutes a rich source of food for the large herds of pigs, whereas the Plantaginetum maritimae and the other plant communities of a more moderate salt resistance are grazed by cattle or used as meadows for mowing (H. BOYKO, 1932).

In India, the succession of grass species from highly saline to least saline conditions consists of *Sporobolus* spp. → *Cynodon dactylon* → *Echinochloa* spp. → *Aristida* spp. → *Elusine compressa* → *Dactyloctenium sindicum* → *Eragrostis tremula* → *Dichanthium annulatum*. The cultivated species can be introduced at different stages of reclamation of salty land.

*3. Vegetable crops*

Tests on relative salinity tolerance of vegetable crops have been carried out at various Institutes. The tests on table beets, spinach, asparagus, and tomatoes indicated that these have good tolerance to salinity and cabbage, cauliflower, broccoli, lettuce, onions, potatoes, and some of the Cucurbitaceae have moderate to fair tolerance. Beans and radishes have only a poor tolerance in this respect. (U.S. Salinity Laboratory, 1954; BERNSTEIN et al., 1951). Though the varieties of green beans showed wide variations, and differences were significant, they had the lowest degree of salt tolerance of any of the vegetable crops tested (BERNSTEIN & AYERS, 1951). Of the spinach types Swiss chard (sada palak) has a higher salinity tolerance than Indian spinach and New Zealand spinach.

*4. Fruit crops*

Most of the fruit trees are sensitive to high salinity. One of the few exceptions in this respect known since ancient times as such is the date palm which flourishes under highly saline conditions both near the sea coast and in the oases. HAYWARD et al. (1946) observed that Alberta peach can tolerate only a moderate amount of salinity. The yields tend to decline over a period of years when ECe of the substrate exceeds 2 mhos. Pears and apples, although being adapted to a more humid climate, are more salt tolerant than stone-fruits such as peaches, plums and almonds (HILGRAD, 1906). Citrus fruits belong to the most salt sensitive category, and of these lemons are the most sensitive, grape fruit the least sensitive. Oranges occupy an intermediate position (KELLEY & THOMAS, 1920; LOUGHRIDGE, 1906). Other salt sensitive fruit trees cultivated also in the semi-arid and arid belts are Persian walnut, mulberry and avocado (AYERS, 1950; KEARNEY & SCOFIELD, 1936). Stone-fruit trees, avocado and citrus accumulate both sodium and chloride in their leaves with subsequent injury. Grape varieties exhibit a wide variation in chloride accumulation. If orange, plum, apricot, and almond trees absorb toxic amounts of salts, defoliation occurs due to this high concentration in the cell sap. In coastal areas, coconut stands temporary inundation with saline water. Like the date palm *Zizyphus mauritiana* is exclusively a plant of arid and semi-arid regions. Even the seeds germinate readily in saline areas, and seedlings make satisfactory progress. Pomegranate has a moderate to fair salinity tolerance and seedlings once established make satisfactory progress. When the proportion of $Na_2CO_3$ in the total salt concentration in the soil is high, it becomes, however, necessary to irrigate frequently at the fruiting stage. In Indian Malta orange raised on Jatti Khatti rootstock showed good results on soils recently reclaimed from salinity. But the plantations tend to deteriorate with a high cationic ratio. The leached soils must have sufficient amounts of active lime and exchangeable potassium to offset the effect of sodium in the soil. Other fruit trees less sensitive to salinity are guava, mango and jack fruit. The grafted varieties of mango and litchi are relatively more sensitive than the ungrafted ones. Extremely sensitive plants to salinity are papaya, coffee and cocoa, of which papaya and coffee are now also cultivated in sub-tropical, semi-arid regions.

*5. Forest trees*

In general, conifers are relatively more sensitive than deciduous trees. Under temperate environment, the trees which showed good tolerance were silver poplar, weeping willow, Russian olive and Siberian elm (SNYDER et al., 1940). Both oak and ash could withstand Cl injury better than the sensitive species of pecan, elm and hickory. TSING et al. (1956) studied the germination and seedling growth of 20 tree and shrub species in pot experiments using soil with a sodium chloride content up to 0.6%. *Azadirachta indica*, *Ailanthus altissima*, and *Sapium sebiferum* showed the least sensitivity to higher salinity levels. The stem cuttings of *Tamarix chinensis* and root cuttings of *Ulmus pumila* also exhibited a high salt tolerance. On calcareous saline soils birch, willow and aspen grow spontaneously (KURCHERENKO, 1957). In solonetzic soils, *Ulmus pinnato-ramosa*, *Robinia pseudacacia* and to a lesser degree *Acer negundo* develop best in the restricted A Horizon and the roots of *Elaeagnus angustifolia* were able to penetrate the dangerous B Horizon (POVETEV 1957; AGAFONOVA 1957). The observations of POPOVA (1957) have indicated that the tree species enumerated below are tolerant to salinity in the following descending order: *Tamarix pallasii*, *Tamarix ramosissima*, narrow leaved olive, honey locust, elm, green ash, Canadian poplar, white mulberry, apricot oak and ash-leafed maple.

For the tropical region the Indian Forest Research Institute, Dehra Dun, has summarised its observations as follows:

1. Saline soils containing over 0.16% soluble salts, even of a low pH value, but possessing a hard pan of stiff clay or Kankar do not support trees or shrub species.

2. Saline soils containing less than 0.16% soluble salts up to a depth of 3 feet with no clay or kankar pan, even though they have a pH value above 9.0 in lower horizons have a scattered growth of *Acacia arabica, Salvadora oleoides, Tamarix* spp., *Capparis decidua, Zizyphus* spp., *Azadirachta indica, Albizzia lebbek, Calotropis procera* and *Prosopis spicigera.*

3. Saline sodic soils with an efflorescence of salts do not bear any tree growth.

4. Saline soils with a high water-table but without a pan or a hard pan lower than 2½ feet depth, carry tree species of *Prosopis juliflora, Acacia arabica, Acacia leucophloea, Capparis horrida, Salvadora oleoides, Capparis decidua.*

5. When the salt content in the surface horizon is less than 0.16% and a clay or kanker pan is below 2½ feet depth, the tree species sustained are *Acacia arabica, Salvadora oleoides, Zizyphus* spp. and *Prosopis spicigera.* Quite generally salty lands in India are more suitable for raising fuel plantations than timber species. *Dalbergia sisso,* a timber species, once established, makes, however, a good growth and withstands a moderate salinity. After preliminary leaching and lister ploughing of Bari soils, which have a high content of carbonates and bicarbonates, *D. sissoo* exhibited the best performance (SINGH, 1963).

## Effect of Aridity and Salinity on Vegetation

Halophytes have a close resemblance to xerophytes due to their frequently very similar structural features. Quite a large number of them are succulents, but non-succulents have also adopted the property to withstand a high salt concentration in the soil or even submersion in sea-water. Succulence is commonly facultative. The sap of succulent species has often a high osmotic concentration which enables them to absorb water from soil solution of high salt concentration. As the seeds of many species are unable to germinate due to a low imbibition pressure developed during the process of germination, their seeds germinate on the parent plant (viviparous habit). This can be observed on many Mangrove plants. Some of the plants show an extensive development of aeration tissues for the conduction and storage of oxygen (DRAR, 1955).

Halophytes as distinct from xerophytes have to endure a much greater stress because the "total moisture stress" in saline soils at the same moisture level is much higher than in salt-free soils. Besides, the halophytes can withstand salt injury much easier due to the accumulation of sodium in the plant sap or due to specific ionic effects resulting from absorption and accumulation of Cl, $SO_4$, $HCO_3$ or $CO_3$. These differences in their make-up also distinguish them from the xeromorphic plants. In fact, halophytes by and large are xerophytes except those which are adapted to aquatic habitat. REPP (1958) has noted that in succulent plants the moisture percentage is high and this dilutes the salt absorbed and defers the harmful effect of salt levels. According to REPP et al. (1959) salt tolerance is primarily to ascribe to the degree of resistance of the protoplasm to salt. HENCKEL & ANTIPOV (1956), therefore, divided the halophytes into "halo-succulents" and "haloxeritic". The former have a low daily and residual moisture deficit and a high water content and the latter have a high moisture deficit and a low water content.

Most of the sclerophyllous plants are adapted to arid or semi-arid conditions but selected ones have adapted themselves in addition to this also to saline habitats. Their rate of assimilation is commonly lower than that of mesophytic plants per unit of area. The leaves have thickened cell walls and great lignification derived from sugars through tannins and condensation of fatty acids to form a thick epidermal cuticle (WOOD, 1934). They are able to endure water loss up to 20% of their water content without injury. The high hydration capacity is due to the presence of hemicelluloses. High water loss is prevented by their lignification, and their thick cell walls in any case prevent a collapse even if the water loss is very high (DAVIES, 1955).

The soft succulent leaved shrubs of *Atriplex vesicaria, Kochia sedifolia* and *Bassia* spp. are able to withstand prolonged drought and show halophytic characters. These are halosucculents. Their mesophyll cells have a high water storage capacity and the leaves have a dense felt or vesicular hairs on the epidermis, but have a less thick cuticle. The cells contain pentosans and show a high protein content, leading to a high hydration capacity. This enables the plants to reduce transpiration and to retain a certain stability of their water content (DAVIES, 1955). BEADLE et al. (1957) have observed that *Atriplex* species are definitely stimulated by addition of salts to a base nutrient (50 meq/l appeared as optimum); NaCl stimulates the most, $Na_2SO_4$ somewhat less and KCl the least. Other examples of halosucculents are *Suaeda prostrata* and *Salicornia herbacea;* examples of haloxeritic plants are *Petrosimonia triandra, P. litwinowii, Echinopsilon hyssopifolium* and *Salsola soda* (HENCKEL & ANTIPOV, 1956).

When the plant experiences a water stress, the viscosity of the protoplasm at first decreases (reaction phase) and then starts to increase (restitution phase), but when the xerophytic conditions are developed slowly, the viscosity increases slowly without the initial decrease. On the other hand the permeability during the "reaction phase" decreases (STOCKER, 1960). LEVITT (1956), however, has not observed these two phases. According to him on dehydration the protoplasmic viscosity increases and permeability is reduced. STOCKER (1960) further observed that the osmotic value of the cells increases due not only to the decrease in water content but also to the formation of osmotically active substances. Thus xeric plants have a manostatic system in the cells which helps them to regulate their osmotic concentrations (THIMANN, 1960).

The rate of respiration of plants when subjected to conditions of drought increases at first and then falls off below the normal rate (STOCKER, 1960). HENCKEL (1954) observed a similar reduction in respiratory activity as an adjustment to salinity, on the other hand, THIMAN (EVENARI, 1953) observed a marked increase in the respiratory rate of leaves at salinity levels that affect the growth of sensitive and moderately tolerant crops. For the more tolerant crops, respiration increased at the higher levels of salinity compared to the moderate levels which varied from 0.17 to 2.9% salt in the soil. POKROVSKAIA (1957), and SARIN & RAO (1958) found a very close correlation between hydration (water content) and respiration of seedlings when these were maintained at 0.6% $Na_2SO_4$ solutions. The effect of salt on respiration levelled off at 60 hours while the effect on hydration continued to increase in wheat seedlings up to 96 hours. The obvious conclusion is that beyond a critical level hydration may no longer have been a limiting factor for respiration. Glyco-halophytes such as *Artemisia salina* and *Atriplex tartarica* tend to have higher respiratory rates. Their respiratory rate decreases markedly with increasing salinity and they restrict also the accumulation of salt in the cells (HENCKEL, 1954).

In xerophytes photosynthetic behaviour has been studied by STOCKER (1960). Under natural conditions the xerophytes have two peaks, the photosynthetic activity decreasing during noon hours. When the water stress is great, the curve of photosynthesis shows one peak in the morning hours and no recovery during the afternoon. Sometimes the deepest point of the curve lies below the compensation point. Under conditions of extreme water deficiency, when the stomata are completely closed, the whole curve is below the compensation point. The decline occurs even when stomata are open and transpiration still continues (SAAD, 1954). The accumulation of carbohydrates depends upon the water stress rate of photosynthesis. Under conditions of water stress, xerophytic plants accumulate largely disaccharides which increase the osmotic pressure of the sap. Normally the monosaccharides tend to disappear as soon as xeric conditions set in. Under saline conditions the plants exhibit frequently higher concentrations of carbohydrates. There is, however, one general exception to this. In *Goebelia alopecuroides*, AKO-

PIAN (1957 and 1958) observed that at a salinity level of 1.3% (with approximately 0.3% Cl and 0.5% $SO_4$) there occurred a decrease in monosaccharides and a corresponding increase in the disaccharides concentration in the leaves. The decrease in total carbohydrate accounted for 2—4%.

Under dry conditions the plants show an increase in protein content. The wilting changes the protein-amino acid equilibrium in favour of the amino acids and causes proteolysis. The synthesizing activity of the proteoses decreases (STOCKER 1960). "Regarding hydrolytic enzymes, it has been generally reported that wilting intensifies their hydrolytic activity to the detriment of their synthetic activity. The shift of equilibrium towards hydrolysis under conditions of water deficiency is more pronounced with drought sensitive than with drought resistant varieties" (EVENARI, 1961). In *Goebelia* plants, under saline conditions, certain amino acids disappeared. The amino acids lysine, threonine, and glycine were present in leaves from trees grown on non-saline soil (TER-KARAPETIAN, 1957). Similar increase in amino acid content from 10% to 15 and 18% (chloride and sulphate salinity respectively) in dry matter of plant was observed by STROGONOV et al., 1956).

In arid regions, salt accumulation occurs particularly in the lowlands or in identical geomorphic situations such as flood-plains, deltas, troughs, low river terraces, lake or coastal terraces. All these are locations where evaporation and transpiration losses are the highest whereas any compensatory precipitation is very low, so that the capillary rise of the groundwater, particularly in soils rich in clay particles, favours the deposition of salts on the surface. In consequence, salinization progresses at full rate. The rate of evaporation is also very much enhanced when the ground-water level is 2—3 m or less and with this intensity of evaporation under arid climatic conditions, the salt accumulation process attains maximum values. Here the evaporation of ground-water far exceeds the precipitation.

Near the sea cost, such lands may in addition rather frequently be flooded by high tides and hurricanes. The accumulated water evaporates and leaves the residue of salts on the ground and in the soil. A third and main case of salt accumulation is caused when irrigation waters have a high content of total soluble salts and there is inadequate precipitation to leach the salts down. In all these cases the soil is soon impregnated with salts in an arid environment (KOVDA, 1961b).

## Reclamation Methods

Against this calamity, various reclamation measures are adopted in different countries. Some of them may be discussed below and, again—because of the particular importance of this chapter in general—a more detailed regional example is presented by describing the measures used in India.

1. Australia

In Australia there occur two types of salty lands. The soils which have primary salinization extensively exist in the vast arid regions of Western Australia, South Australia, North-West N.S. Wales, Victoria and Western Queensland. Here the rainfall is low and felling of great parts of the natural vegetation has accentuated the salinization of basins and slopes. In the sand dune areas of South Australia salinity develops at the base of the dunes. Here too the hydrological equilibrium is disturbed by removal of native vegetation. An application of gypsum and contour furrowing of these lands restores the physico-chemical balance. The following greater storage of water in the soil and leaching out of the salt restores to a certain degree the native vegetation and seeding with forage species may turn such areas even into good grazing lands. In the initial stages, native salt tolerant perennials such as *Kochia* spp., and *Atriplex* spp. are seeded and after a period buffalo grass is introduced (SMITH, 1956; SMITH & MALCOLM, 1959; STONEMAN, 1958a; 1958b). In Western Australia, the hydro-chemical balance in the wheat sheep belt of the arid zone has been restored by adopting a system of ley farming. This consists of planting a ley of Woomera rye grass and subterranean clover which is maintained for 3—4 years with sheep grazing on it. It allows the leaching of the salt, and increases the organic matter and the nitrogen content. After this part of the reclamation work, wheat is taken for a year or two as the crop yielding plant.

Secondary salinization has occurred largely in the irrigated sections of Queensland. Here the water-table has risen very close to the soil surface and the experimentors are confronted with solonetz soils covering an area of 10,000 acres south-west of Brisbane. The soil has a sharply defined prismatic $B_2$ Horizon at 6 to 18 inches depth overlain by loam and clay loam. The exchangeable sodium ranges from 14 to 33% in the A Horizon and 34 to 45% in the B Horizon (ISBELL, 1958). Similar secondary salinization occurred in the Murrumbidgee irrigated areas of New South Wales where 20% of the irrigated area has been salinized. In all these areas production of rice and irrigated mixed pastures has been taken up to restore the hydrochemical balance of the soil. The mixed pastures of *Phalaris tuberosa,* Woomera rye grass and subterranean clover are raised under irrigation for 3—5 years and are followed by rice for one to two years. In 60% of the horticultural areas tile drainage has been installed to keep the water-table down. In other saline areas leaching of salts with open drainage has effectively reduced the salinity hazard.

2. Egypt

For the reclamation of salinized irrigated land, the Department of Agriculture recognises, on a pedological basis, six classes of land. The characters taken into account are soil texture, structure, $CaCO_3$ content, cations and anions in the saturation extract, organic matter, mechanical analysis, electrical conductivity of the saturation extract and total soluble salts. Class I soils are free of salt and are capable of producing all crops; class II soils are slightly salty; class III moderately salty; class IV very salty; in the soils of class V drainage is impaired; and the soils of class VI are unsuitable for cultivation.

The slightly salty lands have 0.1 to 0.2% salt content and their conductivity ranges from 2 to 4 millimhos/cm at 25° C. The rotation adopted on such lands is *Cyperus*-cotton-maize. Their relative tolerance to salinity is in the order shown by this rotation. After maize the test crop sown is *Vicia faba*. Then the rotation is repeated to keep the soil free of salts.

The moderately salty lands have salt concentrations of 0.2 to 0.5% of total soluble salts and the ECe of the saturation extract ranges from 4 to 8 millimhos/cm. Such lands require prior leaching before rice can be planted. They require application of gypsum at the rate of 2 tons per acre per annum for three years. After the first year's leaching, rice is planted in order to further leaching of the land and to render possible the growing of *Cyperus* and subsequently of cotton for reclamation.

Usually a 80 hectare piece is taken up at a time for reclamation. The drains are 3 m deep and are laid out 200 m apart. The subsidiary drains, 25 m apart, join the main drains. The land is flooded to a depth of 15 cm each time and the leachate is drained out. Gypsum is applied at the rate of 5 tons per hectare. It is mixed up with the soil by ploughing up the fields. Within a year already very appreciable improvement is effected in the physical condition of the soil and its productivity. Usually a rice crop can be established (Hilmy, 1944).

3. Hungary

The salty lands of Hungary are in different stages of deterioration. The reclamation measures include mixing of solonetzic and calcareous horizons, application of powdered lime, and lime and gypsum together, and press mud and molasses from sugar refineries. In the initial stage of salinization, leaching with excess water reclaims the land. In the second and third phase of alkalinization, when the sodium ions have been adsorbed by the clay particles to form sodiumized clay, treatment with gypsum is indicated; gypsum and lime and leaching with excess water reclaims the land. Lime suffices when the pH is less than 8. The solods show a neutral or weakly acidic reaction in their A Horizon, and the addition of chalk and organic matter together with a lowering of the water-table reclaims them. In the meadow solonchak like solonetz soil, the water-table is to be found at shallow depths of 3—5 m and seldom at 8 m. The ground-waters are slightly or moderately mineralized. Every summer, the water rises by capillarity and periodically introduces exchangeable sodium into the absorbing complex of the solonetsy soil (SZABOLCZ, 1956a). The lowering of the water-table by this process is essential and we can observe considerable seasonal fluctuations as a consequence, with a maximum height of the ground-water table in spring and a minimum in late summer and early autumn.

4. India

In India numerous and diversified methods are used in the fight against salinity according to the various soil types. In the Punjab, an extensive field-survey of salty lands was carried out and on the basis of the physico-chemical characteristics, their soils were classified into the following four categories:

Category A: Total soluble salt content not exceeding 0.5% and pH value less than 9.0.

Category B: Total soluble salts exceed 0.5% and pH value varies from 9.0 to 9.2.

Category C: Total soluble salts exceed 0.5% and pH value ranges from 9.2 and 9.5.

Category D: Total soluble salt content in the soil exceeds 0.5% and pH value higher than 9.5 (MCKENZIE-TAYLOR, 1940).

More recently the factor of clay content has also been taken into consideration in categorizing such soils (UPPAL, 1963).

The soils of category A are in their initial stages of salinization and can be reclaimed by growing a rice crop for one year with a prior leaching. Such initial leaching helps to increase the water penetration and reduces the salt content considerably. This first rice crop is followed by berseem which too has a high water requirement. Those soils where normal rotation of crops can be introduced contain less than 0.2% total soluble salts and their pH does not exceed 8.5. When, however, the salt content varies between 0.2 and 0.5% and pH between 9.0 and 9.2 two crops of rice in succession with two intervening berseem crops are used to reclaim the land.

Soils of category B require prolonged leaching before a coarse variety of rice can be transplanted on the land. Three or more crops of rice are grown to reduce the salt content and also the pH. The hydrological regime has to be such that the capillarity rise during the summer season is cut off. Much larger quantities of water are required to reclaim these soils than those of category A. The cost of reclamation is compensated by the returns from rice and berseem crops, and where the water costs are not too high it is economical to reclaim them.

The soils of category C can only be reclaimed if the costs of irrigation water are low such as for instance by spring flows, artesian water or canal water allowed at special low rates. Once the leaching process, by flooding the fields, has set in, application of gypsum is quite beneficial. Dressing of farmyard manure improves the structure of these soils. Growing of Dhanicha crop, proceeding the rice, is extremely helpful in increasing the permeability of such soils (McKenzie-Taylor, 1940).

*Sesbania aculeata* (Dhanicha) has a high calcium content and a very acidic sap compared to other green manure crops (Sandhu, 1955; Dhawan et al. 1958) and it fixes the largest amount of nitrogen in saline-sodic soils (Uppal, 1955). In a factorial experiment conducted in the Punjab, green manuring with Dhanicha coupled with application of NPK produced significantly higher yields of paddy and subsequent wheat crop as compared to crops where neither green manuring nor NPK fertilizer were applied. The green manuring with Dhanicha had a marked beneficial influence on the reclamation of saline-sodic soils. Its extensive root system opened up the compact sub-soil and improved its permeability which facilitated the leaching of salts (Kanwar, 1962b). When neither sufficient water was available to grow the Dhanicha crop nor enough water for pre-leaching of the salts, an application of 2 tons of gypsum to the soil at the time of transplanting the rice seedlings and an application of 50 lb. N and of 25 lb. $P_2O_5$ per acre served to reduce the conductivity of the saturation extract from 10.43 to 0.98 in the A Horizon. The yields of paddy and barley crops were comparable to pre-leaching with three feet of water (Kanwar, 1962a). 30—50% of the gypsum requirement of the soil as determined by Schoonover's methods seems to be the adequate dose of gypsum for saline sodic soils with a pH of 8.4 and ECe of 10.43 millimhos/cm at 25° C (Kanwar, 1962b).

Another system of reclaiming salinized soils evolved and adapted in the Punjab was by establishing *Cynodon dactylon* grass after scraping the whole salt efflorescence from the ground surface. Such soils do not have a compact hard pan in the B Horizon and the water-table is usually below 10 feet. With abundant and frequent irrigation and a good basal dressing of farmyard manure the grass establishes itself and opens up the soil. After its establishment it is readily grazed by sheep and cattle. Their droppings add humus to the soil. These soils require three years of rest under the grass for reclamation (Cole, 1940).

Category D soils are difficult to reclaim economically. They require large investment in application of amendments (McKenzie-Taylor, 1940). However, recent work indicates that before long it may become possible to reclaim such land by judicial use of amendments, fertilizers, leaching and growing of Dhanicha crop in rotation with rice.

In Uttar Pradesh, the Usar land Reclamation Committee (1939) classified the saline soils from the aspect of reclamation as follows:

Group I: Mildly alkaline—pH 7.6 to 8.7; easily reclamable.

Group II: Moderately alkaline—pH 8.7 to 9.4; more difficult to reclaim than those in group I.

Group III: Highly alkaline—pH 9.4 to 10.4; difficult to reclaim.

The group I soils can be reclaimed by leaching down the salts simply by the application of heavy doses of water. After this preleaching, the green manure crop of Dhanicha is raised, which increases the permeability of the soil, accelerates the leaching process and reduces the pH. This step is followed by transplanting paddy in the next Kharif season. DHAR & MUKERJEE (1936) treated mildly salinized soils with molasses, molasses gypsum, molasses powdered sulphur and press mud, the byproduct of the carbonation process in sugar factories. He observed that soluble calcium salts replaced the sodium in the base exchange complex and that sodium clay was turned into calcium clay, which improved the physical condition of the soils considerably. Such soils could then be planted with sugar-cane after preliminary leaching.

For the saline-sodic soils of group II an addition of gypsum was essential to reduce the exchangeable sodium by 60 to 70% in the base exchange complex. The quantity of calcium added through gypsum was almost equivalent to the exchangeable sodium. Thereafter leaching and flushing was necessary to drain out the salts from the profile. Growing of green manure crops of Dhanicha followed by paddy in Kharif, ameliorated the condition of such a soil in two years to a state where the salt tolerant crop of barley could be raised (YADAV & AGARWAL, 1959).

In several areas of Usar lands the A horizon, with a thickness of up to 18 inches, has a very high content of total soluble salts, high in soluble and exchangeable Na, and a very high pH, from 8.5 to 11.3. Such soils have a hard indurated clay pan or kankar pan at a depth of 2—3 feet, with or without a high groundwater table. Leaching of such soils is not feasible as the hard pan prevents water penetration. Unless the pan is broken by deep listing, such soils are difficult to reclaim (AGARWAL & YADAV, 1954; AGARWAL, MEHROTRA & GUPTA, 1957). The most economical method of reclaiming them is to plant salt tolerant species of trees, such as *Prosopis juliflora, Acacia arabica,* and *Tamarix articulata,* in pits in which the hard pan has been broken and large doses of farmyard manure have been applied together with non-saline soil in order to establish the seedlings. When the forest has started to be established the geohydrological balance of the soil is restored. Grasses such as *Sporobolus marginatus, Cynodon dactylon,* etc. invade the land and are succeeded by more palatable species. Thus, over a long period, such lands can be brought to a state where the pH is reduced and sodium is leached out of the soil or removed by vegetation. Soils like these belong to group III. On such a soil a reclamation project had been started at Chakeri near Kanpur in 1951. The pH of this soil was 11.0; it has a very high salt content and a hard clay pan, but successful reclamation has been achieved by mechanical shattering of the hard clay pan followed by leaching with sewage water. The rotation adopted was paddy-barley-*Mentha.* In 3 years the pH has been reduced to 9.5 and the salt content from 0.59 to 0.2% (RAYCHAUDHURI & DATTA BISWAS, 1958).

In the Khadar lands of Ganga and Jamna, there occur extensive floods from time to time. These flooded lands then remain submerged over long periods. The ground water, being close to the surface, creates a white efflorescence on the surface during the summer season. The drainage is imperfect and the characteristics of the mildly calcareous alluvial soils in these depressions are their alkaline reaction (pH 8.5 to 9.0) and their saline nature. Provision of an adequate drainage system linked to the rivers will be the proper way of reclaiming such areas.

The Kallar soils belong to group III. These soils largely contain $Na_2CO_3$ and $NaHCO_3$ and no other alkali salts. Their pH ranges from 9.4 to 10.4. Since they have a high clay content their permeability is low. These non-saline sodic soils can be reclaimed by addition of gypsum and mechanical breaking of the hard and compact crust, but the cost of their reclamation is prohibitive (LEATHER, 1914; Usar land Reclamation Committee, U.P., 1939).

The extensive area of saline sodic soils in the Little Rann of Kutch and the Saurashtra area of Gujarat can be reclaimed by leaching, provided that a further encroachment of sea-water is shut off by an embankment from the adjoining sea coast. SHAH et al. (1957) have suggested progressive leaching operations by conservation of the flood waters of the river Banas which flows through this area.

The salinized soils of Maharashtra were classified by TALATI (1947) from the aspect of reclamation as follows:

*1. Mixed saline soils:* They have a low percentage of sodium in the base exchange complex and a fairly high percentage of divalent bases. These soils can be reclaimed by simple leaching and after that by growing rice and sugar-cane for their high water requirements.

*2. Saline soils:* For them, a high content of Na in the soil profile is characteristic, having accumulated due to a high groundwater table. The pH value is below 9.0. These soils too can be reclaimed by growing, as a green manure crop, Dhanicha with a basal dressing of 3 tons of gypsum or $\frac{1}{2}$ ton of sulphur per acre. After addition of this soil amendment and the growing of such green manure crops, sugar-cane can be successfully cultivated.

*3. Alkali and strong alkali soils:* These are stiff alkalinized soils, in which exchangeable sodium ranges in the base exchange complex from 20 to 60% and the pH value is over 9.0, and sometimes even 10, with carbonates predominating. These soils, however, are very intractable and require a prolonged leaching and the application of large doses of gypsum or sulphur. In consequence the cost of their reclamation is very high.

In the mixed saline and in the saline soils the addition of fertilizers and organic matter proved beneficial, and still more so the inclusion of irrigated crops such as lucerne and Shevri *(Sesbania aegyptiaca)*, whereas summer fallowing had definitely a harmful effect. In the case of alkali and strongly alkali soils, where the B Horizon has been compacted an addition of $\frac{1}{2}$ ton sulphur and 2 tons F.Y.M. with the provision of efficient drainage proved quite effective (BASU & TAGARE, 1943a).

## 5. Iran

The four projects where a reclamation is in progress are in the Shahan kareh, Gramsar, Ghezelhesser and in the Khuzestan areas. Here too, salinity has developed because of the rise of the ground water table. Now, land classification and drainage surveys are simultaneously conducted there. Since most of these soils have a high content of gypsum, which makes the reclamation work much easier, only leaching operations based on this land classification are carried out there, in order to reclaim these lands (AYAZI, 1958).

## 6. Iraq

Similarly, in Iraq a land classification survey is conducted along with a drainage survey in order to maintain the saline lands in a productive state. This survey includes texture, depth, permeability or hydraulic conductivity, total soluble salts, exchangeable sodium, gypsum, lime, cations and anions and toxic elements. Most of the aggregated soils of the alluvial Mesopotamian Plain in Iraq contain some gypsum, have a high lime content, and an expanding type of clay. Therefore no amendment is applied. 80% of the soluble salts are leached out of 1 m depth with one meter depth of water. To leach out the salts from 2 m depth, the water amount required is 3 m. Soil amendments are applied only where the lime and gypsum content is low. These amendments are incorporated into the soil by ploughing and thereafter the leaching process is carried out. Where sulphur is applied, time is allowed for it to oxidize and to form gypsum in the soil (AWAN, 1958).

### 7. Pakistan

The salinized soils of Pakistan are classified on the basis of their physico-chemical characteristics and are treated accordingly. Initially the soils were classified on the pattern reported by McKenzie-Taylor (1940). Now exchangeable sodium percentage (ESP), ECe of saturation extract, pH value and clay percentage are the chief factors taken into consideration in regulating the reclamation operations. These soils are mostly saline in nature and, therefore, preleaching of salts by large quantities of water and subsequent cultivation of rice in the Kharif season, when monsoon discharge is available in the canals and from the rivers, are the main features of reclamation operations. The reclamation water is supplied at the rate of one cu. m/sec. per block of 45 acres during the Kharif season. The water-table is maintained by providing 3 to $3\frac{1}{2}$ feet deep drains.

In saline sodic soils, the exchangeable sodium content is high and the soil is compact. Their treatment with gypsum has resulted in some improvement, but the growing of *Sesbania aculeata* is particularly beneficial for the improvement of soil permeability. This plant is grown consecutively for two years and this improvement of the permeability also increases the leaching rate of the soil considerably. In this agricultural process, *Sesbania* is followed by rice, but about six years are required to bring the land to full production where the water-table is below 10 feet (Asghar, 1958).

It had been previously observed that stabilization of the salt layer at 8 feet depth enables the production of even the most sensitive crops like gram *(Cicer arietinum)* on reclaimed lands (McKenzie-Taylor & Mehta, 1941). Later research work in this field has indicated that in lands where the zone of salt accumulation is below 6 feet or where the zone of salt accumulation is only at a 4 feet depth, rice planted in alternate years keeps up the normal soil productivity (Asghar & Khan, 1958).

### 8. U.S.A.

A scientific control of salinity in the soil has evolved by systematic investigations conducted at the U.S.D.A. Soil Salinity Laboratory, Riverside, California. This involves detailed surveys of the salinized areas; studies of the sources and causes of salinity; of the quantities and kinds of soluble salts present; investigations of the pH value of saturation extracts; of the nature of exchangeable ions in clay; determination of the quantity of calcium carbonate or gypsum present; of the texture of the soil and its relative permeability; natural drainage; chemical analysis of water for leaching and investigating the quantity of water available for reclamation, the current soil management and irrigation practices in the area. The reclamation procedures for saline soils are somewhat different from those of sodic soils. For the reclamation of the former the first step is to provide drains to reduce the water-table to approximately six feet. The second step in the reclamation process is the leaching of salts by flushing the land with large quantities of water on the basis of quality of water, pH value, ECe of the saturation extract and of the clay content of the soil. At least 10% of the water on the soil must be drained off. The water is ponded in basins with intercepting borders laid out on the contour at short intervals. Where encrustation of salts exists, the area is flooded, and water is drained off dissolving these surface salts. In these cases leaching is not permitted. When the soil has a clayey texture and permeability is low, the land is allowed to crack, which facilitates the penetration of water. When the pH value has been reduced to 8.5 and ECe to 8—10 millimhos per cm at 25° C, a rice crop is usually sown, and is followed by sweet clover mainly used as green manure. After these steps have been taken perennial fodder crops like strawberry clover, bermuda grass, reed canary grass, or western wheat grass are maintained for a number of years and are grazed on the field. Growing of wheat completes this reclamation process of saline soil.

The reclamation of alkali soils involves the neutralization of excess salinity, leaching of soluble salts, replacing exchangeable sodium with calcium and improving the physical condition of the soil. For leaching of the salts, basins with intercepting borders are laid out on the contour and a drainage system is installed. In irrigation districts, where commercial crops are grown, tile drains are installed. Lime, necessary for the replacement of sodium, is made available by an application of lime or gypsum. The following reactions take place in the soil under moist conditions:

$$CaSO_4 + Na_2CO_3 \rightarrow CaCO_3 + Na_2SO_4$$

($Na_2SO_4$ can be readily leached)

$$CaSO_4 + \text{Na clay Na} \rightarrow \text{Ca clay Ca} + Na_2SO_4$$

($Na_2SO_4$ can be readily leached)

Sulphur, although applied in one fourth only of the dose of gypsum is often very costly. KELLEY (1937) found 3600 lb. of sulphur more ameliorative than 12 tons of gypsum on an alkali soil which had a low content of free lime. The reactions after sulphur application are expressed as follows:

$2\,S + 3O_2 + 2H_2O \rightarrow 2H_2SO_4$ (Reaction brought about by sulphur bacteria in soil)

$H_2SO_4 + Na_2CO_3 \rightarrow Na_2SO_4$ ($Na_2SO_4$ can be readily leached)

$H_2SO_4 + Ca\,CO_3 \rightarrow CaSO_4 + H_2O + CO_2$

Thus $CaSO_4$ and $H_2SO_4$ both react in the soil by forming sodium sulphate which is easily leached by excess water. Such leaching removes all the salts from the soil and the leaching process is continued until the soluble salt percentage is close to 0.2 and electrical conductivity is between 4 and 8 millimhos/cm. At that stage, rice is put in, in order to further the reduction of salinity. The soil is completely reclaimed when the salt content up to 3 feet depth is below 0.2% and ECe is below 4 millimhos/cm. It is, however, also essential to reduce the water-table where it is too high in order to bring about a permanent reclamation. Good management of soil and water is, of course, continuously needed in relation to cropping so that no reversion to salinization of the reclaimed soil may occur (THORNE & PETERSON, 1950).

9. U.S.S.R.

For the purpose of reclamation, the solonchaks are broadly classified into soda solonchaks and others. The former have a very low permeability, a high sodiumization of clay, and their structure is prismatic. Leaching of soda solonchaks is, therefore, difficult. To increase their permeability, deep tillage, laying out of shallow drains and application of gypsum or sulphur to neutralize the high alkalinity are essential. In the other types of solonchaks the hydrophysical properties are far superior, and the salt can easily be leached with large quantities of water. With adequate quantities of gypsum or lime in the soils, their structure is not so adverse, and no amendments are needed (KOVDA, 1961b). If the ground water is within the "capillary fringe" the provision of drains is most essential. This applies particularly to the "wet solonchak" soils (ground water at 1.5—1 m) and to the "moistened solonchak" soils (ground water at 2—3 m). In the case of residual "dry solonchak" soils a mechanical removal of the efflorescence by scrapers or graders is essential, and leaching of these soils requires larger quantities of water than that of other categories of solonchaks. In U.S.S.R., the soil characteristics studied for the reclamation of solonchaks are texture, degree of salinity of the soil and the sub-soil, the level of ground water and its salinity. These data determine the quantity of water needed for leaching. After a preliminary leaching with 12,000—15,000 $m^3$, rice is grown to supplement the draining out of the salts. The irrigation requirements of rice are 20,000 to 25,000 $m^3$/ha. Most of the pre-leaching is carried out in the autumn and winter seasons. An efficient system of drainage is of course essential for the desalinization of solonchak soils (KOVDA, 1961b). The meadow-solonchak types in the Crimea have a high water-table and for their reclamation 3.5—4 tons/ha of gypsum is dressed after which the lands are ploughed to 18—20 cm (NOVIKOVA, 1956).

The classification of the solonetz soils for reclamation purposes provides for two classes: Meadow solonchak-like solonetz and steppe solonetz. In the former case, the groundwater level is to be found in 3—5—8 m depth, and the lower parts of the B Horizon and of Horizon C contain sometimes gypsum and always an appreciable quantity of calcium carbonate. In the steppe solonetz soils, the groundwater level is deeper than 20—30 m. These soil types have a compact structure and a low permeability and an addition of gypsum in these soils as amendment is, therefore, essential. Deep ploughing (to 35—50 cm), to mix the gypsum and lime bearing horizon with the alkaline horizon, reduces the need for application of gypsum (KOVDA, 1961b). An amelioration of solonetz soils in Uzbekistan has been attempted by deep ploughing the soil to a depth over 40—50 cm, mixing up the A and B Horizons, and growing *Melilotus alba*. This process was continuously repeated for 3—4 years. $CaCO_3$ and $CaSO_4$ occur in these soils at a depth below 30—40 cm and mixing up of these layers into the top solonetz layer for a number of years improves the soil condition considerably. In one experiment, the initial exchange sodium percentage of 22 of the soil decreased to 10, calcium saturation increased from 30—40% to 50—70%. As a consequence of this reclamation work, the yield of wheat in this area increased from 200 kg/ha after one year of treatment to 1800—2000 kg/ha after 4 years of treatment. The meadow or steppe-like alkali soils are reclaimed by deep ploughing to a depth of 35—40 cm, here too in order to mix up the calcareous horizon with the alkaline layer. Besides, 3 tons of gypsum are added and the soils were rapidly reclaimed (NOVIKOVA, 1956).

Finally, the Takyr soil type has to be mentioned. These soils are barren lands and show the characteristic block pavement. They are non-permeable and turn frequently into low-water lacustrine swamps. Their pH ranges from 9 to 10. In the sub-soil, they have a high residual salinity containing over 2% salts. The reclamation of Takyr soils is very difficult since the ground water occurs already at a depth of 10—20 cm and leaching by irrigation is carried out without the provision of drainage. 1 m of water desalinizes the soil to a depth of 0.7 to 1.0 m. The water is ponded during the winter months in order to leach these soils. Sometimes sand, 5 cm thick, is spread on the surface and then a frequent irrigation is supplied to these lands. When their alkalinity has decreased below pH 9.0, the area is grassed after deep ploughing and the application of organic matter and fertilizers. A number of years of rest under a grass cover are now given to these soils, and when the physical conditions have materially improved and the pH value has been reduced to 8.0, these areas can be cropped (KOVDA & KUZNETSOVA, 1956).

## General Conclusions

The study of the history of land use as well as that of the present-day land use pattern, the study of factors favouring the development of salinity under arid conditions, of the characteristics of the vegetation in the various arid regions, and also of the adaptations among the plant world to haloxeric conditions, have provided a large amount of background information. Several broad conclusions have emerged from this and we can summarize them as follows:

1. A certain natural equilibrium between the environment and the vegetation was maintained in the arid regions of the world as long as the cultural level there did not transgress the food gathering and hunting stages of the human race. With the increase of population and the improved knowledge of agricultural techniques to exploit the available resources, this equilibrium, particularly in the irrigated areas, was seriously disturbed and very often resulted in salinization of the land.

2. The natural vegetation in arid regions is controlled according to BOYKO's Geo-ecological Law of Distribution; further by the degree of xerophylly; the predominance of geo-chemical factors; and by the simplified structure of the vegetation. Identical factors operate also in specifying the halophytic vegetation in an arid environment. In fact, under haloxeric conditions, the factors operating the distribution of vegetation are still more restrictive.

3. The arid zone vegetation is broadly classified into xerophytes and xeromorphics. The halophytic vegetation has been categorized into euhalophytes or salt-accumulating plants; crinohalophytics or salt-secreting plants; and glycohalophytes, which restrict salt accumulation but have higher respiratory rates. The euhalophytes have been further sub-divided into halosucculents and haloxeritics. Most of the xerophytes show a similar character of hardiness as halophytes and can therefore withstand a high "total soil moisture stress". They develop a high leaf moisture deficit and maintain their physiological activity even at a low water content in the plant tissues. Quite a large proportion of the xerophytic vegetation is capable of converting labile substrates such as amino acids, monosaccharides and fatty acids into storage products like proteins, polysaccharides, fats and oils which raise the osmotic concentration of the sap under conditions of soil moisture stress. Halophytes, particularly euhalophytes, show similar characteristics.

4. Salt crusts have been formed over long periods of time over impermeable layers in several depressions, lake beds, lacustrine lagoons, and Khadar or low lying lands along river banks. These crusts contain sometimes 50—60% of salt by weight and their thickness varies from a few centimetres to a few metres. Such crustal blocks are usually barren of any vegetation except at their fringes where halophytic plant communities are to be observed.

5. A secondary salinization occurs under arid environment as a consequence of the predominance of the evaporation process over the small amount of precipitation, leading to an adverse salt balance in the root zone of the soil. This occurs particularly in those areas where the water-table has risen to the "Capillary fringe". The soils thus salinized may be saline, saline alkali or non-saline alkali and

are usually developing where irrigation has been provised without introduction of efficient water management practices. Certain types such as meadow solonchaks and meadow solonetz soils have developed under arid conditions where the ground water has naturally risen in the area because of its relative lowlying topography.

6. For the restoration of the salt balance, provision of drainage to lower the groundwater level and to flush out the salts is a condition sine qua non. Exceptions are only to be found in dune or gravel areas (see p. 131 ff). In irrigated areas, which are used for commercial crops, the tile drainage system has proved very effective. In other areas an open drainage system, draining away at least 10% of the applied water, ameliorates saline and saline sodic soils. In non-saline sodic soils, an application of soil amendments such as gypsum is essential in order to reduce the alkalinity and to improve the permeability of the soil. In solods, where the B Horizon has compacted, mechanical shattering of the sub-soil is a prerequisite in order to restore their salt balance. If the sub-soil does not have an accumulation of gypsum or calcium carbonate, the addition of amendments is a necessary aid in the reclamation work. An extreme example of reclamation possibilities is shown by the Takyr soils. They have a compact clayey dense crust, 3—5 m thick, and a saline-sodic layer below it, but they have a gypsum bearing horizon at 35—50 cm. Reclamation could be achieved by deep ploughing followed by a leaching process to increase the permeability of these soils and to reduce their salinity level.

7. According to the specific salt tolerance of cultivated crops, they have been categorised into good, moderate, and poor. The tolerance of a species to high amounts of adsorbed or exchangeable sodium is modified by the pH of the soil and the accumulation of soluble $CO_2$. The increase of exchangeable sodium decreases the availability of Ca and this breakdown of the calcium regime in the soil has a powerful inhibiting effect on plant growth due to the differential effects of $Na^+$ and $Ca^{++}$ on the hydration of bio-colloids in the plant. Species which are capable of accumulating large quantities of sodium in their body are least sensitive to the salt in the soil. The high concentration of soluble salts in the soil increases also the osmotic pressure of the soil solution, and this inhibits the moisture availability to the plants because of the increase of the "total soil moisture stress." The toxic influence of different anions such as $SO_4$, Cl, $HCO_3$ and $CO_3$ which accumulate in saline soils, is also specific for each plant species and even varieties of crops.

8. This specific tolerance to various ions provides a useful indication for the inclusion of any crop in different cropping patterns which are suitable for maintaining hydrological and salt balance in the soil. Thus, agricultural methods are apt to prevent a secondary salinization.

9. Finally, non-irrigated salinized soils can be reclaimed by restoring their salt balance through lowering the groundwater level, by the application of amendments, by ponding of rain or irrigation water and planting of salt resistant economic plants species. On places, where an impervious hard pan exists, deep ploughing, or listing increases the rate of infiltration of water and aids in the restoration process of the salt balance. The permanent establishment of salt resistant species is thus facilitated and leads in many cases to economically feasible results even in such areas.

## REFERENCES

ABICHANDANI, C. T. & BHATT, P. N., Unpublished.

ACOCKS, J. P. H., 1964. Karoo vegetation in relation to the development of deserts. In: DAVIS, D. H. S. et al., (Eds.), "Ecological Studies in Southern Africa". *Monogr. Biol.* **XIV**, *100—112.*

AGAFONOVA, A. F., 1957. (Procedures for carrying out experimental works on amelioration of solonetz and highly saline soils for shelterbelts) *Zashchitove lesorazvedenie na Zheleznykh dorogakhv, 49—63* (In Russian).

AGARWAL, R. R., MEHROTRA, C. L. & GUPTA, C. P., 1957. Spread and intensity of soil alkalinity with canal irrigation in Gangetic alluvium of Uttar Pradesh. *Indian J. agric. Sci.* **27** (4), *363—373.*

AGARWAL, R. R. & YADAV, J. S. P., 1954. Saline and alkali soils of Indian Gangetic alluvium in Uttar Pradesh. *J. Soil Sci.* **5**, *300—306.*

AGARWAL, R. R. & YADAV, J. S. P., 1956. Salinity and alkalinity scale to evaluate saline alkali soils for crop responses. *Indian Soc. Sci. J.*, **4**, *141—145.*

AKOPIAN, B. A., 1957. (Changes in the composition of monosaccharides and disaccharides in *Goebelia alopecuroides* when growing on saline soil). *Akad. Nauk. Armienskoi S.S.R. Dok.* **35** (3), *121—124* (In Russian).

AKOPIAN, B. A., 1958. (Peculiarities of the carbohydrate metabolism in plants growing on saline soils). *Akad. Nauk. Armienskoi S.S.R. Izw. Biol. i. Sel'skokhoz Nauk* **11** (11), *69—76* (In Russian).

ANDERSON, E., 1950. "El problema de los suelos salinas y de alcali on la costa del Perú". Peru, SCIPA, Lima *(Series Informaciones Técnicas,* **50***).*

ANDRIANOV, B. V., 1951. (On the problems of the geographical changes in the Amu — Darya Delta). "Problems of Geography" (In Russian).

Anonymous, 1939. Report of *Usar*land Reclamation Committee, U.P. Vol. I. 1938—39, Govt. of U.P. Press, Lucknow.

Anonymous, 1949. "The Australian Environment". The Handbook for the British Commonwealth Specialists Agricultural Conference on Plant and Animal Nutrition in Relation to Soil and Climatic Factors held in Australia. 2nd Ed. C.S.I.R.O., Melbourne.

ANSAURY, S. E., 1955. Report on the Foraminiferal fauna from the Upper Eocene of Egypt. *Publ. Inst. du Désert D'Egypt* no. **6.**

APARICIO, F. DE, 1946. The Comechingon and their neighbors of the Sierras de Cordoba. In: "Handbook of South American Indians. Vol. 2: The Andean Civilizations." *Smithsonian Inst., Wash., Bur. American Ethnology Bull.*, no. **143.**

ARANY, S., 1956a. (Contribution to the role of magnesium in the formation of alkali soils). *Cong. Int. de la Sci. du sol Rap.* **6** Vol. B, *655—659.*

ARANY, S., 1956b. (Use of alkaline waters for irrigation). *Cong. Int. de la Sci. du Sol Rap.* **6** Vol. D. Comn. VI, *615—619.*

ARANY, S., 1956c. (Classification des sols salés de la Grande Plaine basse Hongroise). *Pochvovedenie* no. **7,** *1—7* (In Russian with French summary).

ARMILLAS, P., 1961. Land use in pre-Columbian America. In "A History of Land Use in Arid Regions". *Arid Zone Research* **XVII,** UNESCO.

ARMILLAS, P., PALERM, A. & WOLF, E. R., 1956. A small irrigation system in the valley of Teotihuacan. *Amer. Antiquity* **21** (4), *396—399.*

ASGHAR, A. G., 1958. Causes of and remedies for excess salinity and alkalinity in irrigated lands: Formal discussion. *Second Regional Irrig. Practices Leadership Seminar, Tehran, U.S.I.A.C.*

ASGHAR, A. G., 1960. Report on Irrigation Practices of Pakistan. *Third Regional Irrig. Practices Leadership Seminar, 55—64.*

ASGHAR, A. G. & KHAN, M. A., 1958. Field studies in the prevention of soil salinization. *Agron. J.* **50** (11), 667—671.

AUBERT, G., 1961. Arid Zone soils. A study of their formation, characteristics, utilization and conservation. In: "The Problems of the Arid Zone". *Proc. Paris Symp. UNESCO Arid Zone Res.* **XVIII.**

AUDEBEAU, C., 1924—25. Terres du Bas Delta restées fertiles à la suite de l'abandon de la culture dans le nord de l'Egypte. *Bull. Inst. d'Egypt* no. **7,** *219.*

AWAN, H. A., 1958. Causes of and remedies for excess salinity and alkalinity in irrigated lands. *Second Regional Irrg. Practices Leadership Seminar, Tehran, U.S.I.A.C.*

AYAZI, M., 1958. Causes of and remedies for excess salinity and alkalinity in irrigated lands. *Second Regional Irrigation Practices Leadership Seminar, Tehran, U.S.I.A.C.*

AYERS, A. D., 1942. Salt tolerance of birdsfoot trefoil. *J. Amer. Soc. Agron.,* **40,** *331—334.*

AYERS, A. D., 1950. Salt tolerance of avocado trees grown in culture solution. *California Avocado Soc. Year Book, 138—145.*

AYERS, A. D., 1952. Seed germination as affected by soil moisture and salinity. *Agron. J.,* **44,** *82—84.*

AYERS, A. D., BROWN, J. W. & WADLEIGH, C. H., 1952. A method of measuring the effects of soil salinity on seeds. Salt tolerance of barley and wheat in soil plots receiving salinization regimes. *Agron. J.* **44,** *307—310.*

AYERS, A. D. & HAYWARD, H. E., 1948. A method for measuring the effect of soil salinity on seed germination with observations on several crop plants. *Soil Sci. Soc. Amer. Proc.,* **13,** *224—226.*

BARADEZ, J., 1949. "Fossatum Africae". Paris, Arts et métiers graphiques.

BASU, J. K. & TAGARE, V. D., 1943a. Soils of Deccan Canal tract. IV. The alkali soils, their nature and management. *Indian J. agric. Sci.,* **13,** *157—187.*

BASU, J. K. & TAGARE, V. D., 1943b. Soils of Deccan Canal tract. V. Survey of soils particularly on Nira left bank and Godavari Canal. Deterioration in cane soils. *Indian J. agric. Sci.,* **13,** *512—540.*

BASU, J. K. & TAGARE, V. D., 1954. Land Utilization surveys of saline soils of Bombay State. *J. Soil & Water Conserv., India,* **3:** *24—27.*

BAZILEVICH, N. I., KUZNETSOVA, T. V. & SHELIAKINA, O. A., 1956. (Salt profile of the *Takyrs*). *Takyry Zapadnoi Turkmenii i puti ikh sol' Skokhoziaistvennogo osvoeniia. Moskva, Akademiia Nauk S.S.R., 439—458* (In Russian).

BEADLE, N. C. W., WHALLEY, R. D. B. & GIBSON, J. B., 1957. Studies in halophytes II: Analytical data on the mineral constituents of three species of *Atriplex* and their accompanying soils in Australia. *Ecology* **38** (3), *340—344.*

BENNETT, W. C., 1946a. The Andean Highlands: An introduction. In: "Handbook of South American Indians Vol. 2: The Andean Civilizations," *Smithsonian Inst. Wash. Bur. American Ethnology Bull.* no. **143,** *1—60.*

BENNETT, W. C., 1946b. The Atacameno. In: "Handbook of South American Indians" Vol. 2. *Smithsonian Inst. Wash. Bur. American Ethnology Bull.* no. **143,** *599—618.*

BENNETT, W. C. & BIRD, J. B., 1949. "Andean culture history". *American Museum Nat. Hist. Handbook Ser.* no. **15.**

BERNSTEIN, L., 1961. Salt affected soils and plants. In: "Problems of the Arid Zone". *Proc. Paris Symp. UNESCO Arid Zone Research* **XVIII.**

BERNSTEIN, L. & AYERS, A. D., 1951. Salt tolerance of six varieties of green beans. *Proc. Amer. Soc. hort. Sci.*, **57,** *243—248.*

BERNSTEIN, L., AYERS, A. D. & WADLEIGH, C. H., 1952. The salt tolerance of White Rose potatoes. *Proc. Amer. Soc. hort. Sci.,* **57,** *231—246.*

BHARWAJ, O. P., 1961. The arid zone of India & Pakistan. In: "A History of Land use in Arid Regions", *Arid Zone Research* — **XVII,** UNESCO.

BIRD, J. B., 1943. Excavations in northern Chile. *Amer. Museum Nat. Hist. Anthrop. Papers,* **38** (4), *169—318.*

BOYKO, H., 1931. Ein Beitrag zur Oekologie von *Cynodon dactylon* Pers. und *Astragalus exscapus L.S.B. Akad. Wiss. Wien* (Math.-naturwiss. Kl. Abt. 1, **140,** 9. und 10. Heft.

BOYKO, H., 1932. Über die Pflanzengesellschaften im burgenländischen Gebiete östlich vom Neusiedler See. *Burgenländ. Heimatbl.,* **1,** *43—54.*

BOYKO, H., 1947. The geo-ecological law of plant distribution. *J. Ecol.* **35,** *138—157.*

BOYKO, H., 1955. Climatic, eco-climatic and hydrological influences of vegetation. In "Plant Ecology" *Proc. Montpellier Symp. UNESCO Arid Zone Res.* **V.**

BOYKO, H. & BOYKO, E., 1959. Sea water irrigation — A new line of research on a bioclimatological plant soil complex. *Int. J. Bioclimatol. Biometeorol.* **3** (II, B), *1—24.*

BOYKO, H. & BOYKO, E., 1964. Principles and experiments regarding direct irrigation with highly saline and sea water without desalination. *Trans. New York Acad. Sci.,* Suppl. to No. 8, Ser. II, **26,** *1087—1102.*

BRAIDWOOD, R., 1958. Near Eastern Prehistory. *Science,* **127,** *1419—1430.*

BRAIDWOOD, R. & READ, C. A., 1957. The achievement and early consequences of food production: A consideration of the archaeological and natural historical evidence. *Cold Spring Harbor Symp. Quant. Biol.,* **22,** *19—31.*

BREAZEALE, J. F., 1927. A study of the toxicity of salines that occur in black alkali soils. *Univ. Arizona Tech. Bull.* no. **14.**

BREW, J. O., 1946. "Archaeology of Alkali Ridge, southeastern Utah". Peabody Museum, Harvard Univ., Cambridge, Paper no. **31.**

BRYAN, K., 1929. Flood water farming. *Geogr. Rev.* **29** (3), *444—456.*

BUDARIN, 1957. Quoted by KOVDA (1961).

BUTZER, K. W., 1958. Studien zum Vor- und frühgeschichtlichen Landschaftswandel der Sahara. *Akad. Wiss. Litteratur, Bonn* no. **1.**

BUTZER, K. W., 1959. Environment and human ecology in Egypt during pre-dynastic and early dynastic times. *Bull. Soc. Geog., Egypt,* **32.**

CAMPS, G., 1960. Les traces d'un âge du bronze en Afrique du Nord. *Revue Africaine,* Alger, *31—56.*

CAPOT-REY, R., 1953. "Le Sahara Français". Paris, Presses Universitaires.

CASTETTER, E. F. & BELL, W. H., 1942. "Pima and Papago Indian Agriculture". University of New Mexico Press, Albuquerque.

CASTETTER, E. F. & BELL, W. H., 1951. "Yuman Indian Agriculture: primitive subsistence on the lower Colorado and Gila rivers". University of New Mexico Press, Albuquerque.

COLE, E., 1940. Salty land reclaimed by *dhup (Cynodon dactylon). Indian Fmg.*, **1,** *280—282.*

COLTON, H. S., 1932. Sunset Crater: The effect of a volcanic eruption on the ancient Pueblo people. *Geogr. Rev.*, **22** (4), *582—590.*

COON, C. S., 1955. "The History of Man". London, Cafe.

CROOKES, W., 1917. "The Wheat Problem". 3rd ed. London, Longmans.

CUMMINS, A. B. & KELLEY, W. P., 1923. The formation of sodium carbonate in soils. *Calif. Agric. Exp. Sta. Tech. Paper* no. **3.**

DASTUR, R. H. & SINGH, M., 1942. Studies in the periodic partial failure of the Pb. American cotton in the Punjab. VII. *Indian J. agric. Sci.*, **12.**

DAVIDSON, R. L., 1964. An experimental study of succession in the Transvaal highveld. In: DAVIS, D. H. S. et al., (Eds.), "Ecological Studies in Southern Africa". *Monogr. Biol.* **XIV,** *113—125.*

DAVIES, J. G., 1955. Australia In: "Plant Ecology". *Rev. Res. UNESCO Arid Zone Res.* **VI.**

DELHES, P., 1955. Irak, Jordanie, Liban, Arbie Seondete, Syrie et Ye'men. In: "Plant Ecology" *Rev. Research, UNESCO, Arid Zone Res.* **VI.**

DEPOIS, J., 1961. Development of land use in northern Africa with references to Spain. In: "A History of Land Use in Arid Regions", *UNESCO Arid Zone Res.* **XVII.**

DE SITTER, L. U., 1956. "Structural Geology". McGraw-Hill Book Co. Inc., New York.

DHAR, N. R. & MUKERJEE, S. K., 1936. Alkali soils and their reclamation. I. *Agric. & Livestock, India,* **6,** *850.*

DHAWAN, C. L., BHATNAGAR, B. B. L. & GHAEI, P. D., 1958. Role of green manuring in reclamation. I. *Proc. Nat. Acad. Sci., India,* **27A,** *168—176.*

DRAR, M., 1955. A study on the main characteristics of the ecological groups of arid zone vegetation. In: "Plant Ecology" *Proc. Montpellier Symp., UNESCO Arid Zone Res.* **V.**

DROWER, M. S., 1954. Water supply, irrigation and agriculture. In: SINGER et al. (1954) "A History of Technology", vol. I. Oxford, p. *520—557.*

DUNBAR, G., 1949. "A History of India". Nicholson & Watson, London, 2 Vols.

DYER, R. A., 1955. Structural and physiological features of the vegetation of arid and semi arid areas of the Union of South Africa. In: "Plant Ecology" *Proc. Montepellier Symp. UNESCO Arid Zone Res.* **V.**

EATON, F. M., 1942. Toxicity and accumulation of chloride and sulphate salts in plants. *J. agric. Res.* **64,** *357—394.*

EMBERGER, L., 1955. Africa Du Nord — Ouest. In: "Plant Ecology" *Rev. Res. UNESCO Arid Zone* **VI,** *219—249.*

EMBERGER, L. & LEMÉE, G., 1961. Plant Ecology. In: "The Problems of Arid Zone". *Proc. Paris Symp. UNESCO Arid Zone Res.* **XVIII.**

EVANS, O., 1911. Ostrich farming in Cape Province. In: SOMESET PAYNE (ed.): "Cape Colony". Foreign & Colonial Compiling & Publishing Co., London, *53—56.*

EVENARI, M., 1953. The water balance of plants in desert conditions. *Proc. Desert Res. Symp., Israel, 266—274.*

EVENARI, M., 1961. Plant physiology and arid zone research. In: "The Problems of the Arid Zone" — *Proc. Paris Symp. UNESCO Arid Zone Research.* **XVIII.**

FAUTIN, R. M., 1946. Biotic Communities of the northern desert shrub biome in Western Utah. *Ecol. Monogr.*, **16,** *251—310.*

GAUCH, H. G. & WADLEIGH, C. H., 1942. The influence of saline substrates upon the absorption of nutrients by bean plants. *Proc. Amer. Soc. hort. Sci.*, **41,** *365—369.*

GAUCH, H. G. & WADLEIGH, C. H., 1951. Salt tolerance and chemical composition of Rhodes and Dalbis grasses grown in sand culture. *Bot. Gaz.*, **112**, *259—271*.

GEDROIZ, K. K., 1917. (Saline soils and their improvement). *Thur opt. Agron.*, **18**, *122—140* (In Russian translated by S. A. WAKSMAN).

GEDROIZ, K. K., 1926 (Solodization of soils) *Nosovk Agric. Expt. Sta. Bull.*, **44**, *1—67* (In Russian).

GHOSH et al., 1956. "Rice in India". I.C.A.R., New Delhi.

GREGORY, N. C., 1916. The Navajo country: A geographic and hydrographic reconnaisance of parts of Arizona, New Mexico and Utah. *Dept. of Interior, Washington, Geol. Survey, Water Supply Paper*, **380.**

GREKOV, B. D., 1946. (Peasants in Rus from earliest times to the seventeenth century). *Moscow, U.S.S.R. Academy of Sciences.* (In Russian).

GSELL, ST., 1913—1928. "Histoire ancienne de l'Afrique du Nord". Paris, Hachette, 8 vols. (In French).

HACK, J. T., 1942. "The changing physical environment of the Hopi Indians of Arizona". *Cambridge Peabody Museum, Harvard Univ. Papers*, **35** (1).

HAMDAN, G., 1961. Evolution of irrigation agriculture in Egypt. In: "A History of Land Use in Arid Regions". *UNESCO Arid Zone Res.* **XVII.**

HARRIS, F. S., 1915. The effect of alkali salts in soils on germination and growth of crop. *J. agric. Res.* **5**, *1—53*.

HARRIS, F. S., 1920. "Alkali Soils". New York, J. Wiley, pp. 258.

HARRIS, F. S. & PITTMAN, 1919. Relative resistance of various crops to alkali. *Utah Agric. Coll. Exp. Sta. Bull.* no. **168.**

HARSHBERGER, J. W., 1911. Phytogeographic survey of North America. Die Vegetation der Erde Vol. 13. Edited by A. ENGLER and O. DRUDE. New York, Stechert.

HAURY, E. W., 1936. The Snake Town Canal. Symposium on prehistoric agriculture, University of New Mexico, Albuquerque, *48—50*.

HAURY, E. W., 1945. The excavations of Los Muertos and neighbouring ruins in the Salt River valley, Southern Arizona. *Peabody Museum, Harvard Univ., Papers*, **26** (1).

HAURY, E. W., 1956. Speculations on prehistoric settlement patterns in the south-west. "Prehistoric Settlement Patterns in the New World", New York.

HAYWARD, H. E., 1956. Plant growth under saline conditions. In: "Utilization of saline water". *UNESCO Reviews Res., Arid Zone Research*, **IV.**

HAYWARD, H. E., LONG, E. M. & UHVISTS, R., 1946. Effect of chloride and sulphate salts on the growth and development of the Elberta peach on Shalil and Lovell rootstock. *U.S.D.A. Tech. Bull.* no. **922.**

HAYWARD, H. E. & SPURR, W. B., 1944. The tolerance of flax to saline conditions. Effect of sodium chloride, calcium chloride and sodium sulphate. *J. Amer. Soc. Agron.*, **36**, *287—300*.

HEIMANN, H., 1959. The irrigation with saline water and the balance of the ionic environment. *Potash Rev.*, Berne, July, *1—17;* Aug; p. *1—11;* Sept., *1—18*.

HELTER, V. G., HAGEMAN, R. H. & HARTMAN, E. L., 1940. Sand culture studies on the use of saline and alkaline waters in green house. *Plant Physiol.*, **15**, *727—733*.

HENCKEL, P. A., 1954. Sur la résistance des plantes à la séchesse et les moyens de la diagnostiquer et de l' aumenter. *Vsesoiuzowe Botanicheskoe obslichestvo voprosv Botaniki (Essais de Botanique) Moskva, Akad. Nauk, S.S.R.* **2**, *436—453*.

HENCKEL, P. A. & ANTIPOV, N. I., 1956. Water requirements of euhalophyles in natural environment. *Fizol. Rost.*, **3**, *337—342* (In Russian).

HILGRAD, E. W., 1906. "Soils". Macmillian Co., New York.

HILLS, E. S., 1953. The hydrology of arid & semi arid Australia, with special reference to underground water. *UNESCO Arid Zone "Hydrology" Rev. Res. 179—207.*

HILMY, M., 1944. Reclamation of wastelands *(Barari)* in the northern part of delta. *Proc. Middle East Agric. Dev. Conf., 48—57.*

HOLLAND, T. N. & CHRISTE, W. A. K., 1909. The origin of the salt deposits of Rajputana. *Rec. Geol. Surv., India,* **38,** *154—186.*

HOON, R. C. & MEHTA, M. L., 1937. A study of the soil profiles of the Punjab plains with reference to their natural flora. *Punjab Irrig. Res. Inst. Publ.,* **3.**

HUME, W. F., 1924. "Geology of Egypt". Cairo, Vol. 2.

ISBELL, R. F., 1958. The occurrence of highly alkaline solonetz soil in Southern Queensland. *Queensland J. agric. Sci.,* **15** (1), *15—23.*

JACOBSON, T., 1955. Salinity and irrigation agriculture in Antiquity, Daiyala Basin Archaeological Project. Report on essential results. (Mimeographed).

JACOBSON, T. & ADAMS, R. M., 1958. Salt and silt in ancient Mesopotamian Agriculture. *Science,* **128,** *1251—1258.*

JAMES, P. E., 1942. "Latin America". New York, Lottrop, Lee and Shepard, 908 pp.

KANITKAR, N. V., 1944. Dry Farming in India. *Scientific Monograph* no. **15,** *I.C.A.R., New Delhi.*

KANWAR, J. S., 1962a. Reclamation of saline alkali soils. Quantity of water required for leaching. (unpublished).

KANWAR, J. S., 1962b. Reclamation of saline alkali soils: Amendments, kinds and amounts and cost of application (unpublished).

KASSASS, M., 1955. Rainfall and vegetation belts in arid North-East Africa. In: "Plant Ecology" *Proc. Montpellier Symp. UNESCO Arid Zone Res.* **V,** *58—60.*

KEARNEY, T. H. & SCOFIELD, C. S., 1936. The choice of crops for saline lands. *U.S. Dept. Agr. Cir.* no. **404.**

KELLEY, W. P., 1937. The reclamation of alkali soils. *California Agr. Exp. Sta. Bull.* no. **617.**

KELLEY, W. P., 1951. "Alkali soils, their Formation, Properties and Reclamation". Reinhold, New York.

KELLEY, W. P. & CUMMINS, A. B., 1921. Chemical effects of salts on soils. *Soil Sci.,* **11,** *139—159.*

KELLEY, W. P. & THOMAS, E. E., 1920. The effects of alkali on citrus trees. *California Agr. Exp. Sta. Bull.* no. **318.**

KISELEV, S. V., 1949. ("Ancient history of Southern Siberia"). Moscow (In Russian).

KOVDA, V. A., 1961a. Land use development in arid regions of the Russian plain, the Caucasus and Central Asia. In: "A History of Land Use in Arid Regions", *UNESCO Arid Zone Res.* **XVII.**

KOVDA, V. A., 1961b. Principles of the theory and practice of reclamation and utilization of saline soils in the arid zones. In: "Salinity problems in the arid zones", *Proc. Tehran Sympos. U.N.E.S.C.O. Arid Zone Res.* **XIV,** *201—210.*

KOVDA, V. A. & KUZNETSOVA, T. V., 1956. Popular practice and some theoretical basis of *Takyr* reclamation. *Takyr Zapadvoi Turkmenii i puttikh sel'skokoziaistvennogo osvoeniia.* Moskva, *Akademiia Nauk S S.S.R., 711—717* (In Russian).

KRISHNAN, M. S., 1952. Geological history of Rajasthan and its relation to present dry conditions. *Proc. Symp. Rajputana Desert. National Inst. Sci., India, Bull.* no. **1.**

KRISHNAN, M. S., 1960. "Geology of India and Burma". Higginbothams Ltd., Madras 2.

KURCHERENKO, V. D., 1957. Material on the salt resistance of trees in dry eastern areas of Chkalov region. *Moscow U. Vest. Ser. Biol. Pochvoved. Geol. Geog.*, 12 (2), *99—109* (In Russian).

LEATHER, J. W., 1914. "Irrigation of *Usar* land in the United Provinces". Govt. of U.P. Press, Allahabad, *1—88.*

LEVINKIND, L., 1940. Gross evaporation from standard tanks in the Union of South Africa. *S. Afr. Geogr. J.*, **22**, *22—34.*

LEVITT, J., 1956. Significance of hydration to the state of protoplasm. In: RUHLAND, "Encyclopedia of Plant Physiology" **3**, *650—651.*

LIBEROV, P. D., 1952. Kistorci Zemledelija skifskih plomen podneprov'ja epohi rannevo zeleza V. VI—II. *Materialy po istorii zemeldelija S.S.S.R.* (In Russian).

LILLELAND, O., BROWN, J. G. & SWANSON, C., 1945. Research shows sodium may cause leaf-tip burn. *Almond Facts*, **93**, *1—5.*

LOGAN, R. F., 1961a. Post Columbian developments in the arid regions of the United States of America, In: "A History of Land Use in Arid Regions". *UNESCO Arid Zone Research* **XVII.**

LOGAN, R. F., 1961b. Land Utilization in the arid regions of Southern Africa. II: South West Africa. In: "A History of Land use in Arid Regions", *UNESCO Arid Zone Res.* **VI.**

LOUGHRIDGE, R. H., 1906. Tolerance of alkali by various cultures. *California Agr. Exp. Sta. Bull.* no. **133** (Revised).

MAGISTAD, O. C., 1945. Plant growth relations in saline and alkali soils. *Bot. Rev.*, **11**, *181—230.*

MAGISTAD, O. C., AYERS, A. D., WADLEIGH, C. H. & GAUCH, H. G., 1943. Effect of salt concentration, kind of salt and climate on plant growth in sand cultures. *Plant Physiol.*, **18**, *151—166.*

MAGISTAD, O. C. & CHRISTIANSEN, J. E., 1944. Saline soils, their nature and management. *U.S. Dept. Agric. Circ.* no. **707.**

MAJUMDAR, R. C., RAYCHAUDHARI, H. C. & KALIKINKAR, D., 1955. "An advanced history of India. Part I. Ancient India". MacMillan, London.

MALAURIE, J., 1953. Tonareg et Noirs au Hoggar. Aspects de la situation actuelle. *Amer. Geogr.*, July—Sept., *338—346.* (In French).

MARSHALL, J. B., 1942. Some observations on the tolerance of salinity by cereal crops in Saskatchewan. *Sci. Agr.*, **22**, *492—502.*

McKENZIE-TAYLOR, E., 1940. Making land reclamation precise and profitable. *Indian Fmg.*, **1.**

McKENZIE-TAYLOR, E. & MEHTA, M. L., 1941. Some irrigation problems in the Punjab. *Indian J. agric. Sci.*, **11**, *137—163.*

McKENZIE-TAYLOR, E., PURI, A. N. & ASGHAR, A. G., 1935. Soil deterioration in the some irrigated areas of the Punjab. *Pb. Irrig. Res. Inst. Publ.*, **2.**

MEHTA, M. L., 1940. The formation and reclamation of *Thur* lands in the Punjab. *Pb. Irrigation Res. Inst. Publ.*, 3 (4), *1—54.*

MEIG, P., 1952. Distribution of arid zone climates. Maps nos. 392 & 393, United Nations.

MILLIGAN, A. J., BURVILL, G. H. & MARSH, A. B., 1951. Soil salinity investigations. Salt tolerance germination and growth tests under controlled salinity conditions. *Dept. Agr. West Australia Leaflet* no. **1052.**

MOSSÉRI, V., 1922—23. Le sol egyptien sous le regime d'arrosage par inondation. *Bull. Inst. Egypt*, **5**, *24.*

NEUMARK, S. D., 1957. "Economic influences on the South African frontier, 1652—1836". Stanford University Press, Palo Alto.

NOBLE, J., 1875. "Descriptive Handbook of Cape Colony". Standford, London.

NOVIKOVA, A. W., 1956. Ergebnisse und Perspektiven der Melioration von Alkaliböden in der Krim. *Pochvovedenie,* **8,** *31—43* (In Russian with German summary).

OTA, K. & YASUE, T., 1957. Studies on the salt injury to crops. XI. The difference of the salt resistance in young wheat varieties. *Gifu Univ. Fac. Agr. Res. Bull.*, **8,** *14—22* (In Japanese with English summary).

OTA, K. & YASUE, T., 1958. Studies on the salt injury to crops. XII. The influence of sodium chloride solution upon the germinating force in paddy rice seeds. *Proc. Crop. Sci. Soc.*, Japan, **27** (2), *223—225* (In Japanese with English summary).

PASSEK, T. S., 1949. Periadizacija tripol' skik poselenij. *Izvestija Archeologii* no. **10** (In Russian).

PAVER, G. L. & PRETORIUS, D. A., 1954. Report on reconnaissance hydrogeological investigations in the western desert coastal zone. *Publ. Inst. du Désert D'Egypte* no. **5,** *20—22.*

PENMAN, F., HUBBLE, G. D., TAYLOR, J. A., et al., 1940. A soil survey of Mildura irrigation settlement, Victoria, Australia. *C.S.I.R.O.* no. **133.**

PENMAN, F., TAYLOR, J. K., HOOPER, P. D., et al., 1939. A soil survey of the Merbein irrigation district, Victoria, Australia. *C.S.I.R.O., Bull.* no. **123.**

PENSOKOI, L. K., 1956. (The salt resistance of cotton plant and its effect on the seasonal dynamics of salts under the conditions of Kura-Araks plain.) *Pochvovedenie,* **8,** *86—89.* (In Russian).

PERSHINA, H. N., 1956. (On the development of solonetsosity as a genetic and zonal property of chestnut soils) Moskov. *Ordena Lenina Sol'-Skokhoz Akad. im K.A.Timiriazeva Dok. TSKHA* no. **23,** *165—169* (In Russian).

PICHI-SERMOLLI, R. E. G. 1955. The arid vegetation types of tropical countries and their classification. In: "Plant Ecology" *Proc. Montpellier Symp. UNESCO Arid Zone Res.* **V.**

PIGGOT, S., 1950. "Prehistoric India". Harmondsworth, Penguins Books Ltd, London.

POKROVSKAIA, E. I., 1957. (Certain data on oxidation reduction processes in halophytes) *Painiatn Akademika H.A. Maskova Shornik static,* Moskva, *268—274* (In Russian).

PONCET, J., 1956. La mise en valeur de la basse vallée de la Medjerda. *Ann. de Géogr., Paris, 199—222.*

POPAZOV, D. R., 1956. (Certain peculiarities in the development of solonetz, solonchak, and soloth soils). *Moskov Ordena Lenina Sol'Skokhoz Akad. im KA. Timiriazeva Dok TSKHA* no. **25,** *230—236.* (In Russian).

POPOVA, M. P., 1957. (On the salt resistance of tree and shrub species under conditions of irrigation). *Lesn Khoz.* **10** (5), *27—30* (In Russian).

POVETEV, A. A., 1957. (Structure and types of snow protecting forest plantings on solonetz and extremely solonetz soils). *Zashchitnoc lesorazuodenic na Zheleznykh dorogakl, Moskva, 40—48.* (In Russian).

PRESCOTT, J. A., 1944. A Soil map of Australia. *C.S.I.R.O., Bull.* no. **177.**

PURI, A. N., MCKENZIE TAYLOR, E. & ASGHAR, A. G., 1937. Soil deterioration in the canal irrigated tracts of the Punjab. III. Formation and characteristics of soil profiles in alkaline soils. *Punjab Irrig. Res. Inst. Pub.,* **9** (4).

RAYCHAUDHURI, S. P. & DATTA BISWAS, N. R., 1958. Saline and alkali soils of Asia with particular reference to India. *Trans. Fifth int. Congr. Soil Sci.,* **1,** *191—207.*

REITEMEIER, R. F., 1946. Effect of moisture content on the dissolved and exchangable ions of soils of arid regions. *Soil Sci.,* **61,** *195—214.*

REPARAZ, G. DE, 1958. La zone aride du Peron. *Geografiska Annaler,* Stockholm, **40** (1). (In Swedish).

REPP, G., 1958. Die Salztoleranz der Pflanzen. I. Salzhaushalt und Salzresistenz von Marschpflanzen der Nordseeküste Dänemarks in Beziehung zum Standort. *Öst. bot. Z.* **104** (4/5), *454—490.*

REPP, G., MCALLISTER, D. R. & WIEBE, H. N., 1959. Salt resistance of protoplasm as a test for the salt tolerance of agricultural plants. *Agron. J.*, **51** (6), *311—324.*

ROMYN, R. E., 1924. Ranching in the Transvaal. *J. Dept. Agric.* **8,** *508—517.*

ROSENE, H. F., 1941. Control of water transport of attached and isolated roots by means of the osmotic pressure of the external solution. *Amer. J. Bot.*, **28,** *402—410.*

ROWE, J. H., 1946. Inca culture at the time of the Spanish conquest. In: "Handbook of South American Indians Vol 2: The Andean Civilizations". *Smithsonian Inst. Wash Bur. American Ethnolog. Bull.* no. **143.**

RURAL RECONSTRUCTION COMMISSION, 1945. Eighth report on irrigation, water conservation and land drainage, Govt. Printer, Canberra.

RYBAKOV, B. A., 1945. The Antis and Kiev Rüs: In: "Achievements and future prospects for the development of Soviet archaeology, Moscow". (In Russian).

SAAD, S. J., 1954. Studies on the physiology of the cotton plant. IV. *Proc. Egyptian Acad. Sci.*, **10,** *73—88, 94—106.*

SANDHU, J. S., 1955. Punjab Univ. M.Sc. (Agric.) Thesis.

SARIN, M. N. & RAO, I. M., 1958. Physiological studies on salt tolerance in crop plants. III. Influence of sodium sulphate on seedling respiration in wheat and gram. *Indian J. Plant Physiol.*, **1,** *30—38.*

SAUER, C., 1948. "Colima of New Spain in the Sixteenth Century". University of California Press. Berkeley & Los Angeles.

SAUER, CARL O., 1952. "Agricultural Origin and Dispersals," American Geographical Society, N.Y.

SCHAVRYGIN, P. I., 1956a. Soil solutions and elements of salt status of *Takyrs. Takyry Zapadovi Turkmenii i puti ikh sel' skokhaziaistvennogo osvoeniia. Moskva, Akademiia Nauk U.S.S.R. 459—68* (In Russian).

SCHAVRYGIN, P. I., 1956b. Über Solonetz Erscheinungen in Takyren. *Pochvovedenie* **8,** *44—48* (In Russian with German summary).

SHAH, R. K., VORA, J. C. & TRIVEDI, A. M., 1957. Possibilities of reclamation of the alkali soils of the Little Rann of Cutch. *J. Soil & Water Cons., India*, **6,** *133—137.*

SHANTZ, H. L., 1956. History and Problems of Arid Land Development. Chapter I. "The Future of Arid Lands". Edited by G. F. WHITE, American Assoc. Adv. Sci., Washington D.C.

SHANTZ, H. L. & ZON, R., 1924. "Natural vegetation". U.S.D. Agric. Atlas, Amer. Agric, Part I. Sec. **E.,** *1—29.*

SHCHIPANOVA, I. A., 1955. (On salt resistance of cotton under conditions of Shirvan Steppe.) *Akad. Nauk Azerbaidzhansk. S.S.R. Inst. Pochvoved i Agrokhim. Trudy*, **7,** *233—240* (In Russian).

SHUKRI, N. M. & AZER, M., 1952. The minerology of Pliocene and more Recent sediments in the Fayaum. *Desert Res. Inst. Foud I.* **11** (1).

SIGMOND, A. A. J. DE, 1927. The classification of alkali and salty soils. *First int. Congr. Soil Sci. Proc.* **1,** *330—344.*

SIGMOND, A. J. J. DE, 1932. Hungarian alkali soils and methods of reclamation. *Imp. Bur. Soil Sci. Tech. Common.*, **23.**

SINGH, J., Unpublished.

SLAVIK, B. (ed.), 1965. "Water stress in plants". Proc. Symposium held in Prague, 1963. The Hague, Dr. W. Junk and Praha, Czech. Akad. Sci.

SMITH, S. T., 1956. Handling salty land. W. *Australia Dept. Agric. J. Ser.*, **5** (6), *729—731.*

SMITH, S. T. & MALCOLM, C. V., 1959. Bringing wheat belt salt land back into production. *W. Australian Dept. Agric. J. Ser.*, **8** (3), *263—267.*

SNYDER, R. S., KULP, M. R., BAKER, G. O. & MARR, J. C., 1940. Alkali reclamation investigations. *Idaho Agr. Exp. Sta. Bull.* no. **233.**

SPATE, O. H. K., 1960. "India and Pakistan: A General and Regional Geography". London: Methuen & Co. (Reprint of 2nd edition).

STANDER, G. J., 1954. Water bewaring en doeltreffende beplanning vir die maksimum exploitasie van Suid-Africa se naturalike hulpbronne. *Tydskrif vir Wetenskaf en Kuns,* **14,** *162—177* (In Afrikaans).

STEPHENS, C. G., 1961. The soil landscape of Australia. *C.S.I.R.O. Soil Pub.* no. **18.**

STEWART, G. A., 1959. Some aspects of soil ecology. In: KEAST, A. et al., (Eds.), "Biogeography and Ecology in Australia". *Monogr. Biol.* **VIII,** *303—314.*

STEWART, J., 1898. Effect of alkali on seed germination. *Utah Agric. Expt. Sta. 9th Ann. Rept., 26—35.*

STOCKER, O., 1960. Physiological and morphological changes in plants due to water deficiency. In: "Plant Water Relationships in Arid and Semi Arid Conditions". *Rev. Res. U.N.E.S.C.O., Arid Zone Res.* **XV,** *63—104.*

STONEMAN, T. C., 1958a. Salt land programme for autumn. *West Australia Dept. Agric. J. Ser.,* **7** (3), *359—360.*

STONEMAN, T. C., 1958b. Salt movement in soil. *West Australia Dept. Agric. J. Ser.,* **7** (5), *577—79.*

STORIE, R. E. & MATHEWS, C., 1945. Preliminary study of Chilean soils. *Soil Sci. Soc. Amer. Proc.,* **10,** *351—355.*

STORY, R., 1964. Plant lore of the bushmen. In: DAVIS, D. H. S. et al., (Eds.), "Ecological Studies in Southern Africa". *Monogr. Biol.* **XIV,** *87—99.*

STROGONOV, B. P., IVANITSKIA, E. F. & CHERNIADEVE, I. P., 1956. (Effect of high concentrations of salts in plants). *Fiziol. Rost.,* **3,** *319—327* (In Russian).

SZABOLCZ, I., 1956a. (Alkali soils of Hungary: Distribution and properties of alkali soils.) *Pochvovedenie.* **11,** *9—18* (In Russian with English Summary).

SZABOLCZ, I., 1956b. Les sols à alkalis hongrois. *Congr. int. Sci. Sol. Rap.,* **6,** *603—608.*

SZABOLCZ, I. & JASSO, F., 1959. (The classification of szik soils in Hungary). *Agroke'm Talit.,* **1,** *281—300* (In Hungarian with Russian and German summaries).

TAGAWA, T., 1934. The relation between the absorption of water by plant root and the concentration and nature of surrounding solution. *Japan. J. Bot.,* **7,** *33—60.*

TALBOT, W. J., 1961. Land utilization in the arid regions of Southern Africa. Part I: South Africa. In "A History of Land Use in Arid Regions". *UNESCO Arid Zone Res.* **XVII.**

TALATI, R. P., 1947. Field experiments on reclamation of salty lands in Baramati of Bombay-Deccan. *Indian J. agric. Sci.,* **17,** *153—175.*

TEAKLE, L. J. H., 1937. The salt (Sodium chloride) of rain water. *J. Agric. West Australia,* **14,** *115—123.*

TEAKLE, L. J. H., 1950. An interpretation of some solonized (alkali) soils in South-Western Australia. *Trans. int. Cong. Soil Sci.,* **1,** *389—393.*

TER-KARAPETIAN, M. S., 1957. (Variation in amino acid content of the leaves of *Goebelia alopecuroides* under conditions of saline soils). *Akad. Nauk Armianskor. S.S.A. Dok.,* **25** (3), *117—120* (In Russian).

THIMANN, K. V., 1960. Quoted by EVENARI (1961).

THORNE, D. W. & PETERSON, H. B., 1950. "Irrigated Soils their Fertility and Management". Philadelphia and Toronto, The Blakiston Co.

THUNBERG, C. P., 1795. "Travels in Europe, Africa and Asia". E. & C. Livington, London.

TINDALE, N. B., 1959. Ecology of primitive aboriginal man in Australia. In: KEAST, A. et al., (eds.), "Biogeography and Ecology in Australia". *Monogr. Biol.* **VIII**, *36—51*.

TOBIAS, P. V., 1964. Bushman Hunter-gatherers: A study in human ecology. In: DAVIS, D. H. S. et al., (Eds.), "Ecological Studies in Southern Africa". *Monogr. Biol.* **XIV**, *67—86*.

TOLSTOV, S. P., 1945. Osnoye itogi i ocerednye zadci izuce nija istovii i archeologiii Kera-kalopakii i kara-Kalpakov. *Bull. Acad. Sci. Uzbek SSSR Tashkent* nos. **9—10** (In Russian).

TOUSSOUN, O., 1924. Mémoire sur les finances de l'Egypte depuis les Pharaons jusqu'à nos jours. *Mem. Inst. Egypte*. **6**, *71*.

TSING, T., YI-HSIUNG, F. & WAN-LI, W., 1956. (Salt tolerance of some poplar trees in North Kiangsen). *Acta bot. Sinica*, **5**, *153—176* (In Chinese).

Union of South Africa, 1912. Report of the Director of Irrigation for the Year ending 31st December, 1911, Govt. Printer, Cape Town.

Union of South Africa, 1914. Report of the Director of Irrigation for 1912—13, Govt. Printer, Cape Town.

United States Salinity Laboratory Staff, 1954. "Diagnosis and Improvement of Saline and Alkali Soils". Ed. L. A. RICHARDS. *U.S.D.A. Handbook* no. **60**.

United States Salinity Laboratory Board of Collaborators, 1958. Report of Nomenclature Committee. *Proc. Soil Sci. Soc. Amer.*, **22** (3), *270*.

UPPAL, H. L., 1955. Green manuring with special reference to *Sesbania aculeata* for treatment of alkaline soils. *Indian J. agric. Res.*, **25**, *211—235*.

UPPAL, H. L., 1963. Saline and alkali soils of the Punjab. (Personal communication).

*Usar*land Reclamation Committee, 1939. Report on the reclamation of *Usar*land in U.P. Govt. Press, Lucknow.

VON REENEN, R. J., 1925. Development of irrigation in the Union of South Africa. *S. Afr. J. Sci.*, **22**, *20—41*.

VORONIN, 1955. Quoted by KOVDA (1961).

WADHAM, S., 1961. The problem of arid Australia, In: "A History of Land Use in Arid Regions". *UNESCO Arid Zone Res.* **XVII.**

WADIA, D. N., 1949. "Geology of India". London, Macmillan & Co.

WADLEIGH, C. H. & AYERS, A. D., 1945. Growth and biochemical composition of bean plants as conditioned by soil moisture tension and salt concentration. *Plant Physiol.*, **20**, *106—132*.

WADLEIGH, C. H., GAUCH, H. G. & MAGISTAD, O. C., 1946. Growth and rubber accumulation in guayule as conditioned by soil salinity and irrigation regime. *U.S. Dept. Agric. Tech. Bull.* no. **225.**

WADLEIGH, C. H., GAUCH, H. G. & STRONG, D. C., 1947. Root penetration and moisture extraction in saline soil by crop plants. *Soil Sci.*, **63**, *341—349*.

WAHHAL, A., MUHAMMAD, F. & AHMAD, M., 1957. Soil salinity conditions and growth of crops. (Abstr) *Proc. Pakistan Sci. Conf.*, **9** (3), *2—4*.

WALLACE, A., TOTH, S. J., & BEAR, F. E., 1948. Sodium content of some New Jersey plants. *Soil Sci.* **65**, *249—258*.

WALTER, T., 1904. "Yuan Chavang". Oriental Translation Fund, Royal Asiatic Society, London.

WEDEL, W. R., 1953. Some aspects of human ecology in the Central plains. *Amer. Anthropologist*, **55** (4), *499—514*.

WEDEL, W. R., 1956. Changing patterns in the Great Plains. "Prehistoric settlement patterns in the New World". New York. *81—92*.

WHYTE, R. O., 1961. Evolution of land use in South Western Asia. In: "A History of Land Use in Arid Regions". *UNESCO Arid Zone Res.* **XVII.**

WILLCOCKS, W. & CRAIG, J. I., 1913. "Egyptian Irrigation". London.

WILLEY, G. R., 1946. The culture of La Candelaria. In: "Handbook of South American Indians, Vol. 2: The Andean Civilizations." *Smithsonian Inst., Washington, Bureau of Amer. Ethnology Bull.* no. **143.**

WILLEY, G. R., 1953. Prehistoric settlement patterns in the Viru Valley, Peru, *Smithsonian Inst., Washington Bureau of Amer. Ethnology Bull.* no. **155.**

WOLF, E. R. & PALERM, A., 1955. Irrigation in the old Acolhua domain, Mexico. *Southwestern J. Anthrop., Albuquerque,* **11** (3), *265—281.*

WOOD, J. G., 1934. Carbohydrate changes in leaves of sclerophyll plants. *Aust. J. exp. Biol.,* **11,** *237.*

YADAV, J. S. P. & AGARWAL, R. R., 1959. Dynamics of saline alkali soils of Indo-Gangetic alluvium. *J. Indian Soc. Soil Sci.* **7,** *213—222.*

# PART II:

# PRINCIPLES AND EXPERIMENTS

# BASIC ECOLOGICAL PRINCIPLES OF PLANT GROWING BY IRRIGATION WITH HIGHLY SALINE OR SEA-WATER

BY

HUGO BOYKO*)

(with 15 figs.)

## General Remarks

For several decades, the interest of many scientists, engineers and recently even of statesmen has been more and more directed to solutions of a rather new general problem of mankind, the problem of sufficient water supply. Man's water requirements in former centuries were relatively easily satisfied; for the main demands were restricted to potable water and agriculture. Both were adapted to the disposable water amounts and not the other way round. In regions where even the small amounts of necessary potable water became too scarce, man simply left the uninhabitable places and settled anew in other, more convenient surroundings or wandered as nomad from one well to another. Agriculture relied mainly on rainfall and only relatively small parts of the agricultural areas were cultivated by horticultural methods, including irrigation. As a matter of course, such irrigation methods, although used everywhere, played a major rôle for the sustenance of populations in arid and semi-arid regions only. In the arid regions proper, e.g. in the Nile valley, even cereals had to be grown by flooding from rivers or rich wells in the vicinity, presumed that the water was more or less fresh water.

In humid regions, no water problem with regard to amount or quality existed until the 20th century. This situation has now completely changed. On the one hand, population growth and the demands of a rising standard of life, hygiene etc. in all regions, arid and humid alike, expansion of horticultural methods in agriculture, a rapidly growing industrialization, more intensive methods in mining processes, considerably expanded water requirements everywhere. On the other hand, pollution, of rain drops already in the air and still more pollution of the existing surface and underground water resources by the huge amounts of waste water from industrial and urban areas, worsened by the biological effects of accumulated detergents, insecticides, etc. is steadily restricting the available amount of fresh water even in humid regions.

* Address: World Academy of Art and Science, General Secretariate, 1 Ruppin Street, Rehovot, Israel.

The fascinating idea of obtaining usable water from the sea as a permanent solution to these problems had become scientifically feasible centuries ago. During the last two decades, conversion of sea-water on a large scale became also technically possible. The problem of its economical implementation had, however, not yet been solved.

At this stage, successful experiments involving direct irrigation with highly saline and even sea-water (BOYKO, 1957b, 1964; BOYKO & BOYKO, 1959) indicated new ways to a solution for a number of important crops, and it seems that a combination of both lines of research, the conversion methods on the one hand, with methods described in this book on the other hand, may bring the scientific, technical, *and* economic solution nearer, at least for many agricultural branches, and to great parts of the earth's surface.

For the vast areas of dune land, the global total area of which is much larger than the whole of Europe and seven times as large as the agricultural area of the United States, these new experiments may mean a most important eighth alternative in addition to the seven ones enumerated in the official report on the "Desalination Research Conference" convened by the National Academy of Sciences and the National Research Council of the USA at Woods Hole, Massachusetts, 14. June—14. July 1961. In this report the following alternatives to desalination, recommended for extensive research, are listed as means to overcome the threatening shortage of fresh water (on page 85):

"a) Substitution of saline water for fresh water by development of marine products and salt-tolerant crops; assistance to growth of plants in saline soils; municipal and industrial uses.

b) Reclamation of poor but adequately watered soils.

c) Modification of natural phenomena (climate; natural freezing; condensation; cloud seeding; electrical discharge).

d) Improvement of transportation, storage and use."

In the full text many excellent alternatives are dealt with in some detail. Most of the research work necessary for these alternatives recommended in the report is of high value but requires considerable amounts of financial aid to be carried out adequately.

Compared with them, the sums necessary for research on the new alternatives described in this book and indicated by the experiments of plant growing on sand with saline water imply only a negligible financial burden.

Though this new line of research (BOYKO & BOYKO, 1959) could solve only part of the problem, it may possibly become a major part in this struggle for Human Welfare.

Another alternative is to combine both methods as recommended in this book in a former article (H. Boyko, 1966) (see page 19), namely to desalinate sea-water to brackish water and to use this for valuable crops on sand.

In the present article, we are dealing with the theoretical side and shall try to explain the underlying principles whereby the experiments on direct use of highly saline irrigation water up to oceanic concentration could be successful.

Here too, as in many other cases, ecological observations in nature are primarily responsible for the results, and the experiments carried out much later served mainly to verify them.

When such experimental tests under well-controllable and well-controlled conditions lead to the same results then we can rely objectively on the somewhat subjectively influenced observations and measurements in nature. This is frequently not the case the other way around. Experiments in the laboratory are in many instances not reliable sources for general conclusions, simply for the reason that it is humanly impossible to imitate the innumerable combinations of all factors in any natural plant habitat. This applies, of course, also to the infinitely complicated biological problems including the microbiological and biochemical ones caused by aridity and salinity. Laboratory experiments are frequently misleading, particularly in this field of research.

The same questionable reliability is to be ascribed to any dogmatic adherence to mathematical formulae of soil science in general, if they are to be applied to any work with living beings, in our case with plants.

The studies of the author on these problems started with extensive ecological observations in arid and saline regions of various countries from the tropics to the arctic. It may therefore be of some value to recapitulate some of the basic principles found in natural habitats of vegetation on saline sites.

*Patterns of Halophytic Vegetation*

Wherever sea-water or any other water of higher Total Salt Content has direct contact with the vegetation, we are confronted also with the phenomenon of halophytic plant species, adapted to salinity by an inherent higher or lower salt tolerance.

This salt tolerance is very different with different species and we can observe on such sites, wherever they may be, distinct patterns of plant distribution, according to the following few principal factors:

1. quantitative and qualitative salt content of soil and/or water

2. physical soil features

Fig. 1. Halophytic zonations around the "Unterer Stinkersee", Burgenland:
Zonations: 1. Water or white salt cover (in the lake-center)
2. Crypsidetum aculeatae
3. Suaedetum maritimae
4. vegetationless belt covered by white salt
5. loose *Atropis distans* belt with *Lepidium crassifolium*
6. dense *Atropis distans* belt
7. Plantaginetum maritimae with *Aster tripolium* as codominant species
8. Cynodontetum belt with *Ononis spinosa*
Above the Cynodontetum which is still dependent on groundwater, there is already a short grass steppe, the Festucetum pseudovinae (see Fig. 2 and 3). (from H. Boyko, 1932).

3. specific salt tolerance of plant species

4. height above water-table and/or horizontal distance from open water surface

5. periodical fluctuation of water-table or temporary flooding

6. topographical features.

Any thorough ecological study of a specific coastal part or of the vegetation around an inland lake without an outlet, which means in almost all cases a salt lake, results in the recognition of a specific pattern according to these six main features (compare also Chapman, 1962, 1966).

During the last 35 years, the author had the opportunity to investigate these patterns in humid and arid zones alike and in all

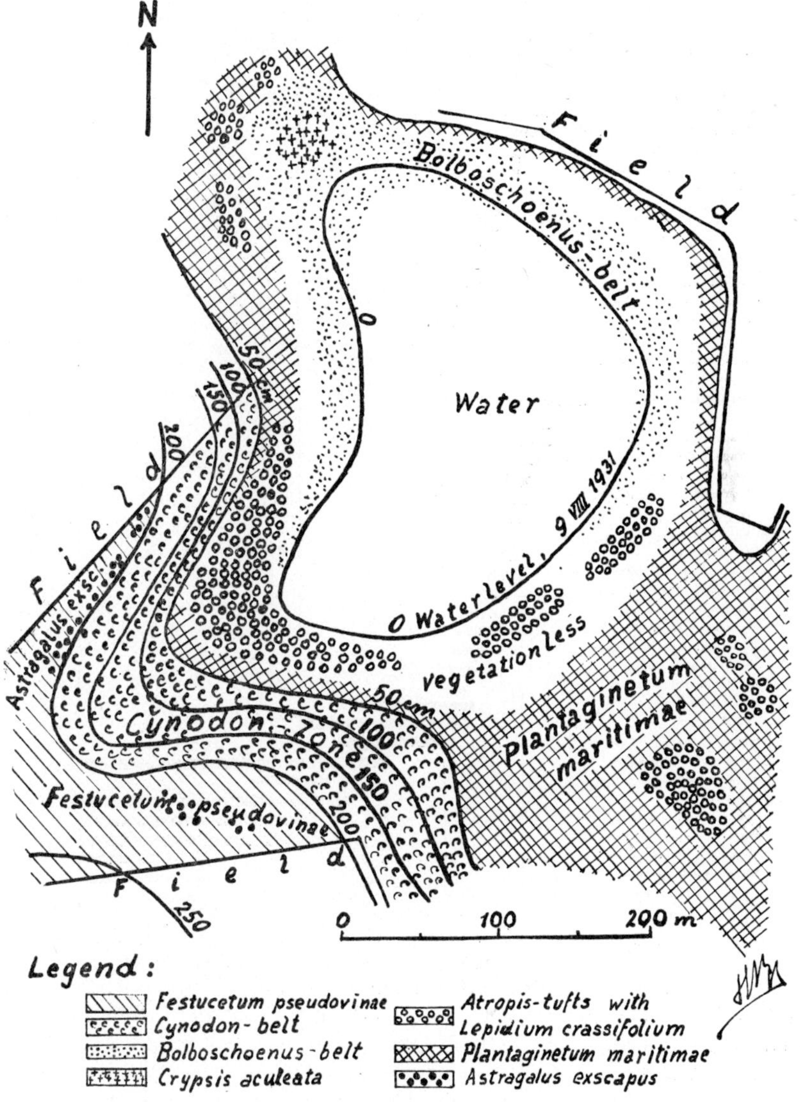


Fig. 2. Zonation in a salt lake in the Seewinkel (Austria).

continents, except in South America and in the Antarctic. A few examples of such halophytic vegetational zonations from a temperate semi-arid Central-European region and from a hot desert region in the Middle East may be added here in order to show likewise several different methods of representation other than photographs or tables.

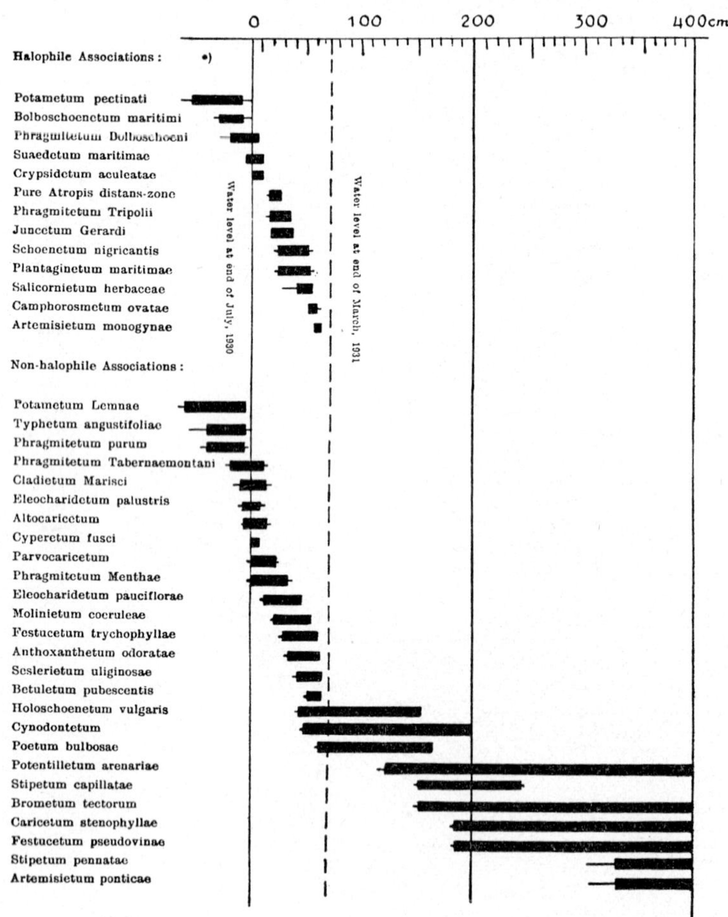


Fig. 3. Relations of zonations east of the Neusiedler See (Austria).

Fig. 1 shows a photograph of such a zonation in and around a salt lake in the Seewinkel in Burgenland, Austria, taken in 1931 (H. Boyko, 1932).

In Fig. 2 such a zonation around another of the numerous salt lakes there is mapped and thus the topographical features and the relative heights of each plant association above the groundwater level are made visible (H. Boyko, 1931).

Fig. 3 shows these relations graphically for the whole region east of the Neusiedlersee (H. Boyko, 1951).

Fig. 4 presents the relations of various zones according to the height above the highly saline groundwater in the oasis Yotvata (Ein Ghadian) in Wadi Araba between the Dead Sea and the Red Sea.

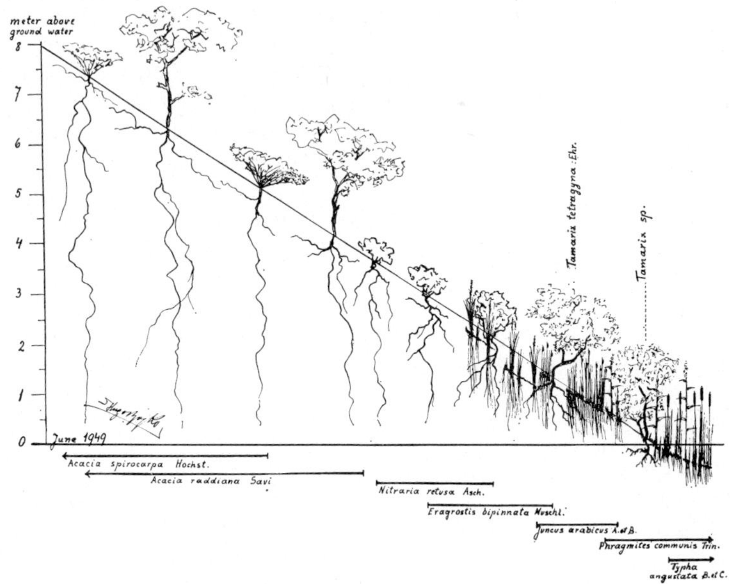


Fig. 4. Zonation in the oasis Yotvata (Wadi Araba), Israel.

The main pattern is always the same: the plant species occur in zonations and each zonation is composed of a specific combination of species building a distinct plant association (the concept "association" is used here in the sense of the Zurich-Montpellier school).

There are, however, certain *exceptions* as for instance:

a) The decisive influence of *seasonal fluctuations of the water table* in flat basins caused by evaporation during the dry season, leaving near the centre a mosaic of extremely halophytic *annuals* (like *Salicornia herbacea, Crypsis aculeata, Suaeda maritima, Suaeda salsa,* etc.). This mosaic is caused by small topographical differences in the bottom of the dried out lake and is as a whole encircling the mostly white salt layer covering the centre of these lakes like a snowfield.

b) Islands of non or less salt tolerant plants and plant communities could be found in the midst of highly salt tolerant plant associations, wherever the soil had a *higher permeability,* e.g. an island of gravel or sand.

c) Sometimes highly salt tolerant and slightly salt tolerant plant species seemed to grow together, but in these cases the slightly salt tolerant species (e.g. the relatively high perennial *Plantago maritima* or *Triglochin maritimum* surrounded by the lower annual

*Salicornia herbacea*) have their deeper-reaching root-system in *cracks;* they have therefore soil and water of a much lower salt content at their disposal than the shallow growing annuals which have the upper layer of the drying soil for their germination and living space, where high salt accumulation is the result of capillarity and evaporation leading to efflorescence on the soil surface.

d) A remarkable exception from the usual zonation pattern was found on several places by the author in the desert region of Wadi Araba. There, the regular order of vegetational zonations was inverted exactly into its contrariness. Soon it could be proved by the author that this was caused by *artesian pressure* indicating much higher amounts of underground water resources than originally expected.

These ecological studies of arid and saline conditions gradually led to three new lines of research (H. Boyko, 1965):

1) *Ecological Climatography:* the investigations of halophytic plant associations in semi-arid regions of Central Europe resulted in the first climate map based on vegetation (H. Boyko, 1934). Another example of a hot desert region is the Climate map of the Sinai Peninsula (H. Boyko & E. Boyko, 1957);

2) *Ecological Hydrology:* using plants and plant communities as qualitative and quantitative indicators of hydrological features, including those of salinity (H. Boyko, 1953);

Ecological climatography and ecological hydrology are combined to form the new scientific branch of "Ecological Geophysics" (H. Boyko, 1965);

3) *Saline and Marine Agriculture,* the main principles of which are dealt with in the following chapters.

Various experiments in different countries have proved that in general a higher soil permeability allows the use of irrigation water with a much higher salt content than soils with a lower permeability. In experiments described elsewhere in this book (see page 214 ff) extremes of permeability on the one hand and of water salinity on the other hand have been used by the author in order to find a practical applicability of this main principle. Dune sand and gravel are the main substrata in which numerous plant species were grown

---

It appears as if salt tolerance may be raised by several times in all plants simply by growing them on sand or gravel. At the request of the author and fostered by UNESCO verifying experiments were carried out in other countries as well and gave the same results. Raised salt tolerance is, however, the result of a very complicated combination of several geophysical, chemical, biochemical and general biological principles, most of which have not up to now received the attention they deserve.

with irrigation water of high salinity. A number of cases were successful even with pure sea-water of oceanic concentration.

The most important principles are:

1) *Quick percolation* of the irrigation water (see page 140) followed by

2) *Good aeration* of the root systems (p. 142)

3) *Easy solubility* of NaCl and $MgCl_2$ where they constitute the main components of the Total Salt Content (T.S.C.) (p. 142)

4) *Lack of sodium adsorption* to sand particles in contrast to the easy adsorption of Na to clay particles (p. 145).

5) *"General Rules of Application"* (p. 145)

6) The principle of *"Partial Root Contact"* (p. 147)

7) *The Viscosity Theory* (p. 152)

8) The condensation of *Subterranean Dew* around the feeder roots (p. 158)

9) *"Biological Desalination"* (p. 165)

10) *Adaptability* of many plant species to the factor of erratics and to *fluctuating osmotic pressure* (p. 168).

11) *The Balance of Ionic Environment* (p. 173).

12) *Raised Vitality* — a newly recognized result of certain biochemical processes (p. 175)

13) *Microbiological influences* (p. 178).

14) The geophysical theory of *"Global Salt Circulation"* (p. 180) including the "Coastal Salt Circulation", the "Magmatic Salt Circulation" and the "Biological Salt Circulation", and finally

15) *The Basic Law of Universal Balance* (p. 195).

Of these 15 principles, the first four were known long ago and the balance of ionic environment (Nr. 11) was dealt with extensively by H. Heimann elsewhere (1958).

In the following, a better understanding of them, of the old principles as well as of the new ones is attempted or at least they are presented with the hope that continuing deeper going research may elucidate them and their complicated interrelations.

## The Principles

### *1. Quick Percolation*

A high permeability of the soil has two very important properties as its consequence: one is a quick percolation possibility of the irrigation water and the second a good aeration of the root systems, decisively influencing the life process of the plant itself and at the same time also the microflora and -fauna in the soil.

This permeability is, of course, particularly high in sand or gravel.

Soil permeability is mainly expressed by the speed of percolation in cm per hour and may reach several thousand cm per day. In the field, this permeability varies very much, even over small areas. An unpublished investigation (by the Hydrological Service of Israel under the direction of Dr. Walter STERN) of an area in Wadi Araba, which superficially seemed to be a rather monotonous stand of single *Nitraria retusa* ASCH. shrubs in an Eragrostidetum bipinnatae, showed as great differences as from 50 to 1500 cm/day percolation speed along a horizontal line of 250 metres.

These differences are mainly caused by different percentages of sand and small stones in the soil. On the other hand, areas of even pure sand with regularly distributed sizes of the sand particles are apt to show great differences in percolation speed. The reason for this is that, in addition to fine sand, silt and clay particles which the coarser sand is mixed with, there exist distinct runoff channels for the percolation stream along the slopes, and holes of small diameters, filled with looser sand or almost hollow, created by decomposed roots, insects, reptiles or rodents. These act as main percolation ways in a more or less vertical direction downward. If the soil in general contains a relatively high amount of salt, then these ways are much less saline or even completely washed out. KARSHON (1956) studied such erosion channels in the gravelly hammada desert north of Eilat in the Negev and found the profiles of the surrounding soils of a relatively high salinity, whereas the profiles beneath them were completely washed out. According to the prevailing opinion of the soil scientists an accumulation was to be expected just there, because of the greater influx of salts from all sides.

Such places are frequently the best suited for germination and establishment and are often the only places in true deserts where the single plant individuals occur. Also sand hills in more humid areas, although in general barren or beset with sparse vegetation

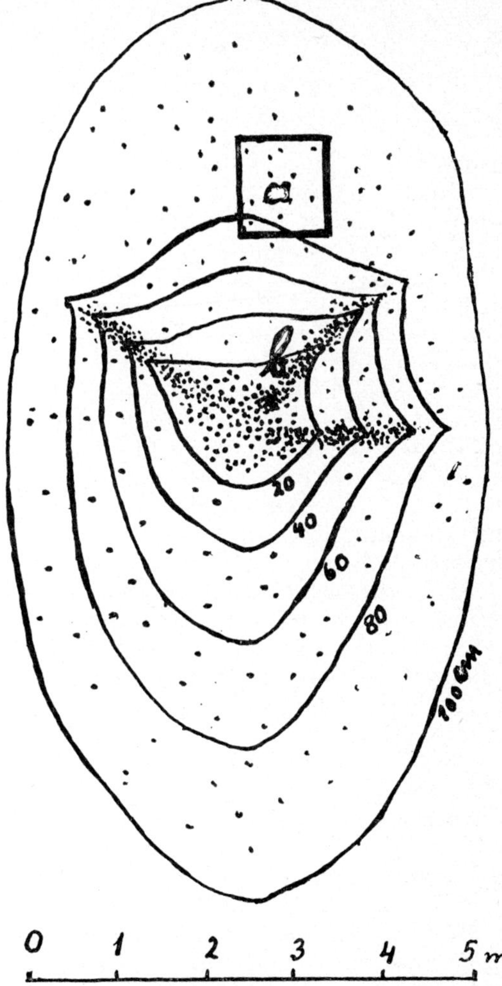


Fig. 5. Plantsociological records in and around a sand trough (BOYKO, 1934): (Place: Dam near Podersdorf, Burgenland; Date: June 4, 1932).

a) Outside the erosion channels:
area: *1 square metre;* medium and fine sand with 15—20% pebbles on the surface; less than 10% vegetation cover, consisting of:
7 individuals of *Bromus tectorum* L. (dry, fruiting)
2 individuals of *Alyssum alyssoides* L. (dry, fruiting)
1 individual of *Convolvulus arvensis* L. (budding)

b) In the channels:
Medium and fine sand with pebbles in 2—4 cm depth only.
Vegetation cover about 90% consisting mainly of *Bromus tectorum* L., growing there with 8—10 individuals per *1 square decimetre* (!), that is more than 100 times as many as in a).

The whole plantsociological record of the channels and in the trough is as follows:

| | *Abundance* | *Dominance* | *Periodicity* |
|---|---|---|---|
| *Bromus tectorum* L. | 5 | 4 | fr. |
| *Camelina microcarpa* ANDRZ. | + | | bud. |
| *Centaurea rhenana* BOR. | + | | dr. (from last year) |
| *Linaria genistifolia* MILL. | + | | bud. |
| *Silene conica* L. | 3 | 2 | fl., fr. |
| *Tunica saxifraga* SCOP. | 1 | 1 | fl. |
| *Reseda lutea* L. | + | | fol. |
| *Alyssum alyssoides* L. | 2 | 1 | fr. |
| (*Plantago indica* L. | + | | seedl. fol.) |
| (*Delphinium consolida* L. | + | | fl.) |

only, on such places show a strikingly denser plant cover mainly constituted by annuals.

A plant-sociological record from the sand and salt steppes of Central Europe may serve as an example (see Fig. 5.).

Under temperate climate conditions, however, sand belongs to the driest habitats (WALTER, 1951), contrary to sand habitats in the hot arid regions. But in both places the larger amount of water available to the plants on these localized places mentioned above is an additional favourable factor. Regarding the occurrence of salinity in the soil or in the water or in both, such places offer the best conditions in their vicinity for plant life in general.

There is, however, also a limit to the percolation speed, above which the plant roots have not enough time for their necessary intake of water and nutrition. This limit varies with the specific physico-chemical properties of each plant species, the individual size of the root system, the distribution pattern of the latter, the intensity and time span of the water stream, the temperature of soil and water, and last not least with the chemical composition and concentration of the solution running along the feeder roots.

*2. Good Aeration*

Closely connected with a good soil permeability, also a good soil aeration is one of the decisive factors for plant life.

Contrary to usual agricultural soils, consisting of a more or less high percentage of clay and silt particles, the space in sandy soils and still more in sand proper, leaves a high air volume between the soil particles. The influence of that good aeration as a consequence is decisive for all living beings which are bound to a permanent or temporary soil habitat. This applies to the aerobe soil bacteria as well as to non-bacterial components of the microflora and microfauna in the soil, and it applies similarly also to all subterraneously living animals and to all subterranean plants or parts of them. The mechanical soil structure may put a certain limit to the use of sand for the larger holes of rodents, etc, and the frequent extreme dryness in the upper layer will restrict bacterial and other micro-life.

But the principle remains always the same. With regard to plant-roots we shall have to deal with this influence later when we are going into the details of the principle of "partial root contact" (page 147).

*3. Easy Solubility of NaCl and $MgCl_2$*

Speaking of marine agriculture and particularly of irrigation with sea-water or dilutions thereof, we are confronted with an irrigation solution containing about 75% NaCl and about 10% $MgCl_2$ of the Total Salt Content (T.S.C.). The concentration may be different but the composition of sea-water is very similar in all oceans and seas, gulfs, fjords etc. connected with them. The concentration can be as low as in the Baltic Sea (6000—8000 p.p.m. T.S.C.) or near the average (about 35,000 p.p.m. T.S.C.), or above it (40,000 p.p.m. T.S.C.), as in the following example:

| *Water from Gulf of Eilat** | (part per million — p.p.m.) |
|---|---|
| Cl | 23,100 |
| $SO_4$ | 3,140 |
| $HCO_3$ | .149 |
| Ca | .547 |
| Mg | 1,490 |
| Na | 12,600 |
| | 41,026 p.p.m. |

pH 8

Always we are confronted with a composition in which the sodium and magnesium ions play a dominant rôle among the cations and chlorine and, to a much lesser degree, sulphate among the anions. It is mainly this fortunate composition, including also all trace elements, which makes irrigation with sea-water more feasible than that with any other saline water of the same concentration.

The two salts NaCl and $MgCl_2$ are also frequent components of saline wells or of under-ground waters especially in arid regions. One of the main characteristics of both is their extreme solubility. Sodium chloride already dissolves at an air humidity of 80% and magnesium chloride at even less than that. Both are among the first to be washed down into deeper layers with irrigation, or — if evaporation and capillarity lead the solution upwards, to build an efflorescence on the soil surface during the irrigation- or rain intervals. Fig. 6 on page 236 may serve as example. There, a sandy field 8 km south of the Dead Sea in Neot Hakikar was irrigated first 3 times, then 2 times and finally once weekly with a desert underground water containing a T.S.C. of about 3000 p.p.m. with about 1000 p.p.m. Cl, throughout the year since autumn 1962. The soil analyses showed after 1½ years of irrigation an accumulation of salt in the uppermost surface layer only (1—2 mm), whereas the salt content in the root layers remained constantly lower than the soil had contained before irrigation with this saline water started, resulting in good yields of various crops (tomatoes, gladioli, Rhodos grass, Juncus).

Soil analyses in an irrigation experiment with sea-water of oceanic concentration in Bet Dagon (Israel) with various crops gave the following results (H. Boyko & E. Boyko, 1959a):

The irrigation was carried out on dune sand during the dry season

* The Gulf of Eilat (also called Gulf of Aqaba) is part of the Red Sea and lies at the southernmost part of Israel.

**Table I.**

Analysis of a soil profile after 100 times irrigation with sea-water of a T.S.C. of 32,700 p.p.m. including 23,780 p.p.m. NaCl.

| Depth of soil samples | Moisture % | Cl p.p.m. | Calculated NaCl in p.p.m. (Coeff. 1.6) | NaCl content in irrigation water | T.S.C. in p.p.m. | T.S.C. of irrigation water |
|---|---|---|---|---|---|---|
| 10 cm | 1.8 | 1,380 | 2,208 | | 2,800 | |
| 30 cm | 1.8 | 1,150 | 1,840 | 23,780 | 2,350 | 32,700 |
| 60 cm | 2.4 | 750 | 1,200 | | 1,450 | |
| 10 cm | 5.0 | 1,600 | 2,560 | | 3,000 | |
| 30 cm | 2.4 | 1,070 | 1,712 | 23,780 | 2,200 | 32,700 |
| 60 cm | 2.9 | 800 | 1,280 | | 2,550 | |

of 1958 with an ineffective rainfall of 1 (one) mm in May. It is unknown if and how far the results were influenced by dew.

This easy solubility has also a negative effect. The respective ions are more readily taken up by certain plant species by their root system and this leads in general to salt accumulation in the aerial parts of the plants. The processes of numerous species in order to overcome this accumulation is best described by CHAPMAN (1962, 1966), the summarizing of the four main principles being:

1) A sodium efflux pump mechanism (e.g. in certain littoral marine algae);

2) Special glands for excretion (e.g. with certain *Tamarix* species, *Frankenia* spp., *Avicennia* spp., *Statice* spp., *Spartina* spp.)

3) Storing by succulence (*Salicornia* spp., *Suaeda* spp., *Arthrocnemum* spp.);

4) Shedding of parts (leaves etc.) (*Calotropis* spp., *Eragrostis bipinnata, Juncus* spp.) and also species of group 3, after accumulation has reached a critical stage.

In many cases of the last mentioned type shedding takes place long after the respective parts have ceased to be included in the process of metabolism. They can remain dry and dead on the plant for years and even decompose there. *Eragrostis bipinnata,* for instance, is one of such examples. We may separate this type as a fifth one from the others.

It is most probable that in some cases we are confronted with a combination of two or more of these processes, as for instance in *Calotropis procera* R.BR. where we can find the succulence and also the shedding of leaves after prolonged irrigation with high concentrations. The same can be observed under natural conditions after exsiccation during a prolonged dry season on a saline habitat, both having the same effect of over-salinization of plant cells in the vegetative parts.

Salt excretion by plants plays an important rôle with certain peoples in their nutrition in regions far from the Sea or from other mineral salt resources. (see page 194).

Specific biological mechanisms for the extraction of salt from sea-water and excretion exist also in the animal kingdom. A current research program of the Harvard Medical School is at present at work to investigate such a mechanism in certain glands of sea gulls. These birds have necessarily to dispose of their surplus of sea salt taken in in great amounts during their diving and fishing actions as well as with their continuous breathing in salt laden air.

In view of our rapidly growing demand of fresh water and the development of desalination techniques, an exact knowledge of such salt processing organs in plants and animals and of their mechanisms may result in most valuable basic information for technological progress in this field and thereby turn out to be of high economic importance. (See also the principle "Biological desalination", page 165 ff.).

*4. Lack of Sodium Adsorption by Sand*

Closely connected with the high permeability of sand as a favourable factor for plants under saline irrigation is the fact that the dangerous phenomenon of deflocculation in normal soils by adsorption of sodium to the clay particles is non-existent or to a negligible degree only in sandy soils, depending on the percentage of clay in the soil texture. Clay particles of a normal agricultural soil, after having been in contact with saline water containing sodium, will be swelling immediately after new irrigation, even with fresh water or rain. The original flocculated structure, so necessary for good plant growth, is lost, and after drying out, the soil becomes a compact structureless mass, frequently with surface cracks in characteristic polyeder form. No crop plant and very few others can thrive in such a habitat.

On sand we are not confronted with this danger. The lack of clay particles or their negligible amount in the soil will help to retain the loose structure and optimal aeration and this fact much enhances the possibility of saline irrigation on sandy soil or on sand.

Contrary to soils containing more clay rainfall here is an advantage, washing down the remainders of salts and the more frequent the rainfall the better.

*5. General Rules of Application*

All these factors together and their influence on plant-growth led to the formulation of the following general rules for irrigation with highly saline water and with sea-water. These rules are mainly based on experiments along the climatic profile from the hot desert region of Eilat, to the cool temperate and humid region of the Frisean Islands and of Stockholm, including the semi-arid mediterranean climate near Tel Aviv and in Southern Italy, and the more humid summer climate of Orinon (near Santander) in Spain, where the experiments are carried out during the drier summer season.

These *General Rules* read:

1. If salt concentration is constant, then

a. the higher and/or the more frequent the rainfalls, the higher can the clay content be;

b. the higher the temperature, the lower must the clay content be.

2. If clay content is constant or lacking at a given concentration from oceanic concentration downwards, then

a. the more frequent or more evenly distributed are effective rainfalls, the higher is the number of potential plant species;

b. the higher the saturation deficit, the smaller is the number of potential plant species.

3. In semi-arid regions annual crops with a growing period during the rainfall season are more promising of economic results than others and need in many cases only additional saline or sea-water irrigation, the latter having at the same time a potential fertilizing effect (see page 175).

4. Seeds bred from parent plants grown with saline water show in general better growth results than those from parent-plants grown with fresh water or with water of low salt content only (see page 173).

5. Agrotechnical details for economic purposes have to be worked out as a matter of course for each species separately.

6. Accumulation of NaCl or $MgCl_2$ is not to be feared in sand or gravel, deeper than the root systems, if at least once a year an effective rainfall occurs.

---

The diameter even of the finest sand particle (0.02 mm) is about twice as large as the thickest root hair (0.01 mm), that of coarse sand particles about 100 times as large. (Remark to Fig. 6.).

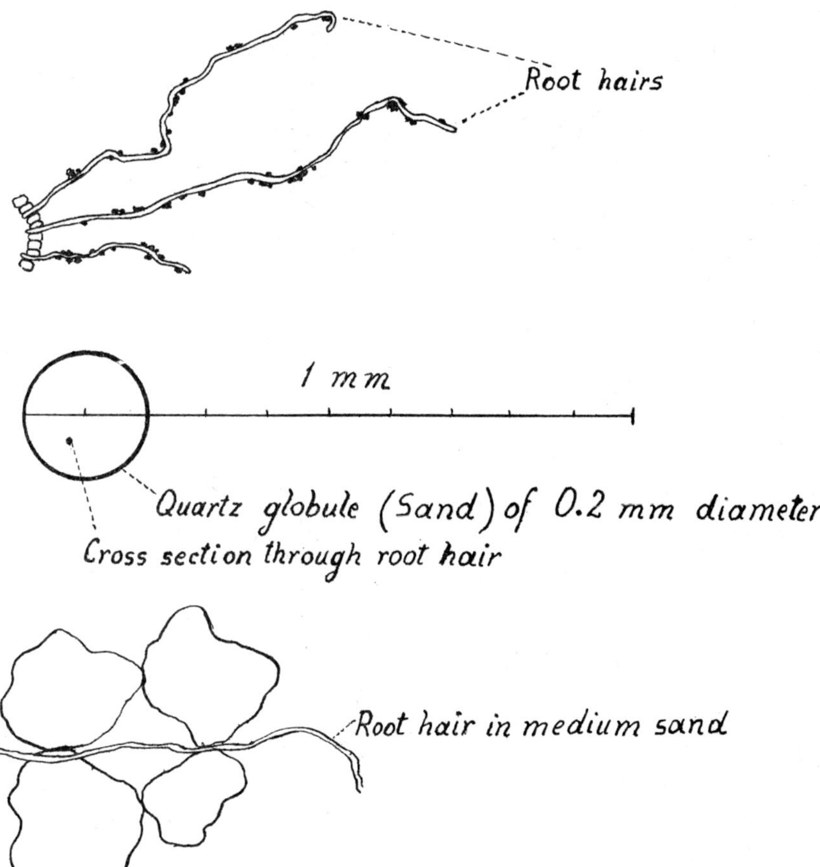


Fig. 6. Relative size of roothairs and of medium sand.

*6. The Principle of "Partial Root Contact"*

This new principle, connected with that of "Subterranean Dew" has first been mentioned in 1957 (H. BOYKO, 1957a) and in more details in a lecture at the New York Academy of Sciences (H. BOYKO & E. BOYKO, 1965). It is based on the fact that in sand the space between the sand particles is usually many times bigger than a feeder root (root hair) can occupy (see fig. 6).

The approximate diameters of feeder roots are from 0.005 to 0.01 mm. Several hundred such feeder roots with a length up to 1 mm have been counted per square mm by several authors and on various species. Two such counts by the present author on *Juncus arabicus* LAM. gave 416 and 524 feeder roots per square mm re-

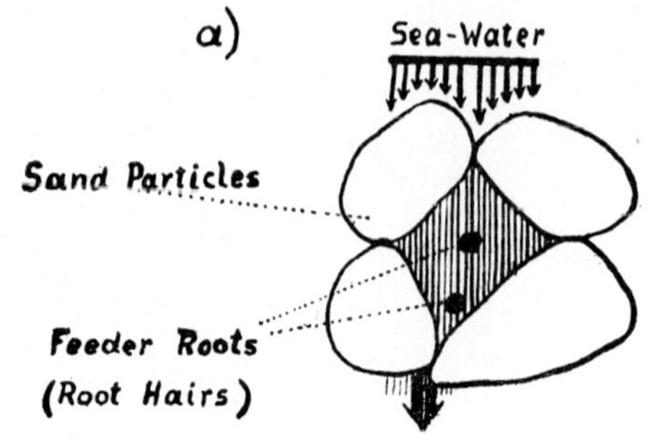


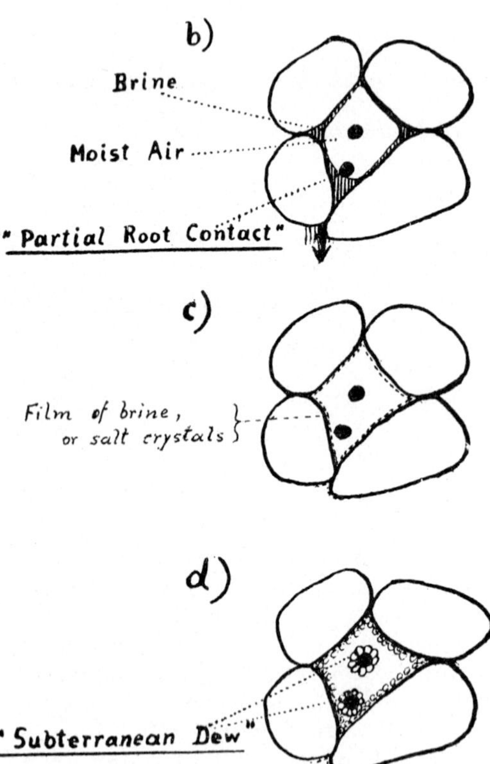


Fig. 7. The principle of "Partial Root Contact".

a) 1st step: Irrigation with highly saline or sea-water
b) 2nd step: "Partial Root Contact"
c) 3rd step: After percolation
d) 4th step: "Subterranean Dew"

Explanation to Fig. 7:

*1st Step:*

*Irrigation with highly saline or sea-water*

The microscopically tiny feeder roots (root hairs) (200—500 per square millimetre on the surface of the freely visible root parts) are surrounded on their whole surface with sea-water, but only (and this is the important point) for a relatively short time. Most plant species can endure this for a short time without damage as long as they have enough life energy to be selective in their uptake of the nutritive components of the sea-water or other saline water and to resist the difference of osmotic value or to adapt themselves to it. (In the drawings the cross sections of the feeder roots are enlarged. In nature they are only 1/10 to 1/100 of the space diameter between the particles of medium sand).

*2nd Step:*

*Partial Root Contact*

The quick percolation in sand has as its most important consequence the fact that the surface of the feeder roots is very soon only partially in contact with the saline solution or the sea-water, and partially or very soon even, as a whole, in contact with air, the oxygen of which raises considerably its life energy and activity, and with it also its selective capacity for the intake of nutritive material from the solution. This applies — but with a certain retardation — also to roothairs pulled to the walls of sand particles during percolation by surface tension.

*3rd Step:*

*After Percolation*

It is assumed up to now (see below on page 154) that a thin film of brine or salt is adhering to the surface of the sand particles. The surface of the feeder roots, however, is at least partly surrounded by moist air*). This air moisture condenses from time to time, mostly in the cool morning hours (particularly in continental deserts with their great temperature differences between day and night), and thus leads to the decisive next step:

*4th Step:*

*Subterranean Dew*

This condensation of air moisture in the soil called in a former paper "subterranean dew" (1957a) supplies the feeder roots with fresh water irrespective of the fact that a chemical analysis of the soil will show that there is still a certain percentage of salt in the soil, for this amount of salt adhering to the surface of the sand particles is in no contact with and therefore of no biological influence on the living feeder roots.

The next irrigation, whether after one day or after several months, washes away any possible accumulation of NaCl and $MgCl_2$ even from these sand particles, and only a certain very low amount of salt remains there constantly as an equilibrium.

* See also the Viscosity Principle, (page 152).

spectively (the plants were taken from their sandy habitat and, after several hours of drying in the air, put in water. 20 hours later, the fresh feeder roots were counted, or, more exactly, calculated from 104 and 131 on 1/4 mm$^2$ each).

If we take for comparison the space between medium sized globule sand particles of 0.2 mm diameter, then the diameter of a feeder root has the approximate proportion as shown in fig. 6. With part of its length equalling the diameter of the air space the feeder root has a proportional volume in the magnitudinal order of about 1/1,000 to 1/100,000 of the air space between the respective sand particles. If in addition we consider the insignificant capillarity

**Table II.**

The relation of surface to particle size (from L. D. Baver, 1940).

| Diameter of sphere | Textural name | Volume per particle $\left(\frac{1}{6}\pi D^3\right)$ | Number of particles in $\frac{\pi}{6}$ cc. | Total surface $\pi D^2$ × number of particles |
|---|---|---|---|---|
| 1 cm | Gravel | $\frac{1}{6}\pi(1)^3$ | 1 | 3.14 sq. cm = 0.49 sq. in. |
| 0.1 cm (1 mm) | Coarse sand | $\frac{1}{6}\pi\left(\frac{1}{10}\right)^3$ | $1 \times 10^3$ | 31.42 sq. cm = 4.87 sq. in. |
| 0.05 cm (0.5 mm or 500 μ) | Medium sand | $\frac{1}{6}\pi\left(\frac{5}{100}\right)^3$ | $8 \times 10^3$ | 62.83 sq. cm = 9.74 sq. in. |
| 0.01 cm (0.1 mm or 100 μ) | Very fine sand | $\frac{1}{6}\pi\left(\frac{1}{100}\right)^3$ | $1 \times 10^6$ | 314.16 sq. cm = 48.67 sq. in. |
| 0.005 cm (0.05 mm or 50 μ) | Coarse silt | $\frac{1}{6}\pi\left(\frac{5}{1000}\right)^3$ | $8 \times 10^6$ | 628.32 sq. cm = 97.34 sq. in. |
| 0.002 cm (0.02 mm or 20 μ) | Silt | $\frac{1}{6}\pi\left(\frac{2}{1000}\right)^3$ | $125 \times 10^6$ | 1,570.8 sq. cm = 1.69 sq. ft. |
| 0.0005 cm (0.005 mm or 5 μ) | Fine silt | $\frac{1}{6}\pi\left(\frac{5}{10,000}\right)^3$ | $8 \times 10^9$ | 6,283.2 sq. cm = 6.76 sq. ft. |

| | | | | |
|---|---|---|---|---|
| 0.0002 cm (0.002 mm or 2 μ) | Clay | $\frac{1}{6}\pi\left(\frac{2}{10,000}\right)^3$ | $125 \times 10^9$ | 15,708 sq. cm = 16.9 sq. ft. |
| 0.0001 cm (0.001 mm or 1 μ) | Clay | $\frac{1}{6}\pi\left(\frac{1}{10,000}\right)^3$ | $1 \times 10^{12}$ | 31,416 sq. cm = 33.8 sq. ft. |
| 0.00005 cm (0.0005 mm or 500 mμ) | Clay | $\frac{1}{6}\pi\left(\frac{5}{100,000}\right)^3$ | $8 \times 10^{12}$ | 62,832 sq. cm = 67.6 sq. ft. |
| 0.00002 cm (0.0002 mm or 200 mμ) | Colloidal clay | $\frac{1}{6}\pi\left(\frac{2}{100,000}\right)^3$ | $125 \times 10^{12}$ | 157,080 sq. cm = 169 sq. ft. |
| 0.00001 cm (0.0001 mm or 100 mμ) | Colloidal clay | $\frac{1}{6}\pi\left(\frac{1}{100,000}\right)^3$ | $1 \times 10^{15}$ | 314,160 sq. cm = 338 sq. ft. |
| 0.000005 cm (0.00005 mm or 50 mμ) | Colloidal clay | $\frac{1}{6}\pi\left(\frac{5}{1,000,000}\right)^3$ | $8 \times 10^{15}$ | 628,320 sq. cm = 676 sq. ft. |

and quick percolation, we can follow the progress of irrigation with saline or sea-water in sand in four steps as shown in figs. 7a to 7d:

Purely physical experiments with Mediterranean sea water, sent through dune sand, which Dr. E. MATZ kindly made at the present author's request in his laboratory at the Negev Institute for Arid Zone Research, also showed that no salt accumulation takes place. Irrespective of the particularly high T.S.C. of about 40,000 mg/l, this seems to be the first percolation and accumulation experiment with natural sea-water on dune sand.

An experiment of Dr. A. WERBER's (1936), which was carried out as early as 1936 with a cooking salt solution of 800 p.p.m. NaCl on a red sandy soil from a citrus plantation also must be regarded as a pioneering step in this direction.

It gave the same result of non-accumulation. Then, WERBER's experiment was of a revolutionary nature, and was disregarded at that time.

From the foregoing explanations, it is clear that any effective rain clears every possible remnant of NaCl and $MgCl_2$ at least in the upper layers. But even without rains, each subsequent application of salt water washes the remnants of former applications to lower levels, so that no *accumulation* of these salt occurs in the root layer, whereas the nutritive salts are selected and taken up by the plants as long as they have enough life energy, the latter being strongly supported by the extremely good aeration (and in consequence the high oxygen supply) they receive in sand and gravel. In the long run, the "Global Salt Circulation" prevents any significant salt accumulation even in deeper layers (see page 180).

### 7. *The Viscosity Principle*

The Viscosity problems are closely connected with all irrigation problems. In our case, i.e. in irrigation with highly saline or sea-water on sand, they are particularly connected with the principle that the feeder roots are in space as well as in time, only partially in contact with the highly concentrated solution of the irrigation water. This is not the case in agricultural soils where the space between the soil particles is of a similar small magnitudinal order as the diameter of the feeder roots, surrounding the latter with its capillary water and/or the water-films adhering to the surface of the soil particles. Factually the "film" of brine, generally assumed to adhere to the surface of the soil particles after percolation of the main mass of saline irrigation water must occur also in sand. But it cannot remain there for a longer period in the form of a continuous uninterrupted film, since the viscosity of these saline waters is much too low.

We do not need, therefore, to go into the details of the complex problem of viscosity, which is of such a particular importance for soils containing a considerable amount of clay particles. In this respect we can simply refer to the existing textbooks dealing with irrigation and soil science. Here, it is sufficient to consider the diameter of such a theoretical film after the irrigation. What, however, we want to stress here is the relation of the *physical* phenomenon of viscosity in the process of highly saline irrigation on sand to the *biological* processes of the plants.

Particularly important physical features in this connection are the non-accumulation of salt, the porosity of a temporarily dry salt layer, and the film diameter.

The chemical analysis (Table I) during an experiment (BOYKO & BOYKO, 1959a) showed that after 100 times of irrigation on sand with sea-water of oceanic concentration during the rainless summer period of 1958, the remaining air-dry and very porous layer of salt adhering to the surface of the sand particles had an approximate thickness of about $10^{-6}$ mm; sand particles with a diameter of 0.2 mm are taken as the basis for the calculation of his figure because they constituted about 90% of the sand experimented with.

The number of particles differ as a matter of course very much with the diameters and the structure. The same applies, although to a much lesser degree, to the total surface of the particles. We can calculate for instance 1000 particles of coarse sand (with 1 mm diameter) per cc against 1,000,000 particles of fine sand with 0.1 mm diameter, that is 1000 times as many particles. The total surface is however only 10 times as large, namely about 30,000 square mm with fine sand of 0.1 mm diameter, against 3,000 square mm with coarse sand of 1 mm diameter.

Table II, taken from BAVER's textbook (1940) on soil physics, showing the relation of surface to particle-size, may be inserted here in order to facilitate similar calculations for conditions of other particle-sizes (p. 150—151).

In all these calculations the sand particles are taken as globules and as put together as if each globule would represent its circumferential cube, and not packed!

The calculated total surface of the globules is therefore always the theoretical minimum.

There is almost no surface activity in sand and therefore no Na-adsorption, in sharp contrast to clay and the great activity of its colloidal particles, the surface of which per cc is thousands of times as large as that of sand particles.

Regarding the classification of soil systems according to particle-size the internationally frequently used ATTERBERG scale* differs somewhat from the scale used by the US Bureau of Soils, as is to be seen from the following contraposition:

* ATTERBERG, A., Die mechanische Bodenanalyse und die Klassifikation der Mineralböden Schwedens. *Int. Mitt. f. Bodenk.* 2, *312—342,* 1912.

Soil Systems according to particle size

| Atterberg Scale in mm (= international scale) | | Scale of the US Bureau of Soils in mm | |
|---|---|---|---|
| Clay | <0.002 | clay | <0.005 |
| Silt | 0.002—0.02 | silt | 0.005—0.05 |
| Fine sand | 0.02—0.2 | very fine sand | 0.05—0.1 |
| | | fine sand | 0.1—0.25 |
| | | medium sand | 0.25—0.5 |
| Coarse sand | 0.2—2.0 | coarse sand | 0.5—1.0 |
| | | fine gravel | 1.0—2.0 |
| Gravel | >2.0 | gravel | >2.0 |

If, as a general basis for our purpose, we take Quartz globules of 0.2 mm diameter, then we have as a minimum 125,000 globules per cc with a minimum surface of 157 square cm, if the globules are placed vertically above one another. Actually the surface will probably be many times (possibly 10 or 100 times) as large, because:

a) the sand particles will be of very different size and shape with many finer sand particles between the coarser ones;

b) they all will be more or less packed; and

c) in addition to the sand particles silt and even some clay particles (e.g. of aeolic origin) are always to be found between.

The total sum of salts remaining in the sand after 100 times irrigation with sea-water of oceanic concentration was in the experiment about 0.200% of the oven-dry weight of the soil (moisture content was 2.5%).

The total minimum volume of sand globules per cc is about $\frac{1}{2}$ cc.

The specific weight of Quartz sand is about 2.7

The specific weight of NaCl sand is about 2.2

The specific weight of $MgCl_2$ sand is about 1.6

Based on these data we come to the following figures for sand (quartz globules) of a particle size of 0.2 mm diameter with the globules placed vertically above one another:

(Remarks by the Physicist Dr. Ph. STOUTJESDIJK, Holland: "1 cc of sand contains 1/2 cc of quartz (1.35 g) globules with radius 0.1 mm. The volume of such a globule is: $4/3 \times 3.14 \times 0.001$ mm$^3$ = 0.00418 mm$^3$. Its weight is $2.7 \times 0.00418$ mg = 0.0113 mg. Hence 1.35 g of sand (1 cc) contains 119,700 sand globules with a total surface of 157 cm$^2$. Salt content = 0.2%, i.e. per cc of soil $0.002 \times 1.35$ g = 0.0027 g with spec. weight of 2.2 this is 0.00135 cc. Thickness of salt film would be 0.00123/157 = 0.000008 cm = $8 \times 10^{-5}$ mm. Of course this is a hypothetical film as capillary forces will concentrate the brine at the points of contact between the sand grains") (and make the film between these points still thinner).

The film evenly covering the surface of unpacked sand of 0.2 mm diameter must therefore have a hypothetical maximal thickness of less than $10^{-4}$ mm. Considering the greatly enlarged surface by packed texture and the numerous fine sand and silt particles, we have to assume a thickness between $10^{-5}$ and $10^{-6}$ mm.

This extreme thinness of a theoretical salt layer adhering to the surface of the sand particles leads to the following conclusions derived from experimental results and theoretical considerations:

If during every year a new salt layer should accumulate in addition to the previous year's layer, it would need many rainless (!) years to build up a very porous layer of, let us say, $10^{-3}$ mm or 1 $\mu$; or that, in other words, in sand areas *with* any efficient rainfall, even after 100 or 1000 years this layer cannot become an obstacle against aeration as long as the adsorption of ions, particularly of sodium-ions, by clay particles does not amount to counteracting proportions.

In normal agricultural soils, such an accumulation actually occurs in the course of a few years. In Italy, the experimenters and agriculturists therefore interrupt their saline irrigation experiments by a fallow year every few years in order to allow the rainfall to restore the former conditions. It may be of interest to mention in this connection the ancient religious law observed even today by orthodox Jewish farmers and settlements in Israel, that every seventh year is to be observed as a so-called "Smittah-Year", during which cultivation of fields has to be interrupted. This law may stem from old Babylonian or even earlier times, when irrigation methods led to recognizable gradual salinization.

It is most probable that the priests of that time who made very careful observations on these vital problems and also kept written records and an exact kind of book-keeping on harvests, recognized the beneficial influence of rainfall also in this respect, after periods of salt accumulation (see page 13).

However, a yearly accumulation of this tiny salt layer is impossible in the root layer in sand for the various reasons mentioned before and of which the main ones may be repeated here:

Quick percolation and absence of capillary movement in sand; high solubility of NaCl and $MgCl_2$; and the yearly rainfall (even in absolute deserts erratic rainfalls occur once in a number of years). The "Global Salt Circulation" (see page 180) is also closely connected with the phenomenon of non-accumulation.

(Prof. P. Dansereau drew my attention to possible differences on non-quartz sand, e.g. coral sand. In my opinion, such differences will be found with high probability and for various reasons: Higher porosity and moisture-retaining power of coral sand in comparison with quartz sand; Ca as antagonist to Na; On the other hand subterranean dew will probably not occur as frequently or not at all, because of the much lower heat-capacity of limestone than Quartz, and in addition for the small differences between day and night temperatures in the humid Tropics. Thereagainst the more frequent and more effective rainfalls in the humid Tropics, compared with the arid deserts, would probably enhance the success of experiments with sea-water on coral sand).

Another beneficial factor in saline irrigation on sand is the porosity of such a thin salt-layer. A small and simple experiment can easily be made by anybody to demonstrate this high porosity after evaporation of the water and drying out of the salt particles.

If we allow sea-water to evaporate on a horizontal surface into air of low humidity, a very porous layer of hygroscopic sea-salt will be left, mostly of a reddish or greyish colour, consisting of various minerals, some of them appearing in small cristals, others more crystalline.

The same occurs when, in the process of sea-water irrigation on sand, — following the quick percolation of the sea-water and the slower evaporation of the $H_2O$ into the air-space between the sand particles —, first a water-film, then a saturated brinefilm and finally an infinitesimally thin, very porous and in addition interrupted layer of salt adheres to the surface of the sand particles.

As is to be seen on the figures 7a—7d, after the above-mentioned irrigation-, percolation- and evaporation-process, there is enough air left in the space between the sand-particles and in contact with the feeder roots to supply these with oxygen and thus with the source of life-energy necessary for them to remain active and selective.

If the air is too dry then the feeder roots will die and new ones will appear at the event of the next contact with water by irrigation or by another cause. The lumps of salt-crystals still adhering to the sand-particles will be washed away into deeper layers.

Again, after each consecutive irrigation with highly saline or sea-water, a new brine film will soon adhere to the surface of the sand particles and will partly also be in contact with the feeder roots. But, as mentioned above, after 100 times of irrigation with water of oceanic concentration or 34,000 p.p.m. (3.4%) T.S.C. only 0.2% salt was found in the soil as a kind of equilibrium. This small amount of salt regularly distributed as a continuous film on the surface of the sand particles would result, as shown before, in a film of about $10^{-6}$ mm thickness; but such a film is physically impossible. Even soap bubbles – and soap water has a much higher viscosity than sea-water – explode usually long before the thickness reaches $10^{-4}$ mm. The calculation of this figure is based on the following consideration:

A column of 10 cc soap water is supposed to be blown up to a soap bubble of 100 mm diameter, the surface of which is then 31,416 square mm (or approximately 30,000). The volume of 10 cmm distributed over an area of 30,000 square mm results in a film of 10/30,000 or $1/3 \times 10^{-3}$ mm thickness.

The brine adhering to the sand particles can therefore only be in the form of an irregular net or in patches, or, if completely dried out, in separated heaps of crystals leaving the other parts of the surface of the sand particles in contact with the air.

The smallness of a film of $10^{-5}$ or $10^{-6}$ mm is of such a magnitudinal order that it is, like the diameter of molecules, best expressed in Angström = Å (1 Angström = $10^{-7}$ mm). This maximum thickness is therefore 10 Å or about the size of one hydrated Li-ion. (The size of a hydrated lithium ion is 10.03 Å; the size of a hydrated Na-ion is 7.90 Å, that of a K-ion 5.32 Å). The diameter of this hypothetical film would therefore be of an approximate magnitudinal order of a few molecules only or partly even a monomolecular film.

These considerations and calculations apply in the same way also to such a hypothetical film on the surface of the feeder roots. Here, the physical factors would be of direct biological importance. *Living* feeder roots are, however, effectively working against such physical obstacles as a salt-brine or salt-cover, threatening to suffocate them.

Such life processes effective as counter-measures are for instance:

Root-exudations or excretions, surrounding of dying root hairs with acid cellsap, expanding the surface of the feeder roots by quick longitudinal growing, and other still unknown physiological and bio-physical or bio-chemical processes.

In addition to this, if there is a dry salt-layer it is, as mentioned above, of a very high porosity, and both, a brine-cover as well as a salt-crystal layer would here too not be a continuous layer owing to the insufficient viscosity of the fluid-solution but would, similarly to that on the surface of the sand particles, rather consist of dispersed small patches of the brine or of clusters of crystals, for the infinitesimal thinness of the original film. In any case, the surface of the feeder roots are in sufficient contact with the surrounding air to fulfill their task of selective plant-nutrition with all necessary efficiency.

Based on the above figures, the maximum thickness of the salt layer on the surface of the sand particles was calculated as of $10^{-5}$ to $10^{-6}$ mm. If the amount of salt found by the chemical analyses has to be distributed not only onto the surface of the sand particles, but also onto the surface of the feeder roots, then the layer is not even as "thick" as that. I counted 416 root-hairs (feeder roots) up to 0.1 mm length and 0.005 to 0.01 mm width per square millimetre of a *Juncus arabicus* individual and on another occasion even 524 could be recorded. The latter figure may represent a particularly high one and the author did not find a higher one reported in the literature. But in all cases the surface per unit is considerably enlarged if we add the surface of the roots and roothairs to the surface area of the sand particles.

The approximate relations of the various diameters are to be seen from Table III.

An irregular, very porous layer of salt crystals full of air between its single, infinitesimally small particles obviously presents no

**Table III.**
Relative Thickness of Salt Film.

| Relation of | Salt Layer (film) | to | Feeder Root | to | Medium Sand Particles |
|---|---|---|---|---|---|
| Diameter in mm | $\frac{1}{1,000,000}$ | : | $\frac{1}{100}$ | : | $\frac{1}{5}$ |
| or | $10^{-6}$ ($10^{-5}$) | : | $10^{-2}$ | : | $2 \cdot 10^{-1}$ |
| The Relation is therefore | 1(10) | : | 10,000 | : | 200,000 |

obstacle to a good aeration and allows the necessary supply of oxygen from the soil air. If such a salt layer on the feeder roots exists and if it has any significance at all in connection with our problems, then it is probably more on the positive side, for its hygroscopicity helps the feeder roots to live and to be active and selective for a much longer period than without it. It does not seem too difficult to study this by a physiological experiment after the successful results of our "dying" experiments mentioned before (see also page 246 ff) have led to these theoretical conclusions.

In the "Dying Experiments" it could be proved that all species irrigated with highly saline solutions (sea-water up to North Sea concentration) showed the same pattern:

The higher the concentration, the more parts of the plant individuals remained green and living after the interruption of irrigation without any rainfall through 9 months, and in the second year through 8 months.

In addition to the physical considerations, it has been found by electron microscopical investigation that the surface of feeder roots is perforated, further that the diameter of these perforations is much wider than the thickness of such a hypothetical film would be, a film which could only exist as a continuous layer if its viscosity were much stronger than that of soap bubbles and not like that of sea-water. All these factors together added to the fact that living roots exude liquids, acids, etc. lead to the conclusion that such a theoretical "film" would not present any obstacle whatsoever for a living plant. The best proof of this, however, is the actual success of the plants that *do* live and grow on sand in spite of the high salinity of the irrigation water.

### 8. *Subterranean Dew*

In addition to the very low viscosity of saline irrigation water and the favourable physical factors of sand there is another phenomenon more nearly connected with the climatic features of the respective area. In many cases it may be the most decisive one for the success of saline irrigation and also for overcoming extremely prolonged dry periods between erratic rainfalls in desert regions. This phenomenon is the "Subterranean Dew", caused by the frequently extensive range of day and night temperatures, particularly in more continental deserts and semi-deserts. Dew in general plays a much more important rôle as a source of water supply for plant life than was until recently accepted by plant physiologists. Collectors of herbarium material, however, have known for a long time that most plants are able to absorb water not only by their root systems but also by their vegetative parts above the soil surface. This was already evident to the present author as a boy at the beginning of this century, when he put his somehow dried out plants after each collection tour in wet newspaper overnight, with the roots outside or without roots at all. In the morning all plants had regained their turgor and could easily be pressed in good shape and prepared for the herbarium.

There is no difference in principle for the plants between the vapours inside the wet newspaper wrapped around the plants and dew or mist in their natural surrounding. CARL TROLL (1935) describes a characteristic vegetation in the "Nebelzone" (fog zone) of the Andes where there is no or almost no rainfall the whole year round but a dense mist supplying the plants with the vital amount of water.

Fog or mist is in general not sufficient to induce seeds to germinate, but one effective rainfall once in decades may bring this about for many species, the seeds of which have an adequate viability. Their further development after germination is then made possible by the dense air moisture.

R. BLOCH (1953) rightly distinguishes two kinds of *dew*, the physical causes of which are very different: "heat capacity dew" on the one hand and "radiation dew" on the other hand. In both cases, the IE factor (insolation-exposure-factor) plays a major rôle (H. BOYKO, 1947). All three phenomena have to be considered in studies of dew condensation on and in the soil as well as on the surface of plant parts, in addition to the features of the specific surface itself (e.g. kind of hairs, body temperatures of the living plant, etc.).

Regarding dew condensation on plant parts above the soil surface and its effect on plants, one of the most authoritative experts on dew problems, S. DUVDEVANI (1953) explained the promoting effect of dew on plants by the fact that *dew occurs mainly during night* which is the best period of plant growth; further dew (or condensation of mist) being absorbed by the growing points of naked buds and growing regions of stems and leaves even in minute quantities will promote meristematic activity and cell growth.

Similar observations are made by F. W. WENT (1953) and presented in his report to UNESCO on the rôle of dew in plant growth in arid regions.

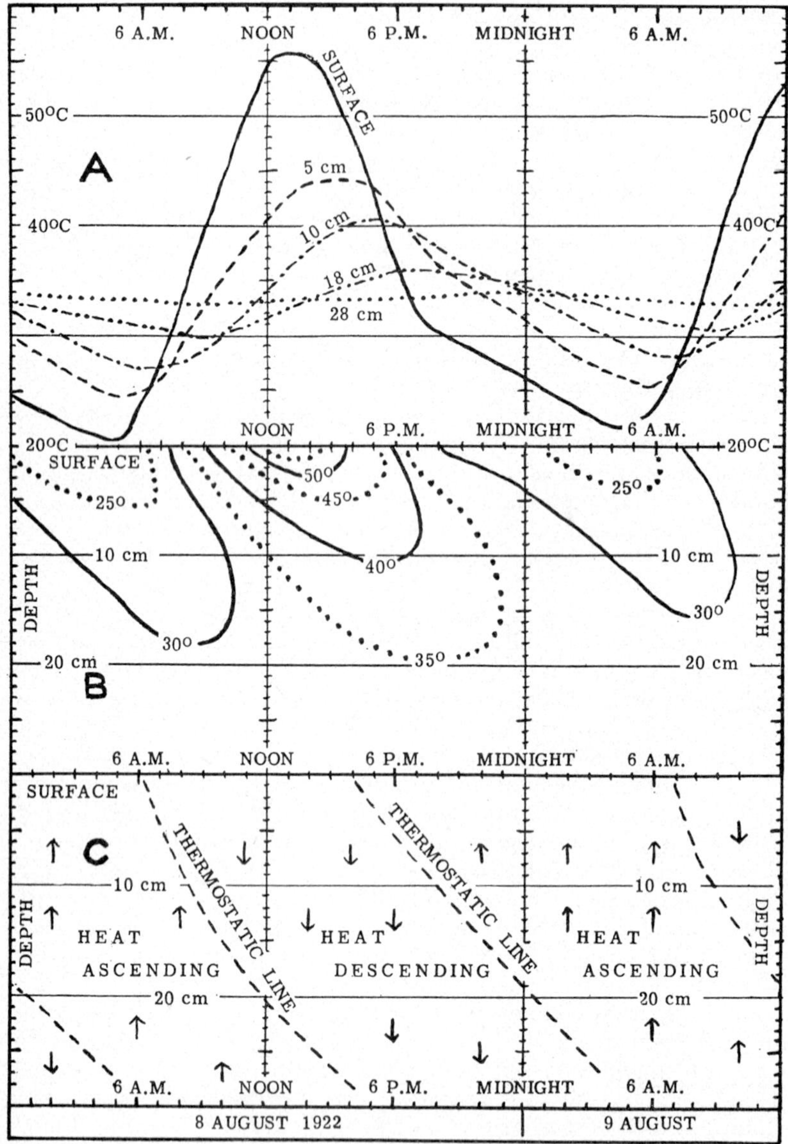


Fig. 8. Changing Temperatures at different levels in sand (after C. B. WILLIAMS, 1954).

In addition to the visible dew formation, there is an invisible condensation of no less favourable effect on plants (H. MASSON, 1951). The difference is, however, only a quantitative one and is irrelevant in connection with our problems. There is no point in subdividing subterranean dew according to these categories. As long as we have no method to make the condensation even of large drops below the surface visible, we cannot well apply this artificial subdivision for condensation water around the root hairs.

The main cause of dew building, the great temperature differences between day and night, and its working operation beneath the surface of sand under subtropic desert conditions are best demonstrated by an example:

Fig. 8 shows such measurements by C. B. WILLIAMS (1954) in Wadi Digla, 12 miles south-east of Cairo, and may also serve as an example of representation of the main relevant temperature data. Fig. 8A shows the gradually changing temperature at different levels in the sand during 36 hours during the hottest season, in August.

Fig. 8B shows the temperature contours at different depths during the day, and Figure 8C shows the movements of heat in the sand at different times, the surface heating during the day and cooling during the night. The lines where the heat movement is momentarily zero, have been called the "thermostatic lines" (MCKENZIE-TAYLOR & WILLIAMS, 1924).

It is, of course, particularly easy for the plants to use water vapour condensing in the soil as subterranean dew around their roots.

Regarding quantitative data of *water uptake by the root hairs*, Hilda F. ROSENE was apparently the first who supplied such data by a microphotometric method devised by her. She and others investigated various crop plants (field pea, tomatoes, radish, wheat, oats, barley, rye, corn, chinese cabbage, Georgis collard, mustard) and a *surprisingly high absorptive capacity* was discovered. The calculated average absorptive capacity of the root-hair system for 24 hours amounted to 55 times the maximum transpiration rate. Hence, on a theoretical basis, maximum transpiration needs could be met if the average rate of intake of the entire root-hair system were reduced to 1/55 or if water entered only 1/55 of the total root-hair surface (ROSENE, 1954).

Field botanists who also include the root-systems in their investigations and records, can frequently find a distinct pattern of root-growth in regions where subterranean dew is one of the main water sources. Two types of root systems were found by the present author as being particularly indicative of this phenomenon:

a) tap-roots growing first vertically downwards until the layer where the subterranean condensations occur most frequently. There, they change their growing direction, bending abruptly into a horizontal one as shown in figure 9.

Fig. 9. Schematic tap root of various desert shrubs, growing in horizontal direction, and thus using the subterranean dew of this layer.

b) a dense subsurface, frequently adventive, root-system utilizing the subterranean dew as to be seen in fig. 11.

This schematic sketch applies, for instance, to various species in the semi-desert of the Central Negev. The critical depth depends on the local climate and soil structure which is decisive for the temperature gradient from the surface downwards. There, in loessy somehow sandy soil the depth where tap roots bend as shown in fig. 9, is about 8—10 cm (Boyko, 1949). These horizontal roots can stretch very far. In the Southern Negev, in gravelly sand, I could follow one of the horizontal roots of *Calligonum comosum* L'Hér. for 17 m without reaching the end. This desert shrub, developing more than one root like this, presents a transition type to that of *Haloxylon persicum* Bge. growing only on sand. This *Haloxylon* species is one of the few true desert trees and its geographical distribution reaches from the extremely wintercold sand deserts of Central Asia to the extremely summerhot sand deserts of Arabia (H. Boyko, 1949).

It manages to obtain its water requirements from two main sources: from underground water by very deep-going roots and by a dense layer of fine and finest roots in the zone of subterranean dew. A profile (figs. 10 and 11) through a sand dune in Wadi Araba shows such a venerable, perhaps 2000-year-old tree growing with the 10 m high more or less stable sand dune.

The profile has been cut into this dune by the author during one of his expeditions, made mainly for the purpose of root investigations and emerged — with difficulty — onto the marked line.

The third water source, direct rainfall, in this region (50 mm yearly average) is too small and too erratic to be of marked significance for this huge plant individual in spite of its covering more than 100 square metres of the dune.

The possibilities of utilizing the principle of subterranean dew for schemes of desert reclamation and the methods to be applied have been presented by the author at the I. Bioclimatological Congress in Vienna (H. Boyko, 1957b).

Subterranean dew may also play an important rôle in the growth

Fig. 10. Excavation of one single venerable *Haloxylon persicum* Bge. tree on a sand-dune in Wadi Araba. The protruding branches are several meters high and give the appearance of many separate trees. The excavation was carried out in June 1949 in order to investigate the true growth relationship of them.

of rock plants in such deserts. The present author found about 70 different species growing in the rock crevices of lime-stone as well as of granit, gneiss and of Nubian sandstone in the almost absolute desert climate of Eilat with not more than 20—30 mm rainfall per year and not infrequently even less.

A. Abraham (see Boyko & Abraham, 1954) investigated the phenomenon of subterranean dew on an annual species (*Amaranthus graecizans* L.) in Jerusalem. There, this species lives its short life cycle from the last winter rains, when it germinates, through the dry summer, relying therefore completely, from an early stage on, on subterranean dew. Abraham constructed similar conditions in the laboratory and was thus able to prove for the first time the building of subterranean dew by an ingenious experiment. This

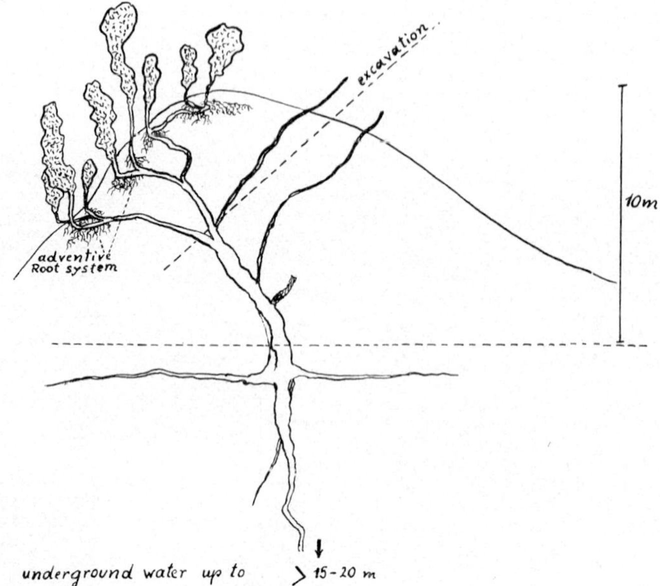


Fig. 11. Profile of *Haloxylon persicum* BGE. through the sand-dune showing its two types of root-systems (see Fig. 10).

phenomenon is of particular importance for desert plants on the one hand and in connection with highly saline irrigation on the other hand, for the reasons explained in fig. 7a—7d.

According to S. DUVDEVANI (1953) and others, the amounts of dew fall can reach a significant percentage of the whole precipitation. If we consider the relatively great part of each rainfall, actually lost for plants by run-off, percolation and evaporation, and on the other hand, if we take ROSENE's figures as a basis for the uptake capacity by root-hairs, added to the uptake of visible and invisible condensation water by plant parts above the soil surface, and if we compare these data with actual measurements of dew per square cm, we then arrive at remarkable potential amounts of fresh water at the disposal of the plants in the form of dew (visible or invisible) under arid and semi-arid conditions.

The observation power of farmers in ancient times was certainly at least as good as it is today and the ancient Jewish prayer for dew seems to be well justified from an agricultural point of view.

The surprising amount of fresh water theoretically at the disposal of the plants by subterranean dew, even in rainless periods and under saline irrigation may be seen from the following investigation:

One root hair of 0.01 mm diameter and a length of 1 mm has approximately 0.03 square mm. By several records, I counted

400—500 root hairs per square mm on *Juncus arabicus* ASCH. et BUCH. (The record was 524, or more exact 131 on 1/4 $mm^2$.)

In the second year, one plant has usually already more than 200 leaves and opposite to each leaf are two about one to one and a half meter long succulent vertical roots and a number of other roots seemingly with nodules. I have found old plants with a basal area of more than one square meter and many thousands of leaves (stalks). If we take a plant with 600 leaves, the basal area of which has a diameter of about 30 cm and if we assume the usually adequate number of roots, i.e. 1200, potentially covered on half a meter of their length with root hairs in the time of water contact, then we arrive at a figure of more than 20 square metres surface of root hairs ready for the intake of irrigation water, dew or rainfall, that is about 200,000 $cm^2$.

Frequently we can find 1 gram (= 1 $cm^3$) of dew on an area of 100 square cm (and sometimes even much more). The result of this theoretical calculation is 2 litres of water or more on one single cool morning for a dew fall available to the root hairs of one single *Juncus* plant, which is a considerable amount, particularly if we consider it compared with rainfall or irrigation. Dew is much more effective because its whole amount or most of it is constantly in direct contact with the absorbing surface of the respective plant-part during the whole process of condensation. The same effect can only be achieved by a multiple amount of percolating water of which a small part only is actually taken in by the plant.

Water uptake of plants is, however, not simply a physical process to be described by a cherished formula, and, if possible still more complicated is the process of salt uptake. The prevailing opinion that the latter is included in the former has long been proved to be wrong. "In cases where a correlation between water uptake and salt uptake is found, it is possible to demonstrate the independence of these two processes by inhibiting one of them without influencing the other one. By means of respiration inhibitors the salt uptake can be reduced while at the same time the water uptake remains unaltered. It appears that in contrast with the salt uptake, water uptake is independent of respiration" (R. BROUWER, 1954). Both vital processes, however, seem to be decisively influenced, at least in many arid regions, by the geophysical phenomenon of subterranean dew which certainly has been too much overlooked up to now.

### 9. *"Biological Desalination"*

One new principle has to be mentioned here, although scientific data about it are yet almost unknown and only the practical experience of the author and some theoretical considerations indicate its important impact. Many exact experiments will have to unveil the quantitative values of its influence on the changing salinity in soil and water.

The principle may be called "Biological Desalination" for it is mainly the biological process of salt accumulation in plants, which renders it workable for utilization by man.

Salt accumulation by specific plant species has already been mentioned in various chapters. Several species are particularly adapted to saline habitats or saline irrigation by accumulating surprisingly high amounts of chlorides. They may excrete the superfluous salt by certain glands like many *Tamarix* species, or they may shed the leaves or stems after salt accumulation has reached the critical state like certain *Calotropis* or *Juncus* species, or they may adapt themselves to high osmotic pressure by other means, e.g. by increasing their succulence.

By using crop plants of these salt accumulating types we are carrying out in principle the same process as with all other crop plants: With each harvest we are diminishing the amount of nutritive elements in the soil. Afterwards we have to substitute this artificially induced loss by adding fertilizers, in order to counterbalance the loss of nutritive elements.

Salt accumulating crop plants, however, extract not only, as a matter of course, the nutritive elements from the soil, but also chlorides and/or other undesirable salts. Each harvest is therefore diminishing the salt content in soil and/or groundwater and in amounts of a magnitudinal order similar to the loss of nutritive elements by crop plants in general or even in higher amounts.

Instead of, however, intentionally substituting the loss of salt in soil and groundwater, as we are doing with the nutritive components, we are substituting the loss by necessary irrigation. By century-long experience and experiments we know approximately the amounts of fertilizers, manure, etc. to keep the correct balance between the amount of nutritive components in soil and soilwater and the loss of them by harvesting the plants. But we have practically very little knowledge only of these relations between the original and/or added salt content on the one hand, and the salt extracted and accumulated by crop plants and carried away by harvesting on the other hand.

A field experiment was carried out by the author together with Dr. Elisabeth Boyko south of the Dead Sea. There, *Juncus maritimus* was planted on 10 dunams (= $2\frac{1}{2}$ acres or 10,000 square metres) and irrigated from Oct. 1962 to August 1965 with saline underground water of about 3000 mg/l T.S.C. with 1000 mg/l chlorine. In spite of the very low rainfall (about 50 mm yearly) the soil profile up to 2 m depth showed after the harvest less Na and Cl content than it had shown before this saline irrigation had started (see graph on page 236).

Only on the surface there appeared a tiny interrupted white salt layer, a fraction of one millimetre thick, as the result of surface evaporation during the intervals between irrigation supplies (about 2 to 3 times a week with a yearly amount of 2000 cubic metre water per 1000 square metre (= 1 dunam)).

The NaCl content through this profile was also less than that of the control profile made through the adjacent, non-irrigated plot, of the same sandy soil type. (The whole area had a uniform soil structure). The two profiles showed no differences of any significance before planting and irrigation started.

A simple theoretical consideration leads to the conclusion that this kind of biological desalination may reach considerable proportions. This artificially created loss is additional to the natural loss of salt through percolation and by leaching. This diminishing process of the salt content takes place in the upper soil layers, except the uppermost one, and in the underground water which is in direct or capillary contact with the root systems of these plants.

Figures of 30—50 kg chlorides per ton dry weight seem to be no extreme estimates for certain halophytic crop plants grown on saline soil or irrigated with saline water.

Killian & Faurel (1935) investigated the succulent species *Halocnemum* and *Arthrocnemum* in Algier and found 2% chlorides of the air dry weights of plants from habitats even poor in salt.

Walter & Steiner (1936) analyzed the cell sap of Mangrove plants grown with fresh water (!) in warmhouses in Germany and received the surprising results that chlorides of the total amount of osmotically effective substances constituted 15—45%.

The possibility to diminish the salt content of water for drinking purposes by biological methods seems to be known since centuries and in various parts of the world. Thus for instance the Japanese are growing salt accumulating algae for this purpose. This knowledge led to numerous experiments with various green algae, "to find a suitable photosynthetic desalination organism". And it lead the participants in the Desalination Research Conference at Woods Hole, Mass., USA (June 19—July 14, 1961) further to recommend that "blue-green algae should also be studied, because they resemble bacteria physiologically in terms of salt tolerance and genetic constitution. Thus in the blue-green algae we have great genetic flexibility".

"While accumulation of sodium chloride, calcium, magnesium and sulfate would be desirable, organisms that would accumulate only calcium and magnesium would be highly beneficial, as the sodium chloride could then be removed far more easily and inexpensively".

Here thoughts are expressed parallel to ours, but only with a view to desalination of drinking water by salt accumulating micro-organisms. In principle, it is, however, the same line of research as we have in mind with crop plants grown in fields and with which we have already some little practical experience.

Something must have been known already in biblical times, and obviously not with micro-organisms but with certain highly developed plant species. We find in Exodus, 15, 22—25:

"And Moses caused Israel to depart from the Read Sea, and they went out into the wilderness of Shur, and they went three days in the wilderness, and found no water.

And they came to Marah, but they could not drink the waters of Marah, for they were bitter (i.e. salty); therefore they called its name Marah.

And the people murmured against Moses, saying, What shall we drink.

And he cried unto the Lord, and the Lord showed him a tree, which he cast into the waters, and the waters were made sweet."

In view of the high reliability of the bible in the description of natural phenomena, no miracle is hinted at between the lines. The text is very clear and leaves no doubts with regard to the actual facts.

My attention was drawn to this most interesting quotation from the Bible by Eng. V. Salkind, a Fellow Member of the World Academy and Hydro-Engineering Consultant of UNO, during a discussion on desalination in general.

These few remarks may suffice to indicate that very wide and to a great part completely new fields are open for research on biological desalination. They include biochemical, ecological, physiological and hydro-engineering fields. As far as economic field crops are concerned, we have for each species to find out particularly:

1) which irrigation amounts are adequate to the maximum potential of salt accumulation without diminishing the plant quality;

2) which is the optimal time for harvesting from both points of view, namely when is the highest amount of salt accumulation to be expected and when has the plant reached the stage of highest economic value;

3) how can we achieve the best combination of desalination and plant quality by selecting and breeding.

*10. Adaptability to the factor of erratics and to fluctuations of osmotic pressure*

Many plants show a striking adaptability to large fluctuations of vital environmental factors. In connection with the subject of this chapter, we are particularly concerned with the adaptability to prolonged drought and to suddenly changing osmotic pressures.

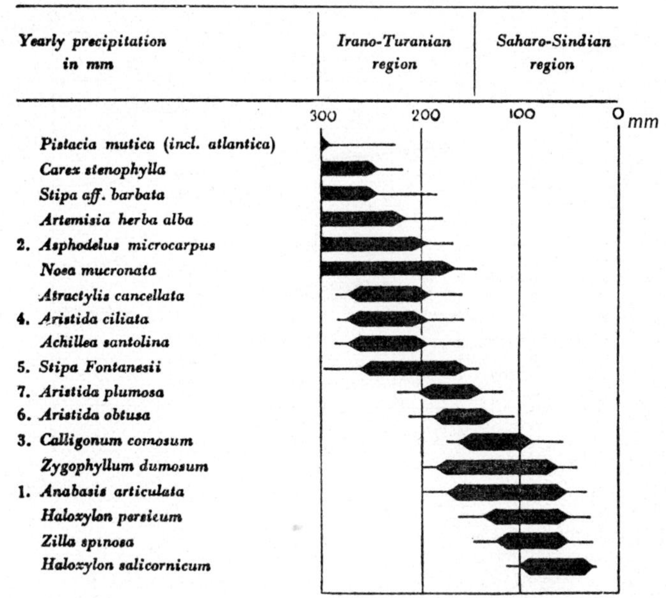


Fig. 12. Method of overlapping of amplitudes.

Legend: ■ = Species growing in optimum conditions. — = Species growing only under favourable local conditions (e.g. southern or northern slope, watercourse, etc.). (Boyko, 1949).

Explanation: Aridity amplitudes of 18 species of the Negev (southern desert part of Israel).

These amplitudes represent a scale of plant indicators for aridity.

The adaptability to the factor of erratics with regard to rainfall or otherwise obtainable water in arid regions has been dealt with by the present author in an earlier paper (1965). There a number of morphological types is described as basis for classification and further research. The morphological features have, however, only a secondary significance compared with the biochemical and biophysical properties and life processes enabling the plants to endure erratic fluctuations of extreme magnitude.

With regard to drought resistance, many plant species are known of a particularly wide ecological amplitude and in consequence, in many cases, of an extremely wide geographical distribution. Frequently this is achieved by the development of ecotypes or strains each with much smaller ecological amplitudes and often with different numbers of chromosomes. A good example is for instance *Dactylis glomerata* L. Plant-sociological records of plant communities in which *Dactylis glomerata* occurs, could be made by the present

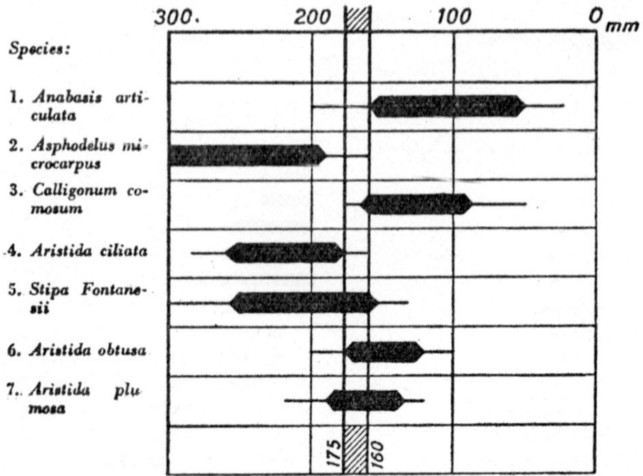


Fig. 13. Practical application of the method of overlapping of amplitudes for determination of the average precipitation on a certain spot.
Explanation: These seven species are taken from the scale of plant indicators for aridity (see Figure 12), and were recorded in one square of 20 by 20 m in Tureibe (Negev). Here their overlapping shows an average yearly rainfall of 160 to 175 mm. From this, the empiric C.S.C. (climatic soil-coefficient) of 30 mm for sand has to be deducted. The result over many years is an average of 130 to 145 mm. (For comparison: rain-gauge measurements in Kurnub show an average of 136.1 mm. Kurnub is 2 km west of our spot and 50 m higher. (BOYKO, 1953).

author in the hot arid Negev (BOYKO & TADMOR, 1954) as well as in the humid arctic of Swedish Lappland.

A similar example is *Agropyrum junceum* BEAUV. from the sand dunes on the North Sea coast with 28 chromosomes and the mediterranean ecotype from the coastal dunes of Israel with 42 chromosomes.

One has, however, to distinguish between this type of adaptability i.e. the development of adequate ecotypes and that of frequent *sudden* changes of environmental factors. Sometimes we can find both in one species and its single individuals. Thus, the cosmopolitic species *Phragmites communis* L., the common reed, is not only extremely versatile in respect to climatic features, occurring in almost all climatic zones of the globe, but also living in the same climatic environment in and near fresh water as well as under highly saline conditions. South of the Dead Sea in S'dom, we can find it under leakages of fresh water pipes and nearby on the extremely salty shore of the Dead Sea itself.

It would be highly desirable to compare the various ecological amplitudes of different species and sub-units of species. Methods to determine such amplitudes should be used more frequently and thereby developed. The International Commission of Applied Ecology worked on this subject during its existence (1947—1961) and numerous comparable values were the results (see BOYKO, 1951). This work has been continued by the International Committee of Ecological Climatography (1958) with regard to climatic factors. In general we can follow a certain pattern if we investigate the border lines of plant associations. There the topographical overlapping in plant-sociological records indicates the various amplitudinal limits of the single species. Fig. 12 shows such a scale of 18 climatic amplitudes with regard to aridity, worked out by the author by numerous plantsociological records. These first amplitudes were the result of about 12 years of work, but once such a basic work has been done, the amplitudes of most of the other species in the respective region can easily be found in one or a few excursions. For there is always a certain pattern of overlapping, and once we know the amplitudes of a certain number of characteristic plants with regard to a specific factor-complex, then the amplitudes of almost all others are to be found quickly by comparison. With regard to the IE (Insolation-Exposure) factor, for instance, this can be done by going around a hill and comparing the topographical distribution according to the angle of slope and compass direction with that of those species of which we have already the basic knowledge. How to use the method of overlapping amplitudes is to be seen from Fig. 13, where the yearly precipitation average is taken as a yardstick of aridity.

Similarly, results by the method of "Overlapping Amplitudes" are to be achieved with regard to salt resistance, where we are confronted with the pattern of zonations according to salinity.

In natural habitats, however, plants are exposed to a sudden raise in salinity on rare occasions only, as for instance by flooding by sea-water. Sudden changes are mainly in the direction of decreased salinity, e.g. by rainfall. Restitution of the formerly higher salt content in the soil is a relatively slow process and more or less parallel to evaporation. But in case of irrigation with highly saline water the plant is actually confronted with such a very quick environmental change. In addition this extreme and sudden fluctuation is connected with a factor, decisive for many vital processes in plant life: the osmotic pressure exercised by the highly concentrated solution first on the cells of the root system and after that on the cell-sap and fluids in the whole plant.

We are still far from full understanding of the complicated osmotic processes, including building of sugar, developing of succulence, and other protective responses, but certain general lines can be observed. Thus many species of high drought resistance show also a good adaptability to fluctuations of osmotic pressure by irrigation with saline water. On the other hand, the amplitude between the optimal osmotic value and the maximal osmotic value is particularly wide in most cases of drought-resistant species, whereas many halophytes show in comparison a significantly small amplitude in this respect.

It seems that the mechanism of adaptation is very specific. Thus the osmotic value of the cell sap in *Juncus arabicus* ASCH. et BUCH. from a saline oasis in Wadi Araba, irrigated throughout the dry season, i.e. 7—8 month yearly, solely (!) with water of oceanic concentration (34,000 p.p.m. T.S.C.) having an osmotic value of about 20 atm., rose to 4.8 atm. only, but all plants did not only survive, but grew steadily and developed in $2\frac{1}{2}$ years from seedlings with 3 to 4 leaves of 4—7 cm height to plants with about 200 leaves and stems of about 100 cm height. Control plants of *Juncus*, irrigated with fresh water had an osmotic value of 2 atm. in their cell-sap but grew much quicker to the same height. On the other hand, the coastal dune plant *Agropyrum junceum* BEAUV., the control plants of which showed an osmotic value of 5.4 atm. under fresh-water irrigation, had already under irrigation with diluted sea-water of only 11,000 p.p.m. T.S.C. (i.e. Caspian Sea water type, with 6.6 atm.) an osmotic value of 7.1 atm., that is a higher osmotic value than that of the irrigation water. It must be added that also *Agropyrum* grew well under those conditions, but also that the osmotic values were determined cryoscopically.

The reliability of the cryoscopic method is doubtful. Thus in his excellent paper on drought resistance presented to the VIII. International Botanical Congress in Paris, O. STOCKER (1954) has shown with great clearness that neither transpiration alone nor cryoscopically measured osmotic values are reliable yardsticks for the water balance in general.

Another phenomenon generally to be observed is that seeds even of euhalophytes need fresh water or water of much lower salinity for germination than the plant can endure when mature. The same applies to seedlings and young plants, and even to parts of grown up plants as long as these parts are in their earlier stages of development.

This adaptation is understandable when we consider the natural conditions with their low or lacking salinity after effective rainfalls or floods, when germination and the first growing stages take place. Closely connected with these observations is the fact that the osmotic value is particularly low in seedlings and young plants (WALTER, 1951).

Numerous investigations indicate that osmotic cellsap values and their smaller or wider amplitudes are specific for each species. WALTER (1951) presents in his monumental work on Phytology a number of examples where certain species brought up under very different climatic conditions retained their osmotic values.

On the other hand we know that an adaptation to higher salinity can be brought about simply by selection. Seedlings grown from seeds, the parent plants of which were already irrigated with highly saline water, resulted already in the second generation in more vigorous plants than those subjected to the same saline irrigation for the first time in their generation chain. These results were achieved in experiments made by the present author with *Hordeum vulgare* L. (Beduin strain) (two generations), *Juncus arabicus* (three generations), *Juncus punctorius* L. (two generations), *Agropyrum* P.B. (two generations), and *Calotropis procera* R.Br. (two generations).

This quick response to selection and the low osmotic value in young plants are important facts to be considered when wild halophytes are to be cultivated with the intention to convert them into cultivated plants. But the knowledge of these facts will further the success also in raising non-halophytic plant species under irrigation with saline water. Evaluation of species or ecotypes with the highest osmotic adaptability in the individual plant and breeding of ever better adapted generations by selection and cross-breeding may considerably help to lead the first prospective results of experiments to results of final economic success.

### *11. The Balance of Ionic Environment*

Already more than half a century ago, the Dutch scientist W. J. v. Osterhout (1909) concluded from his experiments on plants growing with highly saline water that

"1) Each of the salts of sea-water is poisonous where it alone is present in solution.

2) In a mixture of these salts (in the proper proportions) the toxic effects are mutually counteracted. The mixture so formed is a physiologically balanced solution.

3) Such physiologically balanced solutions have the same fundamental importance for plants as for animals."

v. Osterhout's important findings might have been lost, were it not for H. Heimann's recent work (1958) on the balance of ionic environment as one of the main principles of saline irrigation.

Independent from both, the experiments of Boyko & Boyko (1959a, b, 1965) lead to the same conclusions. Direct irrigation with sea-water without desalination was used by them in order to grow ten different plant species on sand, irrigating a number of plants of each with four different types of sea-water and, as a matter of course, a number of control plants of each with fresh water for comparison.

These experiments indicate clearly that the specific mixture of natural sea-water which is percentually about the same for all oceans and seas connected with them, had a decisive influence on the success. The actual concentration varied from about 40,000 p.p.m. T.S.C. (Red Sea and East Mediterranean type) to 11,000 p.p.m. T.S.C. (Caspian Sea type).

Most of the plants developed satisfactorily up to oceanic concentration (34,000 p.p.m.), showing retarded length-growth decreasing with rising salt concentration, but otherwise showing no principal difference in their life cycle. (For details see p. 214 ff).

These experiments and various others with highly saline water made it clear that during the last 30 years the danger of raised osmotic pressure has been far overestimated in its damaging effect on plants; further that the method to use the electrical conductivity of a solution may be good enough to indicate its total salt concentration and thereby its osmotic value, but it is of no practical value for indicating the suitability of any irrigation water ascribed to it up to now.

The opinion of HAYWARD (1956) that "retardation of plant-growth is virtually linear with an increase of osmotic pressure of the soil solution and is, in most cases, largely independent of the kinds of salt present" seems to be misleading. The basis of this widely spread error is to be found in the lay-out of the experiments leading to it. These experiments were not made with a well balanced mixture such as exists in the natural sea-water, but by adding cooking salt or certain other single components to a solution of basic nutrients.

Not the disturbed ionic balance but the raised osmotic pressure was then correlated with the damage done to plant-growth.

However, we know, for instance, that by adding potassium or calcium as an antagonist to any excess of sodium in saline water, although we are raising the osmotic pressure of the irrigation water, we counteract at the same time the damaging influence of its sodium to a soil containing a high percentage of clay and thereby the damaging influence of this irrigation water to plant growth.

There are relatively many well known cases of successful irrigation schemes using irrigation water of a Total Salt Concentration above 2000 p.p.m., some of them in use for centuries, successfully growing various crops. In his paper "The irrigation with saline water and the balance of the ionic environment" HEIMANN (1958) enumerates a number of such examples. Thus a report from Texas (BLISS, 1942) informs us that large areas are irrigated there with water from the Pecos River with about 4520 p.p.m. Total Salt Concentration with a sodium content of 52%. Farming has been successful there since many years. In field experiments near Taranto, in the semi-arid area of Southern Italy, carried out by PANTANELLI & BLANCHEDI (1929), beets, tomatoes, egg plants, celery, carrots, lettuce, sorghum, cotton, soybeans and castor beans gave excellent results, irrigated with water of a T.S.C. of 5400 p.p.m.

Many saline wells of Southern Italy are successfully used for irrigation since centuries. They were investigated by BOTTINI & LISANTI (1955) and showed from lower concentrations up to 7450 p.p.m. total dissolved solids. Similar records of successfully used highly saline waters since many hundreds of years are to be found in Algeria and Tunis, as for instance in the Igli Oasis in the Sahara desert, where cereals, vegetables and cotton are grown with an irrigation water of 4800 p.p.m. total dissolved solids. Such cases are all the more significant as we are confronted in these areas with erratic and only rarely with effective rainfall preventing accumulation of salts in the soil by dissolving and leaching them.

In the end it is primarily the specific capacity of each plant species and of each plant individual to select or to reject the various components of the solution in the soil under the given circumstances, which enables the plants to survive and to develop, or causes their death; the soil factors are of secondary importance only as life fostering factors compared with the former one.

It seems necessary to revive basic research in this field by leading it into yet another direction than until now. Our whole knowledge of the plant-water-soil relation should be put on a more biological, biophysical and biochemical basis rather than on the purely physical and chemical basis as is generally done by soil scientists. It has become almost dogmatic to judge the suitability of any saline water mainly on the chlorine content. Experiments of the present author as well as many other experiments and experiences have shown that this is not justifiable. In spite of the high chlorine content in sea-water, dried sea-salt can even be and is used under certain conditions as a fertilizer, as for instance in fields of the coastal regions south of the Caspian Sea, where a high and frequent rainfall favours this unusual fertilizing method.

It is particularly the recent work of H. HEIMANN (1958) who has stressed time and again that a new and better evaluation of the effect of cations and anions has to substitute the former one-sided approach. The entire ionic environment has to be taken into consideration, and its balance to be aimed at as the primary goal in saline irrigation.

### *12. The Principle of "Raised Vitality"**

In the course of the irrigation experiments with sea-water and a series of dilutions of it in comparison with fresh water, a new principle was found by "dying experiments", apparently the first

* The concept "Vitality" is used here in the sense of vigour (see PIERRE DANSEREAU's book: Biogeography, p. 219).

of its kind. The experiments were carried out by the present author in the Negev Institute for Arid Zone Research in Beersheba, with the following five species:

1) *Agropyrum junceum* BEAUV. *var. mediterraneum,*
a perennial coastal dune plant (seeds collected from dunes near Rehovot)

2) *Calotropis procera* R.BR.,
a small desert tree with succulent leaves (seeds collected near S'dom, south of the Dead Sea)

3) *Juncus arabicus* ASCH. et BUCH.,
a perennial rush, the seeds of which were taken from parent plants growing in the oasis of Yotvata in Wadi Araba

4) *Juncus punctorius* L.,
parent plants collected by Dr. Elisabeth BOYKO in 1956 from a freshwater course near the Monastery of St. Katherina on Mt. Sinai at about 1450 m altitude

5) *Rottboellia fasciculata L.,*
a grass collected in Erythrea by two former Assistants of the author, A. ABRAHAM and N. TADMOR.

Of each of these species, a number of plants were left to die simply by interrupting further irrigation and by exposing them to the following rainless seasons which in 1962 and in 1963 were particularly long: about 9 months in the summer of 1962 and 8 months in 1963. Details of this experiment are described in a later article of this book (see page 214 ff). Here, the underlying principles only may be shown by means of these examples.

It turned out that in the first case, i.e. with *Agropyrum junceum,* all plants of the control experiment, namely alle plants irrigated with fresh water until the interruption, died already during the first summer, whereas the plants irrigated until the interruption with sea-water survived and developed a number of new green shoot when the winter rains set in.

Further, as with the other four species, the percentage of surviving green parts was the higher, the higher the total salt content of the irrigation water had been prior to the beginning of the dying experiment.

The principal difference between *Agropyrum* and the other species is that the four other species are all better adapted to desert conditions and the factor of erratics. With these four species, almost all individuals survived the two long dry periods, but again those formerly irrigated with sea-water of relatively high concentrations used in this new kind of experiments, had after both summers the highest number of surviving green leaves (shoots), or percentage of surviving green parts respectively. The same could be observed with *new* shoots in all *Juncus* individuals. Again, the higher the total salt content of the formerly supplied irrigation water, the higher was also the number of new shoots. In order to achieve a statistical basis for these results, regular periodic measurements were made carried out throughout the two years of the "dying experiment" (July 1961—October 1963).

During the international discussions following these experiments, a number of similar observations could be brought forward. Thus, a better resistance against smoke damage was mentioned for plants growing under saline conditions. P. B. SEARS wrote in a letter to the author about his own observations in U.S.A., according to which, though floods from the sea destroyed the standing crops, they had a beneficial fertilizing effect afterwards. These flooded fields and gardens gave significantly better crops in the next cultivation period than those of the neighbouring non-flooded areas.

Interesting information of a similar kind came from Iran. There, on the southern coast of the Caspian Sea, farmers near the sea-shore fertilize their fields with small amounts of sea-water, brought into the fields with donkeys. This method is transmitted from generation to generation and fields cultivated by this method since old times show much better results than those without it.

E. PANTANELLI (1953) found that under the influence of highly saline waters a number of vegetables and fruits achieved a higher quality: "The plant tissues are more turgid and stronger, richer in sugar, minerals, nitrogen and phosphorus compound, and fuller in aroma. The fruits and vegetables show better preserving qualities and stand better the requirements of transport" (quoted from H. HEIMANN, 1958).

The effect of saline irrigation on the biochemical and physiological life processes in plants is certainly much more complicated and diversified than we know. It is of course much simpler to investigate the effect on the soil and from there to draw conclusions to the plants. But this method is frequently misleading. Thus, for instance, an Indian team of scientists, working with sea-water irrigation, found that with sea-water irrigation, the nicotine content in Tobacco was raised (KURIAN et al., 1966). We have no idea whether this is beneficial, or damaging, or insignificant for the life of this specific plant. Our judgment is based more on whether this is damaging or beneficial for human beings, economically or otherwise. Economically, it could be either. It depends, in this case, on whether we grow these plants for smoking or for the production of nicotine sulphate as an Insecticide.

In other instances, it is the height of the plant which is of importance to us. Then, we have in many cases an obviously damaging influence before us, and we speak of a lower vitality. Actually, saline irrigation with this effect does involve damage only to us and not to the plants. The plants may be and in fact are, as the dying experiments clearly showed, of a higher vigour than those irrigated with fresh water. The reason for raised vitality may be that they had a higher amount of nutritive material at their disposal than in the sand irrigated with fresh water, all the more so since intentionally no fertilizer was used, in order to ensure conclusive results; possibly higher hygroscopicity of the sand particles was responsible for binding a greater amount of subterranean dew, or a biochemical process, or all combined.

Investigations by a scientific team at the Oceanographic Institute in La Jolla of the University of California even discovered and separated a growth stimulating enzyme built in the plants by treatment with highly saline water. (Oral information by I. D. ISAACS and H. C. UREY).

All these experiments and observations indicate that the influence of irrigation with sea-water or dilutions of its with high T.S.C. result in changes of form, or size, or number of specific parts (e.g. fruits) of the respective plants, and they may change also their chemism. These changes can be desirable for us or not from an anthropocentric point of view; this depends on the species as well as on our purpose.

However, whatever they are changing, and how significant or insignificant these changes may be for us, human beings, the result is a surprising adaptation of the plant individual to the changed new environmental conditions, including the changed raw material for their metabolism. The innate vigour and the longevity is not negatively influenced, on the contrary: the above-mentioned experiments and observations indicate an even higher vitality rising with the salt concentration in the sea-water used for irrigation up to surprisingly high critical or lethal limits.

It will be an important task for the future to find out the optimum and the limits for each species and ecotype (with cultivated plants, for each variety) on the one hand, and for the various salt concentrations and climatic conditions on the other hand. But the best and comparable results will be achieved when a cooperative and transnational teamwork of combined pure and applied research workers take the lead.*

## *13. Microbiological Problems*

A very wide and new field is open for research on microbiological

* Institutions interested in cooperation may write to the General Secretariate of the World Academy of Art and Science.

problems in connection with saline irrigation in general and with its application on sand in particular.

This chapter must wait for a more authoritative writer than the present one. Since the aim of this article on principles is more the opening of new fields than a compilation of work done in the past, a few general words may suffice here. In general, one can say that this sector of the entire complex problem has been the most neglected one. Research work on the microfauna and microflora of soils has been mostly done in agricultural, horticultural and forest soils. There, the conditions for bacterial life are very different from those in sand and still more from those in sand irrigated with highly saline water.

Here, like in all other habitats, plant development depends to a great part on nitrogen fixation by soil bacteria. The specific conditions may, however, not be very favourable for such soil-inhabiting N-fixing bacteria like *Azetobacter* or for the nitrifying ones like *Nitrosomonas* or *Nitrobacter*, but we know that many dune-plants and not only Leguminosae are living in symbiosis with root-bacteria, thus developing root nodules and profiting from their mutual symbiotic nitrogen fixation.

On the other hand we know that salinity is no handicap for bacterial life in general. The relatively numerous bacteria living in the almost saturated brine of the Great Salt Lake in Utah is the best proof of this (FLOWERS & EVANS, 1966) (see page 373). Further, on flooded or otherwise irrigated fields of arid regions under tropical temperature conditions, pioneering blue-green algae may play the same role in nitrogen fixation as the nitrogen-fixing bacteria do under other conditions in a temperate, humid climate, thus maintaining the soil fertility. A. BROOK (1964) brings the following example:

"During the rice-growing season in India species of Cyanophyta that have been shown by experiment to be able to fix atmospheric nitrogen are present in great numbers. In such ricefields, crops have been grown year after year without the addition of manure to the soil. There is considerable evidence to indicate that the blue green algae are largely responsible for maintaining the level of soil fertility, the part played by bacteria being negligible".

Most of the Cyanophyta occur only in fresh water, but there are also others living in the Sea, e.g. *Trichodesmium erythraeum* in the highly saline waters of the Red Sea with more than 40,000 p.p.m. T.S.C.

Although life of micro-organisms in general may be the poorest in sand areas of hot arid regions, this is completely changed when these areas showed a higher moisture supply. A relatively rich microflora is described by KILLIAN & FEHER (1935, 1938) and others for such habitats. According to our own studies in Wadi Araba, one of the driest and hottest parts of the world, dead plants and plant parts do not normally decay. They dry out completely above the soil surface or become charred beneath it. Where, however, any source of water keeps the site moist, there are heaps or layers of decayed plant material. This moisture can be supplied also by a hygroscopical salt crust or exposed salt content of shedded leaves and twigs. On such places the shrubs of *Nitraria retusa* ASCH., for instance, or *Statice pruinosa* L. and other desert species are frequently to be found with heaps of moist and decaying plant parts surrounding their basis (I frequently measured heaps of 70 cm height beneath the canopy of old *Nitraria retusa* bushes). This would not be possible without the presence of various kinds of decomposing fungi and bacteria. The actual process, including that of putrefaction, i.e. the breakdown to simpler ammonium compounds, and still less any nitrogen fixing activity by bacteria or any other micro-organisms has not yet been studied there.

Investigations of this kind on such decaying heaps in all their stages and in their natural surroundings are greatly to be recommended, for from their progressed and last stages we may learn how to cultivate those nitrogen fixing micro-organisms most adequate for our purpose. Cultivating and inoculating them into the soil-plant system or introducing them into the irrigation water may become an important side-line in research as well as in the practice of saline irrigation.

## *14. The Phenomenon of "Global Salt Circulation"*

Introductory Remarks:

In the following pages, the author attempts to explain a fundamental geophysical process of vital biological importance for all animal and plant life, which has not been sufficiently evaluated up to now. This process is connected with almost all geophysical branches of science (Geology, Hydrology, Meteorology and others) and also with all biological branches in the same way as the fundamental natural process commonly known as "Global Water Circulation". We are speaking of the similar and simultaneous circulation of certain easily soluble minerals, in particular sodium chloride and magnesium chloride. This circulation may therefore be called "Global Salt Circulation".

This Global Salt Circulation (or to be more exact, global NaCl and $MgCl_2$ circulation) constitutes a specific and distinct part of the well-known general mineral cycle. It is best comparable in its movement with the global water circulation and not with that of the other minerals because a major part of the circulation of the other minerals depends on the movement of the earth crust (elevation of sediments, mountain building etc.,) and the subsequent erosion.

The speed of the movement of NaCl or $MgCl_2$ in solution back to the ocean is approximately the same as that of water and certainly not of the same order of time that is required to level mountains by wind and water erosion.

Our ecological observations on this subject through three decades (since 1929) may perhaps also serve to a small degree as an additional ecological and bio-meteorological contribution to the solution of two old mineralogical and geophysical problems: the rôle of rock-salt deposits in the phenomenon of sodium and chlorine content in river waters, as well as its rôle in the salt content of the oceans.

In connection with our sea-water experiments, however, the most important point is that our observations and experiments give us a simultaneous reasonable explanation for the surprising fact that *no dangerous salt accumulation is to be feared* also in the long run, if water with a high NaCl and $MgCl_2$ content is used for irrigation of economic crops under natural conditions, provided that it is carried out in accordance with the principles and methods described here.

These results have now been proved by laboratory experiments, as well as in field plots at various locations and under various climatic conditions with numerous plant species. Inadequate experiments, e.g. with incomplete drainage or other handicaps in laboratories or in the fields, were the main reason for the misconception that a dangerous salt accumulation would occur.

Coastal or Local Salt Circulation:

Before we deal with the problem of the invisible and only partly measurable global salt circulation, we want to deal in further details with the much more conspicuous phenomenon which has often been observed and has also been measured at several locations during the last two decades, namely the coastal salt spray.

It is a well-known fact that large amounts of salts are blown inland by this spray along all coasts, probably without any exception. The amount decreases rapidly with the distance from the shore, but even in the heart of the continents, salt is continually deposited, as it is brought by air movements and precipitation, not only from the comparatively very few and small salt-covered spots on terra firma, but from the oceans as well.

From the numerous measurements of J. T. Hutton (1958) from the shore to 200 miles inland in Victoria (Australia), it seems probable that a part of the *sodium chloride content in rain-water* stems from other locally conditioned sources, but by far the greater part there is also of oceanic origin.

Another example of particularly high chlorine content of the air in the steppe and forest-steppe region north of the Black Sea is explained by E. Eriksson (1958) by the presence of $SO_2$ in the air as a consequence of specific regional conditions. This $SO_2$ stems from oxydized $H_2S$ and is brought to earth again as $SO_3$. Release of HCl from the ground can occur only in arid regions with the help of the very hygroscopic sulphur trioxyd. In salt waste regions rich in sodium chloride, the presence of $SO_2$ in the air will invariably lead to a loss of HCl. As $SO_3$ is formed and added to condensation nuclei, it reacts with NaCl and water, forming $Na_2SO_4$ and HCl.

V. V. BURKSER (1951) measured the total chlorine content in the air of this region and found the following very high figures: below 46° N., 531.8 g/m³, 47° N., 146.8 g/m³. It is clear that such a high amount of chlorine particles lead ERIKSSON and probably everybody else to doubt that it could be blown over the area as salt particles from the Black Sea. But such high figures of Cl-content in the air are exceptions for inland areas and are mainly to be found nearer to the shore under the direct influence of the coastal salt spray.

Ecological remarks:

The amount of salt spray along the coasts of the sea depends on various factors:

1) on the foam-formation on the tops of the waves;

2) on the strength and frequency of winds sweeping inland from the sea;

3) on the topography of the coastal area, varying between the two extremes of coasts, i.e. slowly ascending ones and vertical or overhanging cliffs. In dune-areas, on the height of the foredune and that of the main dunes; the dune pattern, etc.;

4) on the conditions for evaporation (wind velocity, humidity, air and water temperatures, size of drops, IE-factor (insolation-exposure), etc.;

5) on the amount, the frequency and the intensity of precipitation;

6) on the physical structure of the coastal soils, its permeability, hygroscopicity, etc.;

7) on the salt content of the sea-water which can be, for instance, 6,000 p.p.m. as in the Baltic Sea or 270.000 p.p.m. Total Salt Content as in the Great Salt Lake of Utah.

In flat areas, the amount of spray reaching the surface declines gradually with the distance from the shore. In dune-areas, however, the amount is the highest on the foredune, then declines and rises again at a certain distance. The whole distance of significant salt spray varies according to the topography and reaches inland for about 200—300 m in the coastal dune-area of Israel. This is also likely to be the approximate distance in other countries with dunes of about 5—10 m height and with a medium wind velocity.

Under specific conditions this coastal salt spray may greatly influence soil or vegetation or both relatively far from the seashore, where it originated. Thus for instance, the minute and therefore very light water particles of the frequent mist (up to 200 days yearly) in the Namib desert of the south west african coastal region have a high salt content which is brought by the mist up to 50 km inland (WALTER, 1936).

The smallest coastal zone covered by salt spray is of course to be observed when the coast consists of cliffs. In this case, the glycophilous, i.e. non salt tolerant, vegetation zone may reach the upper border of the cliff itself.

The individual plant species and still more the plant communities (zonations) are excellent indicators of the relative amount of salt brought by the spray.

In Israel, for example, high amounts of salt spray on dunes are indicated by the zone where, for instance, one or more of the following species dominate:

*Cakile maritima* SCOP.
*Cyperus mucronatus* (L.) MAB.
*Eryngium maritimum* L.
*Euphorbia paralias* L.
*Pancratium maritimum* L.
*Silene succulenta* FORSK.,

whereas the beginning of the zone with *Artemisia monosperma* DEL. as the dominant plant species shows the border where the daily significant coastal salt spray ends except on stormy days (BOYKO, BOYKO & TSURIEL, 1957).

H. J. OOSTING was, it seems, the first who measured the amount of salt brought onto the Atlantic coastal dunes in the U.S.A. by spray, and since then such measurements have been carried out in other countries with similar results.

Using the measurements by H. J. OOSTING and W. D. BILLINGS (1942) a calculation derived from the data published by these authors gives the following average amount of windborne salt per vertical square meter under certain wind conditions:

On a single calm day, as much as 0.6 g were caught on this vertical screen; on a moderately windy day, "when breakers were rolling", as much as 2.5 g. 40—50% of this considerable amount is caught by the windward side of the foredunes, whereas the next highest amount is to be found on the crest of the rear dunes, with lower amounts between. These figures give an idea of the order of magnitude of the amount of windborne NaCl in coastal zones.

In areas with long dry seasons and frequent winds from the sea-side, the accumulation can theoretically reach a considerable degree, particularly so in warm climates where the moist particles evaporate immediately upon touching a warm soil surface (like, for instance, hot sand) or where they even evaporate in the air before touching the ground.

### Comparatively Insignificant Effect of Salt Accumulation

Notwithstanding the quantity of windborne salt continually being brought into coastal regions (and of salt brought in occasionally e.g. by springfloods), meteorological precipitation, runoff, percolation and groundwater flow constantly work against the accumulation of these salts, not only in the root-layer but in the whole soil profile.

Irrespective, therefore, of the topographical type of a specific coastal region and irrespective of the amount of salt continuously sprayed onto the soil surface of all coastal regions of our globe and — still more startling — irrespective of the permeability of the soil, we find surprisingly small amounts or no traces at all of the high salt layers which should have accumulated according to a very simple calculation: (The possibly smaller amount of salt precipitation on dune slopes compared with that on the vertical screen is taken as outbalanced by the higher amounts on stormy days).

Specific weight of NaCl = 2.1—2.2

2.5 g NaCl per $m^2$ and day would therefore create a layer of about 0.001 $mm/m^2/day$, i.e. in 3 years for approximately 1000 days a layer of about 1 mm, if compact, and if loose a still higher one.

This would already mean the accumulation of a layer of rock-salt of approximately one meter thickness along all coastal regions in the course of a relatively short geological period, let us say of 3,000 years.

I do not know of any example where such a layer or even any layer at all, conceivably of this origin, can be found near the shore, with the occasional exception of an ephemeral one of no significance.

For this surprising geological phenomenon of "Non-Salt-accumulation" there is only one reasonable explanation which presents itself as a matter of course — the high solubility of NaCl and $MgCl_2$ and, as its self-evident consequence: the "Coastal Salt Circulation" (see "Coastal Salt Circulation" in Fig. 15)*.

Even in coastal regions where the desert is bordering on the sea and rainfall is a rare occurrence, we do not find such a layer of rock-salt, because even the scanty rainfall in these arid regions is sufficient to wash the NaCl and $MgCl_2$ into the sea again. Those rare exceptions, such as Har S'dom (Djebel Usdum) on the shore of the Dead Sea, are obviously submarine deposits left in basins

* The components other than NaCl and $MgCl_2$, constituting altogether only about 15% of the solids in the spray, are mostly taken up by the vegetation acting as excellent fertilizers.

without an outlet and now exposed to complete evaporation. Only the theory of coastal salt circulation explains in a completely satisfactory way the fact that there is no such permanent salt-layer to be observed in coastal regions, although most of them have been in approximately the same correlation to the sea for many thousands of years. In most of the coastal regions, rising or falling of the coast or the sea level respectively of 2—3 metres or thereabouts does not make much difference in this calculation, although these processes may have influenced decisively the history of towns and people there (Bloch, 1963a, b).

Further: Completely calm days are very rare in most of the coastal regions and on most days the white tops of breakers are to be seen, i.e. the amount of spray salt is to be expected to be about 2.5 g per $m^2$/day. In addition to this, it may be emphasized that in almost all coastal regions storms are not rare but occur several times every year, thereby increasing manifold the daily average of spray salt brought in. However, in spite of this, even in coastal regions which have not changed significantly in the last 1,000,000 years or more, there is no known layer of salt of other origin than that of temporary salt-pans (natural or artificial).

As a matter of course, erosion is at work everywhere and has been so through all the ages since the surface of our globe became solid, but erosion of rock is generally a very slow process compared with the quick dissolution and washing away of any NaCl and $MgCl_2$-fallout. This is also the case in comparison with erosion of rock-salt deposits which disappear much more slowly once they are exposed to wind and weather.

The most suitable regions for investigating this phenomenon are probably those where an extremely arid climate and in consequence a climatic desert borders the sea. Together with my wife and co-worker, Dr. Elisabeth Boyko (1952), we measured the NaCl content

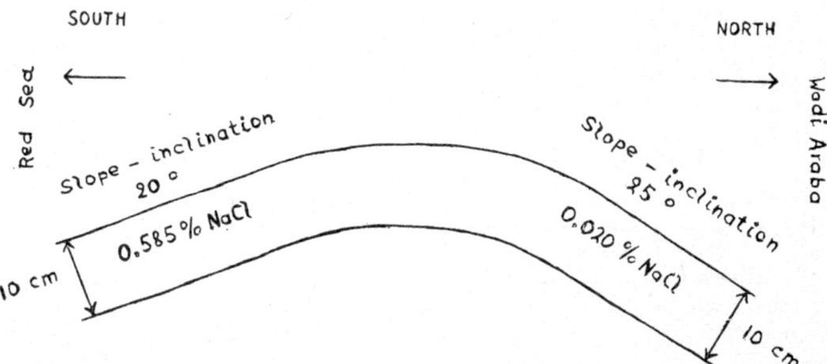


Fig. 14. Comparative distribution of sodium chloride on southern (spray side) and northern (spray shadow) slopes on the gravel hills in Eilat, about 500 m from the shore at the end of the dry season (1950).

**Table V.**

Wind Conditions in Eilat
Taken from "Meteorological Notes",
Meteorological Service, State of Israel, 1956.

| F | N | NE | E | SE | S | SW | W | NW | |
|---|---|---|---|---|---|---|---|---|---|
| | | | | | Time 08 | | | | |
| 0 | | | | | | | | | 556 |
| 1 | 273 | 267 | 48 | 12 | 14 | 39 | 34 | 28 | 715 |
| 2 | 587 | 458 | 20 | | 33 | 21 | 15 | 20 | 1154 |
| 3 | 1871 | 1068 | 9 | 3 | 34 | 6 | 18 | 84 | 3093 |
| 4 | 2441 | 952 | 3 | 3 | 53 | 14 | 21 | 43 | 3530 |
| 5 | 1487 | 470 | | | 3 | 9 | | 19 | 1988 |
| 6 | 638 | 138 | | | 6 | | | 3 | 785 |
| 7 | 117 | 30 | | | 5 | 5 | | | 157 |
| 8 | 8 | 14 | | | | | | | 22 |
| | 7422 | 3397 | 80 | 18 | 148 | 94 | 88 | 197 | 12000 |
| | | | | | Time 14 | | | | |
| 0 | | | | | | | | | 286 |
| 1 | 161 | 125 | 63 | 35 | 127 | 66 | 3 | 20 | 600 |
| 2 | 281 | 241 | 45 | 84 | 222 | 157 | 9 | 32 | 1071 |
| 3 | 733 | 533 | 93 | 96 | 706 | 388 | 46 | 78 | 2673 |
| 4 | 1503 | 701 | 40 | 26 | 418 | 106 | 64 | 96 | 2954 |
| 5 | 1874 | 796 | 11 | 3 | 58 | 11 | 15 | 54 | 2822 |
| 6 | 1007 | 261 | 5 | 5 | 8 | 9 | 3 | 12 | 1310 |
| 7 | 178 | 49 | 5 | | 17 | | | 3 | 252 |
| 8 | 22 | 5 | | | | | 5 | | 32 |
| | 5759 | 2711 | 262 | 249 | 1556 | 737 | 145 | 295 | 12000 |
| | | | | | Time 20 | | | | |
| 0 | | | | | | | | | 885 |
| 1 | 402 | 274 | 6 | 15 | 101 | 131 | 52 | 66 | 1047 |
| 2 | 707 | 347 | 15 | 5 | 111 | 153 | 46 | 183 | 1567 |
| 3 | 1885 | 404 | 3 | 3 | 153 | 70 | 70 | 403 | 2991 |
| 4 | 2349 | 337 | 9 | 8 | 74 | 23 | 68 | 609 | 3477 |
| 5 | 1026 | 139 | | | 29 | 3 | 50 | 189 | 1436 |
| 6 | 282 | 69 | | | 12 | 8 | 12 | 63 | 446 |
| 7 | 64 | 8 | | | 6 | | 5 | 6 | 89 |
| 8 | 48 | 8 | | | | | 6 | | 62 |
| | 6763 | 1586 | 33 | 31 | 486 | 388 | 309 | 1519 | 12000 |

Note: F = Beaufort's scale of wind force.

A simple calculation from this table shows that the ratio of frequency and force of those winds, which are responsible for the coastal salt-spray in Eilat, i.e. S and SE winds, to the frequency and force of the winds coming from a northerly direction, i.e. N, NE, NW winds is

1:12*

In spite of this fact, the ratio of salt content in the upper layer of the soil on the southern slope to that on the northern slope is about

30:1

* The calculation shows the following totals:

| S | | SE | | | |
|---|---|---|---|---|---|
| 148 | | 18 | | | |
| 1556 | | 249 | | | |
| 486 | | 31 | | | |
| 2190 | plus | 298 | = | | 2488 |

| N | | NE | | NW | | |
|---|---|---|---|---|---|---|
| 7422 | | 3397 | | 197 | | |
| 5759 | | 2711 | | 295 | | |
| 6763 | | 1586 | | 1519 | | |
| 19944 | plus | 7694 | plus | 2011 | = | 29649 |

The ratio of frequency and force of winds bringing coastal salt-spray onto the southern slopes, to that of the winds influencing the northern slopes is therefore

2488 : 29649 = 1 : 11.92 or about 1 : 12

on the seaward side (southern slope) and on the opposite side (northern slope) of a low hill of about 25 m height above the sea level and approximately 500 meters inland from the shore in one of the most arid regions of the globe, in Eilat (Red Sea); the average yearly rainfall there is about 20 mm. At the end of the dry season in October, the NaCl content in the upper soil layer (0—10 cm) was approximately 30 times higher than that in the same soil layer on the southern slope (See fig. 14).

These figures indicate that the southern and southeastern winds bring in to about 500 m from the Red Sea shore several hundred times (the calculation tells us, in this case $12 \times 30 = 360$ times) more salt than the northerly winds, most of which come from the distant Dead Sea region which is 160 km to the north. These differences disappear every year after effective rainfalls or after irrigation of these slopes.

The fallout of NaCl in the centre of continents is, of course, much smaller and it is not surprising that it has not as yet received the attention it deserves, except in certain border regions of salt deserts, for instance in the Aralo-Caspian region, from where considerable amounts of salt dust are reported to be brought down by storms, leading to an accumulation on topographically favourable places (STOCKER, 1929). But it is certainly surprising that there is no permanent layer at all near the shores, notwithstanding the uninterrupted salt-spray through geological times. For this — we have to repeat — there is only one conceivable explanation: the "Coastal Salt Circulation".

It is not only the run-off waters after rains or from melting snow, etc., that wash the salt from the surface back into the sea. The same process occurs with percolation of salt simultaneously with the rain-water into deeper layers from where it flows steadily with the groundwater stream back into the sea. The situation of a few small basins without an outlet above a completely unpermeable soil layer may lead to a small local accumulation in these particular basins; but such cases are rare and are, therefore, of insignificant importance for our problem.

Global Salt Circulation:

The Coastal Salt Circulation which we have dealt with up to now is, however, only a small part of the whole phenomenon, and from the biological point of view not the most important one. The process as a whole is of a far greater global impact and applies to all parts of the earth, from all continents to the smallest islands, and also from all oceans to the smallest lake and rivulet. Vertically, it reaches possibly up to the Mesosphere and down into the magmatic depths.

In general, we can say that the yearly fallout of chlorine as NaCl or in any other form is, of course, very much smaller further inland towards the centre of the continents than near the seashore, but it is large enough to be measurable. The chlorine content in the rain drops is a well-known fact and so is the fact that the ratio of chlorine in river water to that in the lithosphere is equal to 7.5% to 0.05% or 150:1 (see Table VI and text).

**Table VI.**

Percentage of Solids in Solution in Sea-Water and elsewhere, after F. W. CLARKE (1924).

| | In Ocean | In River Water | In Lithosphere | In Halites |
|---|---|---|---|---|
| | (%) | (%) | (%) | (%) |
| Cl | 55 | 7.5 | 0.05 | 60 |
| Na | 31 | 7.5 | 2.75 | 40 |
| $SO_4$ | 8 | 15 | 0.18 | |
| Mg | 4 | 5 | 2.07 | |
| Ca | 1 | 19 | 3.64 | |
| K | 1 | 2 | 2.58 | |
| $CO_2$ | 0.2 | 33 | 0.45 | |
| $SiO_2$ | — | 8 | 59 | |

"The 50/50 relationship between chlorine and sodium in river water instead of the 60/40 proportion in halites or the 64/36 ratio in ocean water is probably due to the presence of additional sodium ions obtained through the decay of sodium-bearing minerals, especially feldspar, in the rocks. The proportion of sodium to chlorine in the lithosphere is 55:1 (!)."

It can also be seen from the table that chlorine, which is by far the most abundant ion in the solution of the oceans, is the least abundant among the minerals of the earth's crust (see quotation below).

The explanations given for this fact up to now do not seem to be very satisfactory. K. K. LANDES (1960), one of the foremost authorities in this field, summarizes them in his excellent survey on our present knowledge about the Geology of Salt Deposits. There he writes about "The Origin of Sodium and Chlorine in Sea Water" (p. 29—30) as follows:

"It was widely held for many years that the oceans were originally fresh-water bodies and that this water became mineralized in geologic times by leaching of the continental rocks and the transport of sodium chlorine, and other ions, in solution to the sea by rivers.

There is, however, no geologic or biologic evidence supporting this theory. On the contrary, the evidence is to the effect that the ancient oceans were similar to those of today in dissolved mineral content, at least as far back as the earliest Cambrian Sea.

Some of the marine forms of life, entombed in the rocks as fossils, are similar to those found in the oceans today. *Water, likewise entombed between grains of sediment, is on the average as saline or even more saline than present ocean water* [*italics* by the present author].

Furthermore, it is difficult to correlate the chemistry of sea-water with that of either the rivers or the lithosphere [see Table VI]. The most abundant ion in solution in the ocean, chlorine, is the least abundant (of the sea-water ions) in the rocks of the earth's crust. It is intermediate in relative abundance in river-water. The probable explanation for the 150:1 ratio of chlorine in river-water to chlorine in the lithosphere is that the chlorine in the rivers comes from ocean-water [remark of the present author: Up to here, I agree completely, but disagree with the following]
which has been trapped in older marine sediments now elevated above sea level and draining into the rivers, and from the leaching of rock-salt deposits, also of marine origin." (end of quotation)

All these discrepancies disappear if we take into account the most simple theory of all, the "Global Salt Circulation", which is to be seen in Fig. 15. This theory does not exclude the small contribution of salt to the river waters stemming from older marine sediments as mentioned above.

Further, this simple explanation is also in full agreement with both theories on the source of the light alkaline chlorides in sea-water, i.e.:

1) the theory of C. H. WHITE (1942) which holds that the soluble compounds, stratified according to their density, "were dissolved by the first rains and were carried as saturated brine into the primeval ocean" on the one hand, and on the other hand:

2) the theory of W. W. RUBEY (1951) explaining that not only chlorine but also all other excess volatiles and water itself are of

magmatic origin; we agree with LANDES' opinion in this regard. He says that if RUBEY's theory is correct, the amount of ocean water must have increased through the geologic ages, but this water has always been mineralized and *the proportion of ions in solution has been approximately the same throughout geologic times* (*italics* by the present author).

Actually, the two main chlorides $MgCl_2$ (sp.w. = 1.6) and NaCl (sp.w. = 2.1—2.2) must have been dominant in the uppermost layers. According to their small specific weight and because of their high solubility, they must have been brought down to the oceans relatively soon, only slightly slower that the waters themselves were collected into the big basins. It may be repeated here, that NaCl and, even more, $MgCl_2$ both dissolve even in moist air! So far, the explanation of the origin of chlorine content in the oceans is in full agreement with WHITE's theory.

With regard to RUBEY's theory of magmatic origin of the circulating salt, this too is acceptable and does not contradict that of WHITE. Probably both processes were and partly still are acting simultaneously. There is no conceivable reason against the assumption that water and with it *also the chlorides* percolate and are trapped from time to time during geological times, dropping deeper for instance through temporary cracks and emerging again in solutions of hot springs or on the occasion of volcanic eruptions, thus constituting also a "Magmatic Salt Circulation" as an integral part of the whole salt circulation process.

Here, a few words may be inserted about the far-reaching influence of volcanic eruptions on the cooking salt problem in general from the point of view of human history. The decisive influence of magmatic emanations on the macroclimate, and on aridity and salinity as well as on the changes of the sea level, is dealt with in several important papers by BLOCH and HESTER (1963). BLOCH's studies (1963a, b) of historical developments in correlation with these problems are also of interest in connection with our problem.

As will be seen from fig. 15, and irrespective of either WHITE's theory of marine origin or RUBEY's theory of magmatic origin being correct, or both of them, the amount of NaCl in the Ocean can be assumed to be the same through the geologic ages, on the condition that the whole amount of salt blown from the oceans onto terra firma is carried back again by the perpetual process called by us "The Global Salt Circulation".

This global principle is also to be regarded as responsible for the success of our restricted, and in comparison infinitesimally small experiments in irrigating economic plants with sea-water. It explains the natural process that prevented the much-feared salt accumulation in the soil here and in all places where similar experiments were correctly carried out *in the open*.

Fig. 15 and the following text illustrate in detail this underlying general principle of global impact.

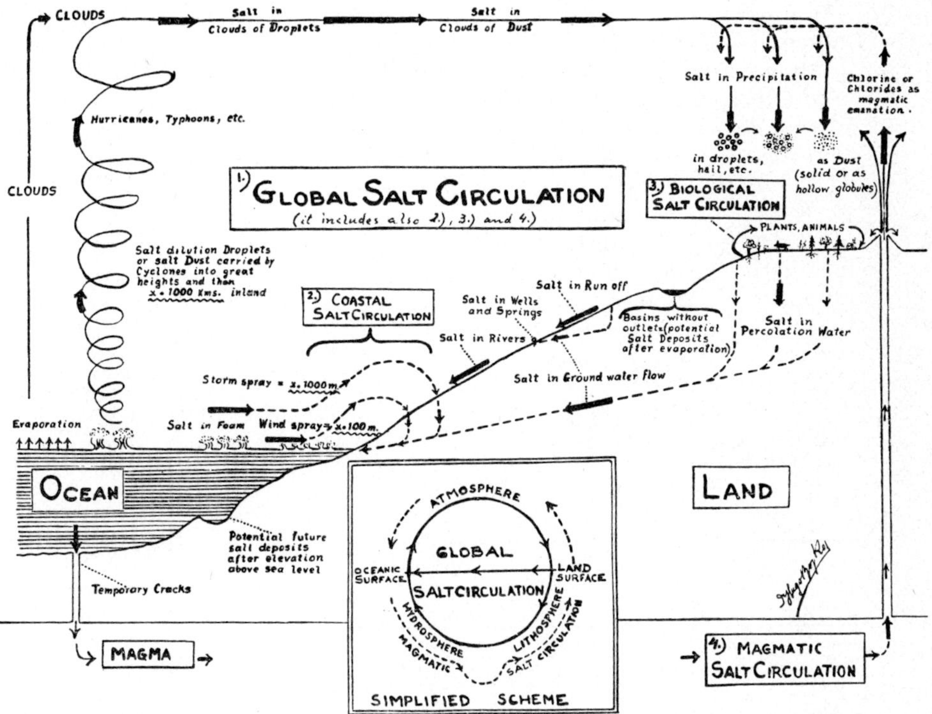


Fig. 15. Schematic presentation of the "Global Salt Circulation" (H. BOYKO, 1962). The easily soluble salts NaCl and $MgCl_2$ have a global salt-circulation similar to and almost simultaneously with that of water.
The different parts of their way during this global circulation, their direction, and the means of transport are explained by arrows. (For more details see the text).

The diagram speaks for itself. In principle, the Global Salt Circulation is the same process as shown before in connection with the Coastal Salt Circulation. The differences are more of a quantitative than of a qualitative nature. The main differences are:

1) the amount of NaCl fallout is much smaller than that near the coasts;

2) the salty droplets are whirled into very great heights by hurricanes and other whirlstorms or similar air movements above the surface of the oceans as well as above land masses, and carried for a long time and over great distances (many thousands of kilometers); against the low coastal winds as salt carrier to a relatively small land strip beyond the shore.

The salt itself in Global Salt Circulation must not necessarily be contained in the solutions of droplets, but may also be carried and fall as dry dust.

About 20 years ago I collected the dust from the terrace of my home in Jerusalem after a remarkable dust storm and brought it to the Meteorological Service of Palestine for closer investigation. It turned out that the dust consisted throughout of hollow globules. These microscopically small globules were often speared by needle-like corpuscules, mainly plant-hairs or parts thereof. The origin of these tiny hollow dust globules is easily explicable by dust adhering to the surface of droplets thus forming a well-cemented shell after the water has evaporated. In this particular case I was told that the dust had been determined as coming from the Lybian desert*.

The different forms in which the ascendent salt as well as the descendent salt may occur in the atmospheric part of the Global Salt Circulation is to be seen from Table VII.

**Table VII.**

Different forms of salt in the Global Salt Circulation.

| A. Airborne Salt | |
|---|---|
| Ascendent | Descendent |
| from the oceans<br>1. dissolved in droplets<br>2. as dust after the water of the droplets has evaporated | to the earth's surface<br>1. dissolved in the droplets of precipitation or with hail, snow, etc.<br>2. in the shell of microscopically small *hollow* dust globules<br>3. as dust in compact particles |
| B. Magmatic Salt Circulation | |
| as magmatic emanation | as in A. or percolating in dissolved state through temporary cracks, etc. into magmatic depths. |
| C. Biological Salt Circulation | |
| Intake by plants and animals in solutions or as solids | return in solutions or as solids |

Much data exist about the ubiquitous content of chlorine in rain-water and in all other kinds of meteoric precipitation. An extensive review on these studies is to be found in the paper of E. ERIKSSON (1953) in the UNESCO series on Arid Zone Research. Basing his calculations on such data, CLARKE (1924) presents an instructive list of the amounts of Cl falling per acre and per year in various countries. Quite generally, the amount declines from the

* The manuscript of the scientific paper on this subject by M. RIM may still be available at the Meteorological Service of Israel.

tropics to the higher latitudes, i.e. the cooler climates, and is, for instance, more than twelve times higher in Ceylon than in England, whereas the rain fall amount is approximately only three times as high.

Biological Salt Circulation:

The continuous intake and return of NaCl by plants and animals and, of course, by man, is not without interest in this connection. The salt balance in living beings is a highly complicated physiological matter and is of vital importance.

Table VIII gives some details about the salt content in the extra-cellular fluids of the human body. This table with its very varied figures serves at the same time also as an example for animal life in general as far as it concerns animals living outside of the oceans.

**Table VIII.**

"Approximate Sodium and Chlorine Content of some Extracellular Fluids of the Human Body" by J. A. Dauphinée (1960).

| Fluid: | Sodium | | Chlorine | |
|---|---|---|---|---|
| | mg% | meq/l | mg % | meq/l |
| Blood plasma | 321—338 | 136—147 | 345—385 | 97—108 |
| Cerebrospinal fluid | 305—340 | 133—148 | 420—450 | 118—127 |
| Gastric Juice | 20—300 | 10—130 | 330—530 | 93—150 |
| Hepatic bile | 280—360 | 122—158 | 330—510 | 93—144 |
| Pancreatic juice | 320 | 140 | 320—355 | 90—100 |
| Jejunal Juice | 320 | 140 | 370 | 104 |
| Sweat | 70—200 | 30— 87 | 99—273 | 28— 77 |
| Saliva | 23— 76 | 10— 33 | 71—150 | 20— 43 |

A healthy body maintains an *exact balance between total- salt intake and output.* This is almost entirely due to the excretion by the kidneys and the amounts excreted by the kidneys are under strict physiological control. These amounts are governed by the body's needs (Dauphinée, 1960).

With regard to terrestrial plant life, it is a well known fact that all plant ashes, as well as animal bodies, contain sodium and chlorine as part of their components. While the relative salt amount in plants is much smaller than that in animals, the absolute amount, however, may exceed the salt content in animals because of the ratio between the general mass of vegetation and that of the animal world. The salt amount varies, of course, very much with the different species, and there are even certain plants as mentioned before, which are able to excrete their surplus salts in solid form. H. Walter (1951) enumerates a number of species and genera from very different families which were found to possess active mechanisms for salt excretion. He mentions *Spartina, Distichlis, Aeluropis* and *Diplachne* among Gramineae, further *Glaux, Armeria, Limonium, Limoniastrum, Frankenia, Cressa cretica,* several Mangrove-trees (*Avicennia, Aegiceras, Acanthus ilicifolius*) and *Tamarix* species.

These salt exudations are sometimes such an amount that they are collected by the population for food. Such particularly high salt amounts are exudated, for instance, by *Tamarix* species in India, and used for this purpose. Similarly, high salt exudations are also found in other *Tamarix* species, e.g. the Mediterranean *Tamarix pseudopallasii* GUTMANN, *Tamarix gallica* L. and others. The morphology of salt excreting glands and the external part of their excretion process is described by John P. DECKER (1961) from *Tamarix pentandra* PALL., a phreatophytic tree in the Southwestern United States.

The numerous chemical studies of plant ashes revealed a great diversity in the amount of their salt content. The well known analyses of food of plant origin in human nutrition may serve as an example for the great variability of NaCl content in plants. Celery or beets, for instance, have a very high salt content which is about 50 to 100 times higher than that of potatoes or beans.

Intake and output of NaCl by plants and animals constitutes the "Biological Salt Circulation" which, too, is a part of the Global Salt Circulation as a whole. It is obvious that this part as well is not restricted to Halophytic Biocoenoses and to saline waters or other saline habitats.

## Saline Soil Formation

With regard to saline soil formation there are areas where salt accumulation occurs — geologically speaking — temporarily on the soil, or — more frequently — in the soil, with a greater speed than the global or coastal salt circulation could equilibrate. This concerns, of course, not only NaCl, but various other compounds too. As far as this specific salt accumulation is concerned, it is described by V. A. KOVDA (1961) and subdivided into various "cycles" depending mainly on the topographical conditions of the specific area. KOVDA distinguishes the following types in the formation of saline soils:

*Continental cycles* (in inland regions that have no run-off).

*Marine cycles* (on coastal plains of dry lands and along the shores of shallow bays).

*Delta cycles* (where accumulation of salt carried by rivers from the inland alternates with accumulation of salts carried in from the sea by the tides).

*Artesian cycles* (carrying the salts upwards from great depths).

*Anthropogenic cycle* (salinization by irrigation, etc.).

These types of saline soil formation may be temporarily restricted but they too are integral parts of the Global Salt Circulation as a whole.

### *15. The Global Salt Circulation and the "Basic Law of Universal Balance"*

Summarizing the main facts, we come to the conclusion that the process of a global salt circulation is a perpetual one:

1) Evidence quoted in this paper shows that the salt amount of the oceans today is the same as it was in very early periods of their existence, and that the salt concentration was already the same at least in the Cambrian.

2) It is a well established meteorological fact that rainfall contains chlorine in measurable amounts. Its origin from the oceans is accepted by meteorologists and is easily conceivable, irrespective of the acceptance of the theory that also magmatic emanations may contribute a part of it.

3) The ratio 150: 1 of chlorine in the rivers to chlorine in the lithosphere is a convincing proof of a perpetual circulation. There is no other way more conceivable than this, of bringing this salt amount perpetually into the run-off, the rivers, etc. and finally back into the sea, than through the NaCl- and $MgCl_2$-fallout by this natural phenomenon. This Global Salt Circulation has to be clearly distinguished from the general mineral cycle through the geological ages in a similar way as the global water circulation is distinguished in its concept from it.

4) Rock-salt deposits are relatively scarce and, still more important, very scattered, and only a very small part of the earth's crust consists of such deposits. They cannot serve as a sound basis to explain the ratio 150:1 in all the river waters running down to the oceans or inland lakes (see Table VI).

5) The yearly NaCl-fallout is, or course, much smaller in the centres of the continents per areal unit than in the coastal regions (compare Fig. 15 and text). In this chapter, where only principles are discussed, absolute figures are not of primary importance; but even if we take only 1/10,000 of the salt amount found in the spray in coastal regions as basis for such a calculation, the result would already be prohibitive for animal and plant life alike outside the oceans or out of water in general because of the high layer of sea-salt, i.e. mainly sodium chloride and magnesium chloride which would cover the earth's surface without the permanent natural process of Global Salt Circulation.

We have to take into account that large areas of all the continents have not been covered by the sea for many hundreds of millions of years and have been exposed to NaCl-fallout throughout these ages. It does not alter the principle if we also take into account wind erosion which too is part, if only a minor part, of the process as a whole.

In the Global Salt Circulation we encounter an interesting and important example of "The Basic Law of Universal Balance" which worked and is working through the ages on a cosmic as well as on a global scale, and we also observe this natural law working on the tiny scale of every single living being in the physiological process of salt balance mentioned above.

In order to describe this process we have tried to explain the full ramifications and connections between the phenomena of *"Coastal Salt Circulation", Biological Salt Circulation"* and *"Magmatic Salt Circulation"*, all of which constitute parts of the *"Global Salt Circulation"*, and we have tried to show its geophysical and biological impact. The whole complex of "Global Salt Circulation", however, is only one of the innumerable proofs of the all-embracing *"Basic Law of Universal Balance"*.

This "Basic Law" reads:

*"Wherever and whenever the Natural Equilibrium is disturbed, re-adjustment takes place until a state of balance is re-achieved"* (H. BOYKO, 1960).

Proof of this law is found in every macro-cosmic event as well as in the microcosmos of every atom, in all geophysical phenomena as well as in the equilibrium of every biocoenosis as a whole and in the chemophysiological equilibrium of every individual living being. In the framework of the Global Salt Circulation the physiological process of salt balance itself thereby takes its place as only a minute part in the inconceivably complicated but harmonic interplay of innumerable processes resulting in all specific life as well as in the bodily decomposition after all specific death.

It is this Basic Natural Law of Universal Balance which embraces all *physical* phenomena for which EINSTEIN found the great mathematical equation $E = mc^2$ — and it embraces as well all *biological* phenomena, the mathematical formulae for which may perhaps be far beyond the range of the surprisingly great but nevertheless definitely restricted mental capacity of man.

## REFERENCES

BAVER, L. D., 1940. Soil Physics. Wiley & Son., New York.

BLISS, L. C., 1942. *Trans Amer. Soc. Civil Eng.* **107,** *1510.*

BLOCH, M. R., 1953. Discussion of S. Duvdevani's lecture on "Dew Gradients in Relation to Climate, Soil and Topography". Unesco — Israel Symposium on Desert Research, 1952, p. 274. Jerusalem.

BLOCH, M. R., 1963a. Some Parallelism of Sea Level Indications and Historic Developments. The Negev Institute for Arid Zone Research, pp. 1—10, Beersheba.

BLOCH, M. R., 1963b. The Social Influence of Salt. *Scient. Amer.* **209,** 1, *89—98.*

BLOCH, M. R. & HESTER, W. B., 1963. Climate and Volcanic Outbreaks. *Bull. Res. Counc. Israel,* Section Geo-Sciences, **11,** 4, *173—174.*

BOTTINI, O. & LISANTI, E., 1955. Richerche e considerazioni sull 'irrigazione con aque salmastre praticata lungo il litorale pugliese. *Ann. Sper. Agr.*, N.S. **IX,**, *401—436.*

BOYKO, ELISABETH, 1952. The Building of a Desert Garden. *J. Royal hortic. Soc.*, 76, *1—8.*

BOYKO, HUGO, 1931. Ein Beitrag zur Oekologie von Cynodon dactylon Pers. und Astragalus exscapus L. *S.B. Akad. Wiss. Wien,* Math.-naturwiss. Kl., **140,** I, 9.u.10. Heft.

BOYKO, HUGO, 1932. Ueber die Pflanzengesellschaften im burgenländischen Gebiete östlich vom Neusiedler See. *Burgenländ. Heimatblätt.,* **I,** 2 (Eisenstadt).

BOYKO, HUGO, 1934. Die Vegetationsverhältnisse in Seewinkel. Versuch einer Pflanzen-soziologischen Monographie des Sand- und Salzsteppengebietes östlich vom Neusiedler See, II. *Beihefte z. Bot. Zbl.*, Prag-Dresden, **51,** II, *600—747.*

BOYKO, HUGO. Die Vegetationsverhältnisse in Seewinkel, I. *Bot. Zbl.*, (destroyed in the Press in Dresden) distributed as Ms.

BOYKO, HUGO, 1947. On the role of plants as quantitative climate indicators and the geo-ecological law of distribution. *J. Ecol.* **35,** *138—157.*

BOYKO, HUGO, 1949. On the Climax Vegetation of the Negev, etc. *Pal. J. Bot.*, Rehovot, Series, VIII, *17—34* (see page 26).

BOYKO, HUGO, 1951. On Regeneration Problems of the Vegetation in Arid Zones. *Proc. VIII. int. Bot. Congr. Congr.,* Stockholm 1950, printed in full: *IUBS,* Series B (Colloques), No. 9: *62—80.* Paris 1951.

BOYKO, HUGO, 1953. Ecological solutions of some hydrological and hydro-engineering problems. Ankara Symposium on Arid Zone Hydrology. Arid Zone Program of UNESCO. Proceedings, Unesco, *247—254,* Paris.

BOYKO, HUGO, 1957a. Ergänzung zum Referat von R. Knapp. (Unterirdischer Tau). *Proc. I. Bioclimat. Congr.* Part. II, Section A., Leiden.

BOYKO, HUGO, 1957b. Circular Letter No. 11, December 1957. Int. Commission of Applied Ecology, IUBS. Rehovot, 1957 (see also Circ. Letter No. 12, Dec. 1958 with additional details).

BOYKO, HUGO, 1960. The Basic Law of Universal Balance. Mimeographed, International Commission of Applied Ecology, IUBS, Presidential Address. Meeting, London, July 1960.

BOYKO, HUGO, 1955. Climatic, ecoclimatic and hydrological influence on vegetation. — Introductory lecture, UNESCO-symposium on Plant-Ecology in Montpellier, UNESCO Series Arid Zone Research, **V,** *41—48,* Paris.

BOYKO, HUGO, 1962. Main Principles of Direct Irrigation with Seawater Without Desalination. Report to UNESCO, mimeographed, Negev Institute for Arid Zone Research, *1—40,* Beersheba.

BOYKO, HUGO, 1965. Some New methods in Ecological Climatography and Ecological Hydrology (Ecological Geophysics). *Proc. III. Biomet. Congr.*, Pau 1963, *924—930*; London.

BOYKO, HUGO, 1966. Introduction, Summary and Outlook in H. Boyko, Ed., Salinity and Aridity — New Approaches to Old Problems. Dr. W. Junk, Publishers, The Hague.

BOYKO, HUGO, & A. ABRAHAM, 1954. Examples of Root Adaptability to Aridity. *VII. Congr. int. Bot.* Rapport et Communications 11 et 12, Paris, *237—239*.

BOYKO, HUGO, & ELISABETH BOYKO, 1957. A climate map of the Sinai peninsula as example of Ecological Climatography. (I. Int. Congr. on Bioclimat. Biometeor. Vienna, 1957). *Int. J. Bioclim. Biomet.* **1,** II, Sect. A.

BOYKO, HUGO & ELISABETH BOYKO, 1959a. Seawater Irrigation — a new line of research on a bioclimatic Plant-Soil-Complex. *Int. J. Bioclim. Biomet.* **III,** II., Sec. B. 1, *1—24* Leiden.

BOYKO, HUGO & ELISABETH BOYKO, 1959b. Seawater- and Saltwater Irrigation. Int. Bot. 1 Congr. Montreal.

BOYKO, HUGO & ELISABETH BOYKO, 1964. Principles and Experiments regarding direct irrigation with Highly Saline and Sea Water without Desalination. *New York Acad. Sci., Trans.*, Ser. II, **26,** Suppl., *1087—1102.*

BOYKO, HUGO, ELISABETH BOYKO & TSURIEL, D., 1957. Ecology of Sand Dunes. Final report to Ford Foundation, Sub-project C — 1e, *353—399*, Jerusalem.

BOYKO, HUGO & N. TADMOR, 1954. An Arid Ecotype of Dactylis glomerata L. (Orchard Grass) found in the Negev (Israel). *Bull. Res. Counc. Israel,* **IV**/3, *241—248.*

BROOK, ALAN, 1964. The Living Plant. University Press, Edinburgh.

BROUWER, R., 1954. The regulating influence of suction tension and transpiration on the water and salt uptake. *VIII. Congr. int. Bot.*, Rapports et Communications, Sections 11 et 12, Paris *221—222.*

BURKSER, V. V., 1951. Aerogeochemical investigations in the steppe and forest steppe region. *Ukrain. Khim.-Zhur.*, **17,** *472—476.*

CHAPMAN, V. J., 1962. Salt Marshes and Salt Deserts of the World. Leonhard Hill, London.

CHAPMAN, V. J., 1966. Vegetation and Salinity. In H. BOYKO, ed. "Salinity and Aridity — New Approaches to old Problems" Dr. W. Junk, Publishers, den Haag.

CLARKE, F. W., 1924. Data of Geochemistry. *U.S. Geol. Survey, Bull.* No. **770,** Washington.

DAUPHINÉE, J. A., 1960. Sodium Chloride in Physiology, Nutrition and Medicine, pp. 409 ff. in DALE W. KAUFMANN: Sodium Chloride, London.

DECKER, JOHN P., 1961. Salt Secretion by Tamarix pentandra Pall. *Forest Sci.*, **7/3,** U.S. Forest Service.

DUVDEVANI, S., 1953. Dew Gradients in Relation to Climate, Soil and Topography. UNESCO-Israel Symposium on Desert Research, *136—152.*

ERIKSSON, E., 1958. The Chemical Climate and Saline Soils in the Arid Zone. UNESCO Series Arid Zone Research; Climatology, **X,** *147—180,* Paris.

FLOWERS S. & F. R. EVANS, 1966. Plant and Animal Life in the Great Salt Lake of Utah. In: Salinity and Aridity — New Approaches to Old Problems (H. BOYKO, ed.) Dr. W. Junk, Publishers, Den Haag.

HAYWARD, H. E., 1956. Plant Growth Under Saline Conditions. UNESCO Series Arid Zone Research, **IV,** *37—71,* Paris.

HEIMANN, H., 1958. Irrigation with Saline Water and the Balance of the

Ionic Environment. Lecture, Int. Potassium Symposium, Madrid.

HUTTON, J. T., 1958. The Chemistry of Rainwater with particular reference to conditions in South Eastern Australia. Canberra Symposium, Climatology and Microclimatology, UNESCO Series Arid Zone Research, **IX**, *285–290*, Paris.

Int. Committee for Ecological Climatography (of the Int. Soc. of Biometeorology) 1958. Committee Report for 1956—1957. *Int. J. Bioclim. Biomet.* **II**, VII, Section A 1b, *1–8*.

KARSHON, R., 1956. Salt Conditions in Erosion channels in Wadi Araba. *Bull. Res. Counc. Israel,*

KILLIAN, CH. & O. FEHER, 1935. Recherches sur les phenomènes microbiologiques des sols sahariens. *Ann. Inst. Pasteur,* **55,** *573—623*.

KILLIAN, CH. & O. FEHRER, 1938. Le rôle et l'importance de l'exploration microbiologique de sols sahariens. *Mem. Soc. Biogéogr.,* **6,** *81–106*.

KOVDA, V. A., 1961. Principles of the Theory and Practice of Reclamation and Utilization of Saline Soils in the Arid Zones, in Proceedings of the Teheran Symposium, October 1958, UNESCO, Arid Zone Research **XIV,** *201–213*, Paris.

KURIAN, T., E. R. R. JYENGAR, M. K. NAYARANA & D. S. DATAR, 1966. Effects of Sea Water Dilutions and its Amendments of Tobacco. (Lecture at the Symposium on the problems of the Indian Arid Zone, Jodhpur, Nov. 1964). in: "Salinity and Aridity – New Approaches to Old Problems (H. BOYKO, ed.) Dr. W. Junk, Publishers, The Hague.

LANDES, K. K., 1960. The Geology of Salt Deposits, pp. *28–69*, in DALE W. KAUFMANN, Sodium Chloride, London.

MASSON, H., 1951. Condensations atmosphériques non-enrégistrables au pluviomètre. L'eau de condensation et la végétation. *Bull. Inst. Franç. d'Afr. Noire,* **10.**

McKENZIE-TAYLOR, E. & C. B. WILLIAMS, 1924. A comparison of Sand and Soil Temperatures in Egypt. *Min. Agr. Egypt., Tech. and Scient. Bull.* **40,** Cairo.

Meteorological Service, State of Israel, 1956. Climatological Normals, Part Two: "Winds", Serie A. Meteorological Notes, No. 15, pp. VII + 77, Jerusalem.

NARAYANA, M. R., V. C. MEHTA, & D. S. DATAR, 1966. Effect of Sea Water and its Dilutions on some Soil Characteristics. (Lecture held at the Symposium on the problems of the Indian Arid Zone, Jodhpur, Nov. 1964) in: Salinity and Aridity - New Approaches to Old Problems (H. BOYKO, ed.), Dr. W. Junk, Publishers, The Hague.

OOSTING, H. J. & W. D. BILLINGS, 1942. Factors affecting vegetational zonation on coastal dunes. *Ecology* **23,** *131–142*.

OSTERHOUT, W. J. v., 1909. *Bot. Gaz.* **48,** *98–104*.

PANTANELLI, E., 1953. Agronomica Generale, pp. *67* and *159*. Bologna.

PANTANELLI, E. & A. BIANCHEDI, 1929. Economica della Capitanata, *67–69*, Foggia.

ROSENE, HILDA F., 1954. The Water Absorption Capacity of Root Hairs. *VIII Congr. int. Bot.,* Rapports et Communications, Section 11 et 12, *217–218*, Paris.

RUBEY, W. W., 1951. Geologic History of Seawater. *Bull. Geol. Soc. Amer.* **62,** *1111–1147*.

STOCKER, O., 1928. Das Halophyten-Problem. *Erg. der Biol.,* **3,** *265–354*.

STOCKER, O., 1929. Ungarische Steppenprobleme. *Naturwissenschaften,* **17,** *189–196*.

STOCKER, O., 1954. Die Trockenresistenz der Pflanzen (La résistance des végétaux à la sechesse). *VIII Congr. int. Bot.,* Rapports et Communication, Sections 11 et 12, *223–232*, Paris.

STOCKER, O., 1958. Morphologische und physiologische Bedingungen der

Dürreresistenz. Int. Kali Inst. Bern, Kalium Symposium, *79—93*, Bern.

TROLL, CARL, 1935. Wüstensteppen und Nebeloasen im südnubischen Küstengebirge. *Z. Ges. f. Erdk. zu Berlin.*

TROLL, CARL, 1956. Das Wasser als pflanzengeographischer Faktor. Hb. Pflanzenphysiol., **III,** *750—786*. Berlin-Göttingen-Heidelberg.

U.S. National Academy of Science - National Research Council: "Desalination Research and the Water Problem" Publ. No. 941, pp. 1—85, Washington, 1962.

WADLEIGH, CECIL H., & MILDRED S. SHERMAN, 1960. Sodium Chloride in Plant Nutrition. *470—483*, in DALE W. KAUFMANN, Sodium Chloride London.

WALTER, H., 1936. Die ökologischen Verhältnisse in der Namib Nebelwüste (Südwest Afrika). *Jb. Wiss. Bot.* **84,** *58—222*.

WALTER, H., 1951. Einführung in die Phytologie, III. III Grundlagen der Pflanzenverbreitung, I. Teil: Standortslehre. Eugen Ulmer, Stuttgart.

WENT, F. W., 1953. Discussion to S. Duvdevani's lecture on "Dew Gradients in Relation to Climate, Soil and Topography. UNESCO-Israel Symposium on Desert Research, p.152. Jerusalem.

WENT, F. W., 1953. Preliminary report on some results of a study of the role of dew in plant growth in arid regions. UNESCO/NS/AZ/128.

WERBER, A., 1936. Is Irrigation with Saline Water possible?. *Hadar,* Tel Aviv, **IX.,** 9, *201—203,* (in Hebrew).

WHITE, C. H., 1942. Why the Sea is Salty. *Amer. J. Sci.,* **240,** *714—724*.

WILLIAMS, C. B., 1954. Some bioclimatic observations in the Egyptian Desert. Biology of Deserts (Symposium), Inst. of Biology, pp.*18—27*, London.

---

# PLANT GROWTH UNDER SALINE CONDITIONS AND THE BALANCE OF THE IONIC ENVIRONMENT

BY

HUGO HEIMANN

(with 4 figs.)

## Introduction

Extension of food production to the land reserves of the arid and semi-arid zones of the world confronts plant science and agronomy with the problem of salinity. The water available for irrigation is very often charged with salts and where rain is lacking, the soil too is impregnated with salts. To put it into figures: Using an irrigation water with 1000 ppm. T.D.S. (total dissolved solids) at an annual application of 10,000 $m^3$ per ha, the total minerals added to the soil in one year amount to 10,000 kg, what is by one order of magnitude more than the minerals applied to the soil by a rather liberal application of chemical fertilizers.

Due to evapo-transpiration salts concentrate in the top layer of the soil and the osmotic pressure of the soil solution may reach very high values.

This paper will not deal with the influence of salinity on soil properties, although compaction of the soil due to high sodium accumulation badly affects the plants growing on such soils, and the discussion will be restricted to the direct effects of salinity on plant growth. Furthermore, "plants" means in this context commercial crop plants and "soils" in general the usual agricultural soils. Dune sand, i.e. the "soil" used in BOYKO's experiments with sea-water (1959, 1964) is not dealt with in this article.

## Osmotic Pressure versus Environment

The generally accepted view is that the excessive osmotic pressure connected with high salt concentrations causes damage to the plant, irrespective of the kind of salts involved. But there are facts often observed in actual farming that contradict this assumption. In many places of the world crops are grown with satisfactory yields using irrigation water with up to 3000 and more p.p.m. of total dissolved solids. The aim of the present research was to bridge the gap between the farmer's experience and the teachings of science. The work started with re-assessing the dominant rôle attributed to osmotic pressure. By scrutinizing all the experimental work published to this point, the following conclusions resulted:

The experimental procedure applied by the scientists who pioneered in this field generally was of the following type: The test plants were grown in solution culture, then salts of all kinds, such as the chlorides and sulphates of sodium, magnesium, calcium, etc. were singly added to the nutrient solution in increasing concentrations. The concentrations and the correlated osmotic pressures were noted at the point where the plants started to suffer, as indicated by leaf scorching and wilting. A comparison of the results proved that this always happened, irrespective of the type of salt used, close to the same value of osmotic pressure. The conclusion was drawn that it is the high osmotic pressure as such that makes the plants suffer, due to the difficulty they have in acquiring the water they need. Unfortunately for many years of salinity research, but fortunately for man who has to grow food under saline conditions, this conclusion proved to be fallacious. The experimental design used in this kind of research was unsound in principle. The application of single salts for raising the salinity of the growth medium fulfilled the requirements of simple and well controlled experimental conditions, but it was not in accordance with a primary biological postulate since long established by Jacques LOEB (1911), v. OSTERHOUT (1909), RUBINSTEIN (1928) and other pioneers in general physiology: The well-being of all organised living cell systems exposed to an ion-charged aqueous environment is conditioned by a well-balanced ratio between these ions. While we know the evolutionary background of this postulate, namely the origin of life in the primeval sea and the adaptation of the functions of life to that environment, we are still lacking its physical explanation. But that does not detract from its fundamental importance and its general validity in animals as well as in plants. Actually, it was proven that plants in all stages of their development can stand much higher osmotic pressures in physiologically balanced solutions than in those of single salts or unbalanced mixtures. Therefore, it is not surprising that crops such as tomatoes and lettuce can be grown in hothouses at salt concentrations, as high as 50,000 kg per ha and at correspondingly high osmotic pressures, as reported by CLAY & HUDSON (1960). Furthermore, BERNSTEIN (1961), LAGERWERFF & EAGLE (1962) and others found that plants are able to counteract high external osmotic pressure by increasing their internal one.

Therefore, within rather broad limits of concentrations and osmotic pressure, the salinity damage is caused by unbalanced ionic ratios and not by osmotic pressure as such.

Hence, by proper balancing the ionic ratios, crops may economically be grown at surprisingly high levels of salinity.

### The Sodium - Potassium Relationships

The cationic environment of plants generally contains: H, K, Ca, Mg and $NH_4$, where ammonium fertilizers are applied. Sodium

(Na), that is normally absent or present only in insignificant concentrations, enters the picture under saline conditions and becomes the predominant cation.

In irrigation water Na-concentrations of 6 to 12 meq. per litre and more are often found and the total quantity of Na entering the irrigated soil may easily reach 3000 to 7000 kg per ha. Furthermore, while the anions are mainly removed from the root zone of the soil by the leaching action of rain and irrigation, at least part of the sodium steadily accumulates in the ion-exchange complex of the soil that is so essential for plant growth. While not entering the pedological aspects of this process, we have to ask: what rôle plays sodium in plant life and crop production? The textbooks of plant physiology and agronomy, including the newest ones, do not give a satisfactory answer. They mainly discuss the following two points:

(1) Is sodium an essential "plant nutrient"?

(2) How far can the cheap sodium chloride or other sodium compounds be substituted for the more expensive potassium fertilizers?

Surprisingly, fundamental and independent research work about the rôle of sodium in plant-life is very scarce. Most of the sodium research was guided and sponsored by the legitimate commercial interests of the industries selling sodium containing fertilizers to the farmers, or by the temporary needs of a war economy lacking supplies of potash. Therefore, the concentrations of sodium applied in the experiments were limited to those obtained with sodium chloride or nitrate as fertilizers. In this application the sodium appears with a few hundred kilos per ha, whereas in salinity ten times as much sodium incidentally enters the soil with the irrigation water.

The situation may be illustrated by the following diagram (Fig. 1), as proposed by HEIMANN (1959):

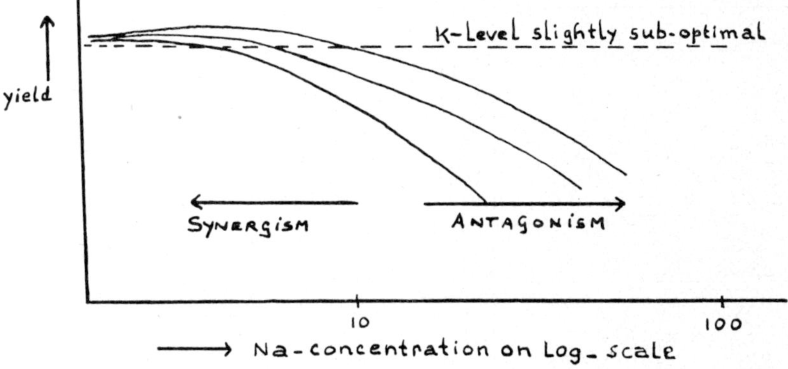


Fig. 1. The influence of sodium in the medium on crop yield.

At the left one finds the field of sodium concentrations obtained by sodium containing fertilizers, and in many cases small additions of sodium produce a slight increase in the crop. Sodium and potassium act in this field as synergists.

On the right side appear the sodium concentrations reached by irrigation with saline water, which are by an order of magnitude higher than those mentioned before. The beneficial action of sodium present in low concentrations is reversed into a detrimental one. This upper range of sodium concentrations, typical for saline conditions, was never fully explored.

In growing food under saline conditions, protecting the plant against sodium becomes a major consideration.

But before dealing with this problem, a question often asked has to be answered:

Why is the sodium accused as the main culprit and not the chloride ion?

The answer is: Firstly, it is a well-based opinion among the biologists that it is mainly the cationic environment of a cell system that matters. The rôle of the anions, although not to be omitted, is of secondary importance. Secondly, those claiming a toxic action of the chloride ion base their claim on the fact that leaves of plants, such as citrus and other fruit trees suffering from salinity, are often high in chloride whereas the sodium appears in them on no alarming level. This is an erroneous conclusion drawn from a too superficial interpretation of foliar diagnosis.

The following table illustrates the situation.

**Table I.**

Alkali Ions in Roots and Leaves.
in milli-equivalents per 100 g of dry matter.

| | | K | Na | Cl | source |
|---|---|---|---|---|---|
| Avocado | roots | 70.3 | 41.4 | — | Martin & Bingham |
| | leaves | 32.8 | 8.4 | — | (1954) |
| Peaches | roots | 17.9 | 15.6 | 21.4 | Hayward et al. |
| | leaves | 76.8 | 2.2 | 11.5 | (1946) |
| Oranges | roots | 26.6 | 13.6 | — | Jones & Pearson |
| | leaves | 30.4 | 4.2 | — | (1952) |
| Groundnuts | roots | 18.0 | 43.0 | — | Heimann & Ratner |
| | stems | 31.5 | 13.4 | — | (1965) |
| | leaves | 11.0 | 4.8 | — | |

It is evident that sodium intruding a plant through the roots is mainly retained by them, and a barrier of a still unexplained kind prevents its transfer to the trunk or stem, and from there into the petioles and the leaves. On the other hand, potassium and chloride ions move freely to the foliage.

So chloride is found in the leaves, whereas the sodium remains hidden in the subterranean parts of the plant less accessible for sampling and analysis, and the wrong conclusion is drawn that the chlorine caused the damage to the tree. Actually, the roots suffered from the intrusion of sodium and their suffering is transmitted to the tree as a whole. There is little undisputable indication of a specific chloride danger, and even cases like that of the avocado tree which is considered specifically sensitive to chloride are still open to questioning.

Therefore, we may safely assume that sodium is the main culprit, and fighting salinity means first of all fighting sodium.

A look into animal physiology will suggest a way how to do that. Contrary to terrestrial plants wherein sodium is normally absent or only present in very low concentrations as a "minor nutrient", in animals sodium and potassium occur in similar concentrations and both are of vital importance. Plants as well as animals evolved from common primitive forms of life that first appeared in the ocean a few thousand millions of years ago, that means in a highly saline environment. So all functions of life in these primitive ancestors of ours were adapted to an environment wherein sodium was the predominant cation. In the course of evolution plants as well as animals left the sea and adapted themselves to terrestrial life, to an environment less heavily charged with ions. And then happened a thing that may be puzzling. Animals including man retained in their body up to our days a considerable concentration of both alkali ions whereas the plants dropped the sodium as a major constituent and retained the potassium only. A somewhat oversimplified answer may be offered: The continuous play of nerves and muscles in animals is activated by sodium and potassium moving in opposite directions. As the nerve is excited or the muscle contracts, potassium temporarily leaves the interior of the cells concerned and goes into the extracellular medium, whereas the sodium enters the cell in equivalent amounts. As the nerve is sedated and the muscle relaxes, both ions move back to their normal location. Plants possess neither nerves nor muscles and, therefore, they are not in need of that interplay between sodium and potassium. Consequently, sodium was dropped in the course of evolution. But why did not the same happen to potassium? The answer may be found in enzyme chemistry. There is a considerable number of vital enzymes, mainly those involved in phosphate transfer, which require the presence of potassium in the medium for their performance, and no other cation of those belonging to life can replace it. In some of these enzymatic reactions sodium exerts an inhibitory function. It is questionable if there are enzymatic processes in plants wherein the presence of sodium is compulsory in the same exclusive way as that of potassium.

It is yet an unsolved riddle what function the potassium fulfils in enzymatic reactions. There are good reasons to assume that the potassium ion influences the structure of the water attached to the enzyme molecules and/or to the molecules of the reactants, which may also lead to a change of structure of the whole complexes involved.

Certainly, more study should be devoted to the ultimate physical background of the actions and interactions of the two alkali ions in the processes of plant life.

The antagonistic relationships between sodium and potassium in animal physiology suggest the question: How far is potassium capable to counteract the detrimental influence of the sodium ion on plants and so to restore the balance in their ionic environment? A study of all the cases of successful farming with highly saline irrigation water reported from Texas, Arizona and other places in U.S.A., from Spain, Southern Italy, Algeria, Tunisia, Russia, Israel and from other countries showed that always potassium was present in the water in concentrations of between 10 and 100 ppm, and the ratio between sodium and potassium was at least that typical for sea-water. In addition, the potassium level of the soil was often found to be very high.

These findings support the assumption of an antagonistic relationship between these two ions in plant life. The textbooks of plant physiology stress the antagonism between calcium and potassium and that between mono- and divalent ions in general, but nowhere is the existence of an antagonism between sodium and potassium suggested and dealt with. A few workers in the past, for instance VAN ITALLIE (1935) and REIFENBERG (1947), have paid attention to the possible existence of such an antagonism and its agricultural

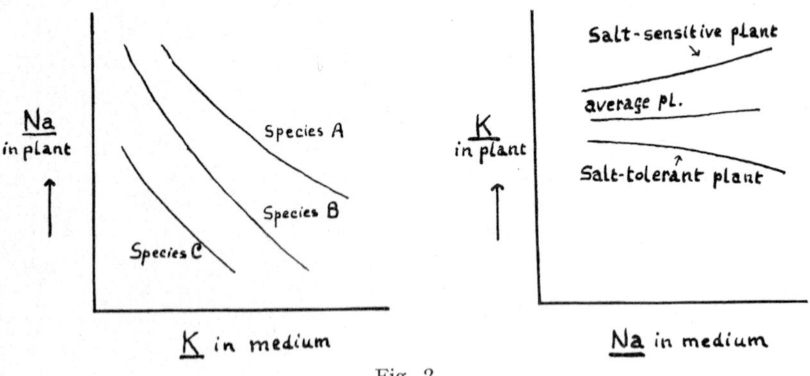


Fig. 2.

implications. But their ideas were not followed up. By experimentation as well as by evaluation of facts incidentally reported in literature the following picture about the interactions of sodium and potassium was obtained (Fig. 2), and this picture seems to be of general validity.

On the left, the influence of potassium in the growth medium on the uptake of sodium by the plant is shown. In all cases studied potassium blocked the uptake of sodium in all plants.

On the other hand, as shown on the right, the effect of sodium added in increasing concentrations to the growth medium generally does not decrease the uptake of potassium. There are cases where a slight depression is noted and these cases concern salt-tolerant plants such as beets or barley. There are cases where the level of potassium was not at all affected, and there are others, one should say the most frequent ones, where the presence of sodium in the medium caused even an increase in the uptake of potassium.

Apparently the salt-sensitive plant reacts to the invasion of sodium by an increased acquisition of potassium in order to restore the balance between these ions in their internal environment.

Obviously, the term of a "normal level of potassium as a nutrient" is not applicable where sodium has invaded the plant and the plant needs protection. Also, the carrier hypothesis of explaining the uptake of the cations by the plant as a process of competition gets into difficulty as the presence of one cation rises the uptake of the other instead of lowering it.

## The Definition of the Environment

Whereas in solution culture the environment of plant roots in respect to the two alkali ions can easily be defined, great difficulties arise where plants are grown in soil. With regard to sodium the situation is rather simple. It appears either in the soil solution or in the ion exchange complex of the soil. But the potassium is present in at least five different states:

1. contained in primary rock minerals,
2. secondarily fixed in the lattice of clay minerals,
3. adsorbed by the ion exchange complex,
4. dissolved in the soil solution,
5. fixed in the bio-phase of the soil.

It is impossible to assign to each of these different states of potassium an activity coefficient for its relative participation in the ionic balance. Furthermore, the situation is very fluid as there is a steady transit between the different states of potassium, actuated by various physical, chemical, and biological factors.

Finally, even if the status of potassium in a certain soil could be determined at a certain moment, that would only concern the bulk of the soil and not the narrow zone of contact between root and soil which matters most for the plant. The following two drawings will illustrate these points:

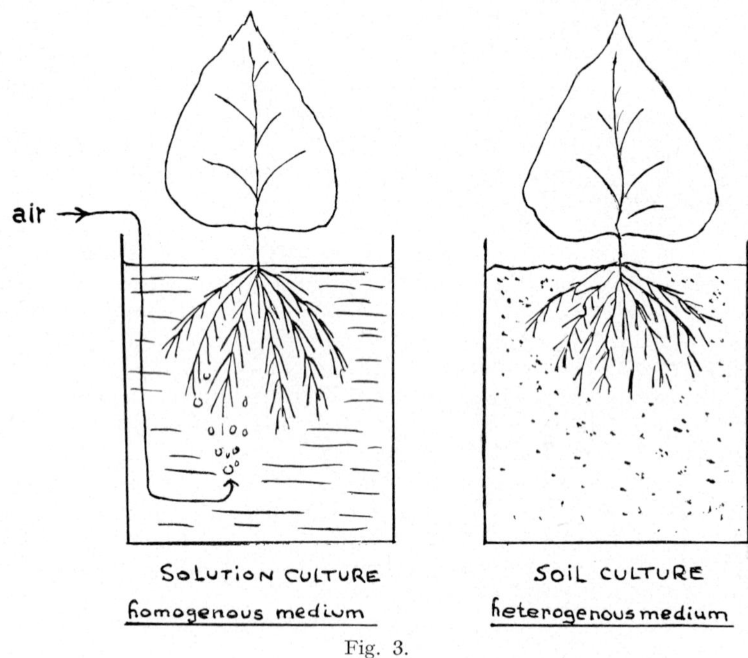


Fig. 3.

In solution culture we have a completely homogeneous root environment. All the cations removed by the root are immediately

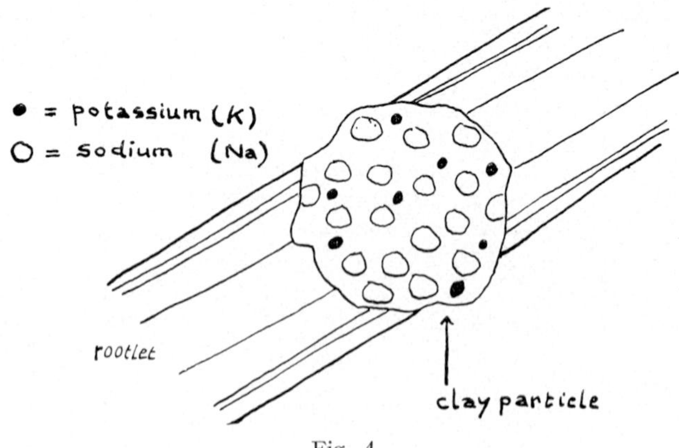


Fig. 4.

replaced from the common pool, assisted by the bubbling of air through the solution, and the periodic renewal of the solution. Not so in soil culture. Soil itself is an extremely heterogeneous medium. The ions needed by the plant, for instance the potassium, are mainly taken from the immediate root environment, and those that are excluded, for instance the sodium, are steadily increasing their share in it. A seriously unbalanced situation is produced, and there are no agents acting quickly restoring the balance.

This state of matter easily explains why plants growing in solution culture generally tolerate much higher salt concentrations than those growing in soil.

The second picture (Fig. 4) illustrates what is going on in the micro-environment, viz. on the points of contact between the active roothairs and the ion-charged clay or organic particles of the soil.

## Root Development and Organic Manure in Respect to Salinity

The above drawings also explain the extreme importance of richly developed and well ramified roots of plants growing under saline conditions.

A large active root surface distributes the loss of potassium from the environment to the growing plant and diminishes the disturbance of the ionic balance on the very points of contact between the root hairs and the ion carrying particles. Therefore, all agents stimulating the development of a densily branched root system assist the plant in tolerating salinity. Farmers working under saline conditions always stress the importance of using organic manure for getting good crop yields. Certainly is root stimulation one of the reasons for it.

Another point is the fact that the rich microflora developing in the soil under the influence of the organic amendments is better capable than the higher plants to attack the potassium confined in the minerals and to put this element finally at the disposal of the crop plant and so to contribute to the balance of the ionic environment.

## The Balanced Ionic Environment in Sprinkler Irrigation

In many countries, the technique of irrigation shifts from flooding the ground to sprinkling the foliage. In Israel, for instance, 90% of all irrigation is already by sprinkler.

In numerous, if not in most plants, the leaves are not less active in the uptake of solutes than the roots. Therefore, the principle of the balanced ionic environment is also valid and of practical importance in respect of the leaves coming into contact with the irrigation water. The cases of scorched leaves observed after overhead irrigation of citrus trees and other plants with saline water have to be explained by the unbalanced ionic composition hitting the leaves rather than by the osmotic pressure exerted by the salts. Again, where sodium is the predominant cation in the water, potassium is the main agent for adjustment of the balance, and for avoiding the damage otherwise caused. A relatively small concentration of potassium in the water, generally from 5—10% of the sodium present, will be sufficient.

On restoring the balance between sodium and potassium, a beneficial action of salinity is often observed in two directions: firstly, a stimulation of growth and a reduction of the growth period for field crops, and secondly, a shifting from vegetative growth to fruit production.

## The Reaction of Plants to a Saline Environment

But plants are not only exposed to an environment, which may be balanced or not, they also actively react to it in different ways. How can the plant be assisted in its struggle with salinity besides by the adjustment of the ionic environment?

### Root Respiration, Trace Elements, and the Organic Factor

Firstly, there is the question of root respiration. The acquisition of water against a high osmotic pressure, and the selective uptake of ions such as potassium, calcium and magnesium in the presence of sodium in great excess, means the expenditure of considerable energy. This energy can only be derived from metabolic processes consuming oxygen. Depriving the roots from oxygen due to soil compaction causes a decreased uptake of potassium and an increased uptake of sodium, and vice versa.

But respiration not only means the easy access of oxygen to the roots. It also involves a long chain of enzymatic reactions taking place within the plants. Many of the steps in this chain need metal activators and catalysts such as Fe, Mn, Zn, Cu, Co and Mo, and the need for the availability of these minor elements within the reach of the roots is greater under saline conditions than under non-saline ones. On the other hand, saline water takes a toll of trace elements, cations as well as anions, from the root zone by ion-exchange. Hereby salinity deprives the soil of trace elements much more than crop removal does. Supply of these elements from outside, or their mobilisation from within the soil, is an important factor for protection of the crop against salinity. Chelates produced by chemical or metabolic processes from the organic phase of the soil may much contribute in supplying the roots with minor elements from the hidden reserves of the soil.

Secondly, as already stressed above, all factors stimulating root development are important in assisting the plant exposed to salinity. Intensive root development means three things:

i. The conquest and exploitation of a larger volume of soil,

ii. A denser network of roots within this volume and hereby a multiplication of points of contact between root and soil, and

iii. A quick renewal of the short-lived active root hairs.

Organic fertilizers such as farmyard manure or a well-prepared compost are known to be very efficient in enhancing root development.

Thirdly, it is a well-known fact that there are varieties within the species of crop plants that are salt-sensitive, and others that are salt-tolerant. Some take up much sodium from a certain environment and others less. These differences are sometimes extremely large. What are the reasons? The conventional answer that this is due to a differential genetic set-up is highly unsatisfactory. It does not explain anything. It only translates the same fact from one language into another. Apparently, one has to assume in plants an autonomous mineral metabolism, in analogy to the situation in animals. That means that plants are capable according to their needs selectively to absorb, to exclude, to retain and to excrete ions present in their environment. The question arises if there are not hormone-like compounds present in the plant which govern the pattern of their mineral metabolism in a similar way the hormones do in animals. The differences between the varieties with regard to their salt resistance could then be explained by a differential hormonal equipment. In this respect more attention should also be given to the microflora living in the rhizosphere, and especially to the mycorrhizal association of fruit trees and field crops. It seems that many fruit trees successfully grown under saline conditions are notoriously infected by ectotrophic as well as endotrophic mycorrhiza. There is some evidence that the pattern of mineral metabolism is changed under the influence of the organic metabolites produced by the microflora and resorbed by the host. It may further be worthwhile to look for synthetic organic compounds applied to the plant and acting in the same direction. This should also include a screening of all the organic substances now experimentally or commercially used to plants as systemic insecticides, dwarfing agents, soil conditioners, osmotic pressure producers and so on. They may incidentally have a more or less strong action on the mineral metabolism of the plants exposed to them, including an increased salt resistance. In this direction a promising new field of research lies ahead.

### Balanced Environment versus "Plant Nutrition"

The concept of the balanced ionic environment (B.I.E.) as shortly outlined in this paper looks attractive from several points of view: Scientifically, the science of soil-plant relationship has become accustomed separately to deal with three categories of ions: Firstly, the nutrients, a convenient designation useful for all who exploit it, secondly, the ions connected with salinity, known as "trouble-makers" and thirdly, the H-ions expressed as pH.

This split of ions into three different categories is only an outcome of the history of thought and method in science and does not befit scientific objectivity. The B.I.E. concept overcomes this artificial division, places all three categories on a common footing, and stresses their interdependence in action.

Consequently, terms like "toxicity" or "deficiency" of ions may need revision and should in many cases be replaced by that of "unbalanced ratios between ions".

From the practical point of view:

The B.I.E. concept brings in line the experience of the farmer with the teachings of science. But, what is more important, the concept proved already its practical usefulness in growing sugar beet, groundnuts, fodder plants and other field crops under conditions of high salinity. It saved tens of thousands of apple and pear trees grown in the semi-arid parts of the Cape Province of South Africa, where the salinity of irrigation water comes close to 3,000 ppm total dissolved solids, and more than 700 ppm of chloride.

Finally, the conventional separation of what is called "plant nutrition" from the environmental factors such as light, temperature, aeration, humidity and so on should be discontinued. The total ionic environment should be dealt with as an integral part of the environment at large.

Environmental conditioning as outlined above, coupled with the selection of salt-resistant varieties and proper soil management, makes it possible to break the salinity barrier and to extend food production into areas to-day excluded from intensive farming. Following the same ecological approach more food can be produced by a given quantity of irrigation water, a commodity of increasingly scarce supply.

## REFERENCES

BERNSTEIN, L., 1961. Osmotic adjustment of plants to saline media. I. Steady state. *Amer. J. Bot.* **48,** *909—918.*

BOYKO, H. & BOYKO, E., 1959. Seawater Irrigation. A new line of research on a bioclimatological plant-soil complex. *Int. J. Bioclimat. Biometeor.*, **III,** II, section B 1, *1—24.*

BOYKO, H. & BOYKO, E., 1964. Principles and Experiments regarding direct Irrigation with highly saline and seawater without desalination. *Trans. N.Y. Acad. Sci.*, Ser. II, **26,** Suppl. to No. 8, *1087—1102.*

CLAY, D. W. T. & HUDSON, J. P., 1960 Effects of high levels of potassium and magnesium sulphates on tomatoes. *J. hort. Sci.* **35,** *85—97.*

HAYWARD, H. E., LONG, E. M. & UHVITS, R., 1946. *U. S. D. A. Techn. Bull.* **922.**

HEIMANN, H., 1959. The irrigation with saline water and the balance of the ionic environment. *Potassium Symposium 1958, 173—220.*

HEIMANN, H. & RATNER, R., 1965. The irrigation with saline water and the ionic environment: field experiments with groundnuts and cow peas. *Oléagineux* **20,** *157—162.*

ITTALIE, T. B. VAN, 1935. *Trans. 3rd int. Congr. Soil Sci.* **1,** *191—194.*

JONES, W. W., PEARSON, H. E., PARKER, E. R. & HUBERTY, M. R., 1952. Effect of sodium in fertilizer and in irrigation water on concentration in leaf and root tissues of Citrus trees. *Proc. Amer. hort. Sci.* **60,** *65—70.*

LAGERWERFF, J. V. & EAGLE, H. E., 1962. Transpiration related to ion uptake by beans from saline substrates. *Soil Sci.* **93,** *420—430.*

LOEB, J. & WASTENEYS, H., 1911. Die Entgiftung von Natriumchlorid durch Kaliumchlorid. *Biochem. Z.* 33, *480—488.*

MARTIN, J. P. & BINGHAM, F. T., 1954. Effect of various exchangeable cation ratios in soils on growth and chemical composition of avocado seedlings. *Soil Sci.* **78,** *349—360.*

OSTERHOUT, W. J. v., 1906. On the importance of physiologically balanced solutions for plants. *Bot. Gaz.* **42,** *127—134.*

RUBINSTEIN, D. L., 1928. Das Problem des physiologischen Ionenantagonismus. *Protoplasma* **4,** *259—314.*

WATSON, D. J. & ORCHARD, E. R., 1960. (Experiments with Poly-ethyleneglycol). *Rothamsted Exp. Sta. Ann. Rep. 1960, 96.*

# EXPERIMENTS OF PLANT GROWING UNDER IRRIGATION WITH SALINE WATERS FROM 2000 MG/LITRE T.D.S. (TOTAL DILUTED SOLIDS) UP TO SEA-WATER OF OCEANIC CONCENTRATION, WITHOUT DESALINATION

BY

HUGO BOYKO AND ELISABETH BOYKO

*Rehovoth, Israel*

## Introductory Remarks

The experiments were based on the results of the author's ecological observations, first in the salt- and sand steppes of Central Europe (BOYKO, 1934) made in the years 1929—1935, and then in South-West Asia and many other countries, where he could study the plantsociological conditions of halophytic vegetation, dune vegetation or both together.

In all cases he found that the salt tolerance was raised considerably on permeable soil and seemingly several times on sand. All experiments described in the following are based on these ecological observations.

Irrigation with *desert underground water* of a relatively high T.D.S. (Total Diluted Solids) was practized by Dr. Elisabeth BOYKO in cooperation with the other author on the completely vegetationless gravel hills in Eilat since 1949. There a flourishing garden was laid out and planted by her under extremely hot desert conditions with about 20 mm yearly average rainfall. The irrigation-water at disposal had 2000—6000 mg/l T.D.S. and 180 flourishing species, most of them non-halophile, proved the theoretical principles to be correct (E. BOYKO, 1952, 1966). Soon similar experiments were made in other countries. A summary of all experiments was published in 1964 (H. BOYKO & E. BOYKO, 1964).

In earlier papers already (H. BOYKO, 1957—1958, H. BOYKO & E. BOYKO, 1959) the authors drew attention to experiments with direct irrigation with *sea-water without desalination*. The results obtained and the significance of this new line of research in various directions induced the first author to ask scientists in other countries to verify these experiments and to enlarge them by using as many different species as possible.

Our first series of experiments with sea-water proper started in June 1957 with *Juncus maritimus* LAM. var. *arabicus* ASCH. et BUCH. and the mediterranean ecotype of *Agropyrum junceum* BEAUV.

These two species were chosen for two main reasons: After all efforts to receive any financial support for such a new line of research proved to be in vain, the outlay of these first experiments had to be restricted to the minimum of species, namely two. They were, however, chosen because of their extremely different ecological requirements in their natural habitats, so that a wide amplitude of applicability could be hoped for.

*Juncus maritimus* var. *arabicus* grows in Israel mainly on heavy swampy soils in the hot desert region of Wadi Araba between the Dead Sea and the Red Sea; *Agropyrum junceum*, however, in the very different climate of the mediterranean coast as a xerophytic perennial grass on the sand dunes (BOYKO & BOYKO, 1959). The experiments indicated, seemingly for the first time, that the two great dangers to agriculture, high salinity on the one hand and shifting dunes on the other, could successfully be fought against by combining them. Since more than a third of the whole of our terra firma is threatened by these two dangers, the significance of these experiments is obvious, particularly in view of the necessity to keep the natural resources for food and industrial raw materials in pace with the growth of the population.

It was this reasoning which induced UNESCO to sponsor the continuation of these experiments, and the World Academy of Art and Science is establishing an international Working Group for their potential global impact on human welfare in general.

Our experiments with saline water of phreatic origin dealt with about 200 different plant species. Our experiments of direct irrigation with sea-water proper (without desalination) used as research material 10 different species.

## Sea-water vice other Saline Waters

The difference between sea-water and saline underground water or other saline waters is a qualitative one and not or to a lesser degree only a quantitative one with regard to the Total Salt Concentration (T.S.C.) or more exact, to the Total amount of Diluted Solids (T.D.S.).

The various types of sea-water have very different concentrations, but all are sea-water proper, for the specific composition of soluble salts and the percentage of the various compounds is in all of them almost the same. It has been stressed by the first author already in the introductory article to this book, as well as in the article on principles that this specific composition presents such a well balanced ionic environment for plants that their salt tolerance with regard to irrigation with diluted sea-water seems to be multiplied in almost all species compared with their tolerance regarding irrigation with saline water of the same concentration but of another composition. The most striking features of sea-water and its dilutions are:

a) the very high chloride content which is always about 85% or so (about 75% sodium chloride and 10% magnesium chloride;

b) it contains all nutritive elements the plants need, including also — and this seems to be of particular importance — the necessary trace elements, and

c) micro-organisms, living or dead, in considerable amounts (their influence is not yet adequately studied).

Experiments with NaCl solutions (WERBER, 1936) or even with synthetic sea-water (BOTTINI, 1961) are not comparable in their biological effect. They are, however, most useful for the solution of many important physical and physiological questions. In the following, experiments with natural sea-water only are described, to the exception of the few field experiments carried out with highly saline desert underground water in the rift valley between the Dea Sea and the Red Sea.

In our experiments four types of sea-water were used.

1) East Mediterranean sea-water taken from the shore south of Tel Aviv. According to the analyses carried out in the chemical laboratory of the Ministry of Agriculture, Israel, it contained:

| | |
|---|---|
| $CaSO_4$ | 2,965 mg/l |
| $MgSO_4$ | 1,713 mg/l |
| MgCl | 3,530 mg/l |
| KCl | 870 mg/l |
| NaCl | 31,700 mg/l |
| Various | 2,822 mg/l |
| | 43,600 mg/l |

These figures are higher than those of other analyses of East Mediterranean sea-water. The reason is probably that the respective place of the shore is rather shallow. East Mediterranean sea-water in general is accounted for as having a T.D.S. of 39,000 to 40,000 mg/l.

In addition to this fact, the container (tank) at disposal on the place of the experiments in Bet Dagon and in Beersheba had an opening of several square decimetres, which was not hermetically closed. Evaporation therefore resulted in a further rise of concentration. The highest concentration was 53,000 mg/l T.D.S., measured by an analysis of irrigation water in use before refilling the tank.

The other sea-water types were simply made by appropriate dilutions (75, 50, 25%) with tap water.

Thus we used, apart from the East Mediterranean sea-water type,

2) sea-water of oceanic concentration with a T.D.S. of 30,000—40,000 mg/l,

3) sea-water of the North Sea type near Holland with a T.D.S. of about 20,000 to 27,000 mg/l,

4) sea-water of the type of the Caspian Sea with a T.D.S. of about 10,000 to 14,000 mg/l.

All control experiments were made with fresh water (tapwater) containing about 700 mg/l T.D.S. with about 200 mg/l chlorine in it for hygienic purposes.

## Principles

The basic ecological principles of irrigation with sea-water or other highly saline waters on sand are dealt with in foregoing articles. Therefore a summarizing enumeration of the old and well known ones as well as of the new ones may suffice here.

They are:

a) Quick percolation (kn);*

b) Good aeration (kn);

c) Easy solubility of NaCl and of $MgCl_2$ (the main components of sea-water) (kn);

d) Lack of sodium adsorption to sand particles in contrast to the easy adsorption to clay particles (kn);

e) the principle of "Partial Root Contact" (H.B.);

f) the Viscosity-principle (H.B.);

g) the condensation of "Subterranean Dew" (H.B.);

h) "Biological Desalination" (H.B.);

i) Adaptability of plants to fluctuating osmotic pressure (m);

j) the "Balance of Ionic Environment" (H.H.);

k) "Raised Vitality" (Vigour) (H.B.);

l) microbiological influences (kn in principle only);

m) the geophysical theory of "Global Salt Circulation" (H.B.);

n) the Basic Law of Universal Balance (H.B.); (H. Boyko, 1966).

## Outline of Experiments

The experiments were made in order to obtain certain quantitative data and tangible results for theoretical and practical applicability in various directions.

Although carried out on a very restricted scale, the results seem to justify similar experiments with many more species and on a much larger scale in view of the diversified possibilities of their application as, for instance, afforestation of coastal dunes, lessening the strain on freshwater resources, particularly in arid and semi-arid countries, productivizing of deserts, and so on.

In this chapter the general outline only of the experiments with sea-water is mentioned. The methods of measuring and some other

* Remarks: the abbreviations in brackets refer to the respective author: kn = known since long (see particularly UNESCO Arid Zone Research Series); H.B. = H. Boyko; H. H. = H. Heimann; m = recognized since several years apart from the present writer by various authors, e.g. V. J. Chapman, O. Stocker, H. Walter and others.

details differ somewhat with the different species and are therefore described in the part dealing with each of them separately.

In the first line the experiments were carried out with the aim to have an exact control:

a) of the amount of saline irrigation water for each individual plant,

b) of the salinity of the irrigation water,

c) of quantitative data of salt accumulation or non-accumulation in the soil,

d) of growth data and specific reactions of the various species, and plant individuals under irrigation with different concentrations of sea-water.

All biochemical and almost all physiological studies had to be left for later research projects, for which laboratory facilities and financial support was to be hoped for.

In order to achieve reliable results in spite of the administrative and financial obstacles, the following methods were used:

### Frames

Circular frames of the same size in diameter and height were made from cleaned asphalt barrels by taking out the bottom. They were put on sand and filled with sand. In order to keep the temperature influence of direct sun radiation or reflex radiation at a minimum, the outside of the frames was painted white. The frames were 90 cm high and had a diameter of 46 cm ($r^2\pi = 1661$ square cm). The sand surface was therefore equivalent to about 1/6 square metre and each liter of irrigation water was equivalent to 5 mm normal rainfall.

The exact amount of sand in each frame varied to a small degree because of the somewhat irregular shape of the asphalt barrels, the sand volume in them being between 1/7 and 1/6.5 cubic metre (the volume of the root layer is dealt with later on and varies with the different species). The sand surface was kept about 5 cm below the fringe of the frame; not lower in order to prevent the irrigation water from overflowing, and not higher in order to avoid too much shadow, thus creating adverse conditions. Further, in the heavier natural soils of the experimental places, slowly inclined furrows of about 50 cm width, i.e. a little broader than the asphalt barrels, were dug out, filled to a depth of 40—50 cm with gravel and sand, and on these furrows the frames were finally put, filled with sand and used for planting. This allowed quick percolation up to 1.30—1.40 and the inclination of the furrows prevented any stagnation of the percolated irrigation or rain waters.

### Lysimeters

Besides these simple constructions, similarly simple and cheap lysimeters were constructed from such bottomless asphalt barrels, but with a deeper and quicker percolation possibility: deeper simply by putting one and a half asphalt barrel together. The whole outlay and the details of these lysimeters are to be seen from fig. 3, 4 and 5. Planting and irrigating is described with each species separately, as are also the measurements of the plant individuals.

Data on Plant growth

As far as possible, plant development was measured on each plant individual separately. The relevant data are comprised in tables according to the respective species. In two places field experiments could also be carried out: South of the Oasis of Yotvata in Wadi Araba and near the Dead Sea. The Desert garden of Eilat also belongs to this group (see the part "Places of experiments").

Meteorological Data

All experiments were accompanied by meteorological measurements on the spot, or the meteorological data of the nearest place were asked for and received by the kind cooperation of the Meteorological Service of Israel*. The main data continuously used were: maximum daily temperature; minimum daily temperature; average daily temperature; rainfall; dew.

Soil and Water Measurements

Soil and water investigations refer to mechanical and chemical analyses, and total percolation amount of the irrigation water. An important part of these geophysical measurements consisted in the determination of the difference between the NaCl contents in the irrigation water and that of the percolated water taken from the containers of the lysimeters after irrigation. Another significant part was the investigation into the salt movement and particularly into the question of salt accumulation in the soil. Sufficient results could be obtained in these first experiments to answer these most important questions, although many and diversified experiments will still have to be carried out under different climatic, soil (sand), and biological conditions in order to enrich our knowledge in these new and partly revolutionary lines of research. However, the main principles as mentioned above could already be proved.

## Places of Experiments

In the following the conditions at the various experimental places in Israel are shortly described, in chronological order:

* Our particular thanks are to be expressed for the collaboration of the meteorological station in the Negev Institute for Arid Zone Research in Beersheba.

## The Desert Garden of Eilat

Eilat lies on the northernmost point of the Red Sea, in the southern part of Israel at a latitude of 29°33′ north of the equator, and is known for its extremely dry and hot desert climate.

Precipitation is about 20 mm (less than one inch) yearly average, occurring erratically during the winter months.

There, in summer 1949 one of us (Dr. ELISABETH BOYKO) was entrusted with the task of planning and laying out a desert garden to make this very desolate and shadeless desert into not only a livable but also attractive dwelling area. The main reason for this project was that the erection of a town and a port were planned here as an outlet for Israel to the Indian Ocean, the Persian Gulf, the Far East and Australia.

Soil: Instead of planting the garden in the depression ("Sabha") with its heavy soil, surface-near saline groundwater and halophytic vegetation, we chose, in accordance with the principles described by H. BOYKO in a former article of this volume, the vegetationless gravel hills, which girdle the foot of the rocky mountains along the greater part of Wadi Araba. The mechanical analysis of this "garden soil" is shown in Table I:

**Table I.**

Desert Garden of Eilat: Mechanical soil analysis

| | |
|---|---|
| Stones | 61.0% |
| Coarse sand | 23 % |
| Fine sand | 12.3% |
| Silt | 2.7% |
| Clay | 1.0% |
| | 100.0% |

The skeleton (stones) represents the rock debris from the surrounding mountains and consists of: granites, gneiss, nubian sandstone, flint, limestone; pieces of pure quartz, feldspar and other minerals are also scattered. At a depth of 20—50 mm we are frequently confronted with a layer rich in gypsum.

Water: In the beginning of the plantation there was only one possibility to provide for the most necessary irrigation water, namely to bring it in barrels by command car from a saline spring (Ber Ora) 18 km north of Eilat. After a few months, however, a pipeline was laid from the oasis Yotvata (Arabic: Ein Ghadian) about 42 km north of Eilat. This water had a total salt content fluctuating between 2000 and 6000 (sometimes up to 10,000) mg/l, the composition of which was approximately as to be seen in Table II.

The amount of irrigation water can only be estimated since no facilities for controlled irrigation could be installed. In view of

**Table II.**

Composition of Solids in the irrigation water for the desert garden of Eilat in % of the T.D.S.

| | |
|---|---|
| Ca | 13.4% |
| Mg | 7.4% |
| K | 0.8% |
| Na | 6.5% |
| $CO_3$ | 4.0% |
| $SO_4$ | 59.6% |
| Cl | 8.0% |
| Fe | (traces) |
| Al | (traces) |
| Si | (traces) |
| Nondetermined | 0.3% |
| | 100.0% |

the high value of every drop in this region, only the minimum necessary for the plants was and still is given. A shrub of 2 m height and 1½ m width, for instance, received about 30 l a week, filled into the circular watering pans provided for each plant according to its size. This kind of irrigation kept the plants living and healthy but also limited their growth.

Nevertheless the purpose aimed at could be achieved as can be seen from the frontispiece of this volume. These two photos show a part of this garden at the beginning of the work in 1949 and the same spot as it looked ten years later in 1959. In 1949 we were confronted there with a desolate, entirely barren gravel desert whereas the lush vegetation visible in Fig. B proved the correctness of the principles and gives new hopes for the productivization of many deserts in the world. The town of Eilat has now about 13,000 inhabitants and is a well visited recreation place.

Plant species: The list of species planted in the desert garden of Eilat is to be found in the complementary volume to this book (Vol. IV of the WAAS-series) (E. Boyko & H. Boyko, 1966). 180 different species were planted: trees, shrubs, perennials and annuals, many of them not only of ornamental value or most useful for their shadow giving crowns, but also of economic value as potential plant raw material as basis for industries. They all flourished except two shallow-rooted succulents (a *Kleinia* sp. and a *Mesembryanthemum* sp.) which died when a breakdown of three weeks of the irrigation system in the hottest month (July) made irrigation impossible; (a pipeline brought the saline water over a long distance (42 km) from saline wells in Wadi Araba).

Growth data and details of these plants in the desert garden of Eilat are also described in the volume mentioned above.

In connection with the studies on soil and water it is noteworthy

**Table III.**

Soil Analysis of Experimental Plot in Yotvata Oasis, after ploughing and before levelling.
(N. H. TADMOR, D. KOLLER & E. RAWITZ, 1958)
(Two representative profiles are given, in percent)

| Depth (cm) | Mechanical composition | | | | | Moisture equiv-alent | Wilting point |
|---|---|---|---|---|---|---|---|
| | Coarse sand 0.2—2.0 mm | Fine sand 0.02—0.2 mm | Silt 0.002—0.02 mm | Clay < 0.002 mm | $CaCO_3$ | | |
| 0— 35 | 2.4 | 56.8 | 7.0 | 6.0 | 27.8 | 22.1 | 12.0 |
| 35— 45 | 1.0 | 68.6 | 8.0 | 7.0 | 15.4 | 26.4 | 14.3 |
| 45— 70 | 1.7 | 67.8 | 9.3 | 6.8 | 14.4 | 27.0 | 14.7 |
| 70— 90 | 2.4 | 69.3 | 5.8 | 4.5 | 18.0 | 27.4 | 14.9 |
| 90—130 | 1.3 | 77.6 | 4.3 | 2.8 | 14.0 | 28.3 | 15.4 |
| 130—170 | 0.2 | 84.5 | 2.5 | 2.8 | 10.0 | 26.4 | 14.3 |
| 170—200 | — | — | — | — | — | — | — |
| Groundwater | | | | | | | |
| 0— 30 | 3.1 | 43.3 | 7.3 | 4.3 | 42.0 | 11.96 | 6.5 |
| 30— 65 | 5.0 | 48.5 | 7.0 | 5.5 | 34.0 | 14.54 | 7.9 |
| 65— 95 | 7.8 | 45.5 | 6.8 | 4.5 | 35.4 | 16.92 | 9.2 |
| 95—115 | 11.7 | 43.4 | 4.5 | 5.0 | 35.4 | 29.10 | 11.5 |
| 115—150 | 16.8 | 44.4 | 1.8 | 6.0 | 31.0 | 26.30 | 14.3 |
| 150—170 | 17.0 | 48.8 | 1.3 | 7.5 | 25.8 | 28.17 | 15.3 |
| 170—210 | 14.8 | 49.8 | 1.5 | 6.3 | 27.6 | 27.40 | 14.9 |
| 210—240 | — | — | — | — | — | — | — |
| 240—250 | 14.4 | 38.4 | 1.5 | 6.3 | 39.4 | 26.18 | 14.2 |
| 250—280 | 2.5 | 22.3 | 5.5 | 9.5 | 60.2 | 23.00 | 12.5 |
| Groundwater | | | | | | | |

to underline here also that no salt-accumulation took place in the soil except some efflorescence on the surface.

It has also to be mentioned that after seven years of successful irrigation with water of an unusually high salt concentration, deeper drilling in the region of the water source considerably improved the quality of the water. It is still not drinkable because of its high content of magnesium sulphate, but has a T.D.S. of only about 2000 mg/l with 600 mg/l chlorine. The first seven years of this field experiment were, however, sufficient to prove that the principles are correct and applicable.

### Yotvata

For two years, in 1953 and 1954, a field experiment was carried out with *Juncus maritimus* at Yotvata, Wadi Araba, with highly saline fluctuating underground water containing, according to the Hydrological Service, 2,000—6,000 mg/l T.D.S. with about 25—

| Water extract | | | | | |
|---|---|---|---|---|---|
| $Cl^-$ | $SO_4^=$ | $CO_3^=$ | $HCO_3^=$ | Total Salts | pH |
| 5.800 | 1.900 | 0.0002 | 0.0035 | 12.400 | 8.4 |
| 1.725 | 1.188 | | 0.0015 | 4.641 | 8.0 |
| 1.600 | 1.038 | | 0.0015 | 4.244 | 7.9 |
| 0.700 | 1.050 | | 0.0013 | 2.738 | 7.8 |
| 0.605 | 1.013 | | 0.0012 | 2.532 | 7.8 |
| 0.205 | 0.838 | | 0.0012 | 1.618 | 7.8 |
| — | — | | — | — | — |
| 1.200 | 1.138 | 0.0002 | 0.0032 | 3.700 | 8.5 |
| 0.500 | 1.050 | 0.0001 | 0.0014 | 2.415 | 8.3 |
| 0.190 | 0.942 | | 0.0019 | 1.883 | 8.1 |
| 0.290 | 0.932 | | 0.0018 | 1.898 | 8.0 |
| 0.183 | 0.913 | | 0.0018 | 1.686 | 7.9 |
| 0.850 | 1.134 | | 0.0018 | 3.114 | 7.9 |
| 0.220 | 0.920 | | 0.0018 | 1.761 | 7.5 |
| — | — | | — | — | — |
| 0.080 | 0.805 | | 0.0018 | 1.362 | 7.5 |
| 0.090 | 0.507 | | 0.0019 | 0.936 | 7.4 |

30% chlorine. The experiment in the field was carried out under the direction of the first author by his coworkers N. H. Tadmor, E. Rawits and Y. Leshem (1958); the soil analyses were obtained from the Laboratory for Chemical Analyses, Ministry of Agriculture.

Climate: hot desert climate with about 50 mm average yearly rainfall.

Soil: loessy sand.

The relatively high groundwater level is characterized by the wild vegetation as it was found before ploughing, namely an Eragrostidetum bipinnatae into which a few individuals from the lower border of the *Nitraria retusa* zonation are already invading. For more details see the soil analyses of Table III (note the sulphate and $CaCO_3$ content).

On the basis of the successful field experiments here and near the Dead Sea, a plantation on an area of 125 acres (50 hectares) has now been started between Yotvata and Eilat as a first step for an economic enterprise on 2000 acres in this region.

### Bet Dagon

Starting in June 1957, controlled experiments with four different types of sea-water (see above) were carried out in Bet Dagon, in the Mediterranean coastal plain of Israel, on dune sand. Measurements including those of the first "dying experiments" were carried out until January 1960. Then the place had to be abandoned.

Fig. 1. Sea-water experiments in Bet Dagon: Part of the set-up.

Climate: warm, semi-arid, mediterranean with rainfall in winter season (October to March). The dry period without any rainfall is usually about 7 months (summer).

Soil: dune sand from coastal dunes. Table IV shows its physical structure.

**Table IV.**

Mechanical analysis: Dune Sand used in Bet Dagon for the sea-water experiments.

| Physical soil-structure analysis in % | | | | | Total % of $CaCO_3$ | Water Solution 1:5 | | % Organic Matter | pH |
|---|---|---|---|---|---|---|---|---|---|
| 1.0 mm | 1.0 — 0.50 mm | 0.50 — 0.25 mm | 0.25 — 0.10 mm | 0.10 mm | | Total Soluble Salts | Cl | | |
| a. 0.0 | 0.1 | 9.3 | 85.7 | 3.7 | 2.3 | 0.011 | 0.002 | 2.2 | 8.3 |
| b. 0.0 | 1.0 | 15.0 | 78.8 | 3.5 | 3.9 | 0.090 | 0.044 | 2.3 | 8.6 |

Plants: Plants experimented with in Bet Dagon:
1) *Agropyrum junceum* L.
2) *Agave fourcroides* LEMAIRE
3) *Agave sisalana* PERRINE
4) *Ammophila arenaria* LINK
5) *Beta vulgaris* L. (sugar-beet)
6) *Calotropis procera* R.BR.
7) *Hordeum* (barley, a beduin strain)
8) *Juncus maritimus* LAM. var. *arabicus* ASCH. et BUCH.
9) *Juncus punctorius* L.
10) *Rottboellia fasciculata* DESF.

Beersheba

Experimenting on the same lines was continued in Beersheba, where the Negev Institute for Arid Zone Research put an adequate space of about 500 square metres at our disposal.

Climate: arid with about 250 mm yearly average winter rains and 8 to 9 months dry period.

Soil: coarse building sand in the Lysimeters (see fig. 5), and dune sand, mainly of medium sized particles in high frames (bottomless asphalt barrels) put on sand and gravel.

Plants experimented with:
1) *Agropyrum junceum* L.
2) *Calotropis procera* R. BR.
3) *Juncus maritimus* LAM. var. *arabicus* ASCH. et BUCH.
4) *Juncus punctorius* L.
5) *Rottboellia fasciculata* DESF.

S'dom

A successful field experiment on 10,000 square metres is carried out with *Juncus maritimus* since 1962—1966, near S'dom, 8 km south of the Dead Sea. Ecological observations on some other species, occurring wild or as intruded weeds in this region, could be combined with these experiments. Irrigation water derived from a saline well with about 3000 mg/l T.D.S. with about 1000 mg/l chlorine.

Climate: The place has a very hot climate throughout the year for its geographical position near the lowest depression of the globe (its altitude is 380 m below sea level). Rainfall is about 75 mm yearly average.

Soil: The very sandy soil shows an even structure down to the relatively high water level (at about 2.50 m below soil surface), and probably deeper. The physical structure consists mainly of sand particles from 0.2 mm to fine sand with about 10% silt and 3% clay. The surface-near groundwater is indicated also by the wild vegetation, characterizing a sandy but saline oasis. The dominating plant community is a Tamaricetum maris mortui. Repeated analyses of soil profiles carried out in connection with these experiments led to the new theory of biological desalination (Boyko, 1965, 1966) (see fig. 6).

## Irrigation and Percolation

The irrigation methods as well as the studies on percolation and accumulation or non-accumulation of salt in the soil had to be adapted to the local conditions of each place and to the means at our disposal. In an earlier paper (Boyko & Boyko, 1959) detailed description of these methods for the first series of experiments with sea-water during the years 1957 and 1958 in Bet Dagon was given. All preparatory work was done in winter 1956/57 and in spring 1957, in order to start with the sea-water irrigation in the midst of the dry season there. This time for starting the experiment was chosen for the following reason:

To our best knowledge, no comparable experiments had been carried out with natural sea-water, nor experiments combining them with dune sand as medium. Not knowing how long plants could survive under such extreme conditions, we assumed on the basis of our theoretical considerations and ecological experience that we could take the risk of a dry period of three months, but not more. It was only later, in the course of the experiments, that we could observe the plants surviving with good results for a 9 months long dry period, with sea-water irrigation in various concentrations as the only water supply.

In general our aim was on the one hand to determine the thresholds of salt tolerance of the various plant species by starting with extremes of salinity and gradually lowering it, and on the other hand, to determine the degree of salt accumulation in the soil, if such an accumulation should occur.

In Bet Dagon and also in the Negev Institute for Arid Zone Research in Beersheba we used four types of sea-water, taking 100% East Mediterranean sea-water as standard and dilutions of it for the other types.

The four types of sea-water (Series Nr. I—IV) used during these experiments are to be seen from Table V.

**Table V.**

Sea-water types used in these experiments.

| Barrels Nr. | Series Nr. | Sea-water and Dilutions | Total Salt Content T.D.S. | NaCl | Sea-water type |
|---|---|---|---|---|---|
| 5 | I | 100% Sea-water | 4.4% | 3.2% | East-Mediterranean |
| 5a | | | | | to Red Sea type |
| 4 | II | 75% Sea-water | 3.3% | 2.4% | Oceanic concen- |
| 4a | | 25% Tap water | | | tration |
| 3 | III | 50% Sea-water | 2.2% | 1.6% | North Sea (1.7-3%) |
| 3a | | 50% Tap water | | | |
| 2 | IV | 25% Sea-water | 1.1% | 0.8% | Aral Sea (1%) |
| 2a | | 75% Tap water | | | Caspian Sea (1.3%) |
| 1 | | Fresh water for | | | |
| | | Control experiment | about 0.04 | | normal drinking- |
| 1a | | (Tap water from | | | water |
| | | the Citrusgrove | | | |
| | | at Bet Dagon) | | | |

Also with regard to the amounts of irrigation water we started with extremes: in 1957 and 1958 during the completely dry season we applied five liters daily to each unit, except on Saturdays and holidays, and reduced this in the rainy season according to meteoric precipitation. From the graph of fig. 2 it is to be seen how the amounts of water application from both sources were recorded.

During the whole time of the first half year of 1959 this irrigation in Bet Dagon was cut to about a quarter of the former amount in order to achieve results on the basis of which economic experiments could be started. The amount of supplied sea-water during these 6 months was about 1 cubic metre per square metre to which the scanty precipitation of this winter season has to be added. This amounted to altogether about 347 mm rainfall, a rather poor amount for this Mediterranean place.

From July 1959 on no irrigation at all could be given, leaving the plants for observation until January 1960, when the place had to be abandoned. These observations lead us to the layout of the "Dying Experiment" in Beersheba, the first of its kind as far as we know. The results are discussed with each species separately in a later chapter.

The main purpose of all these experiments was to obtain an

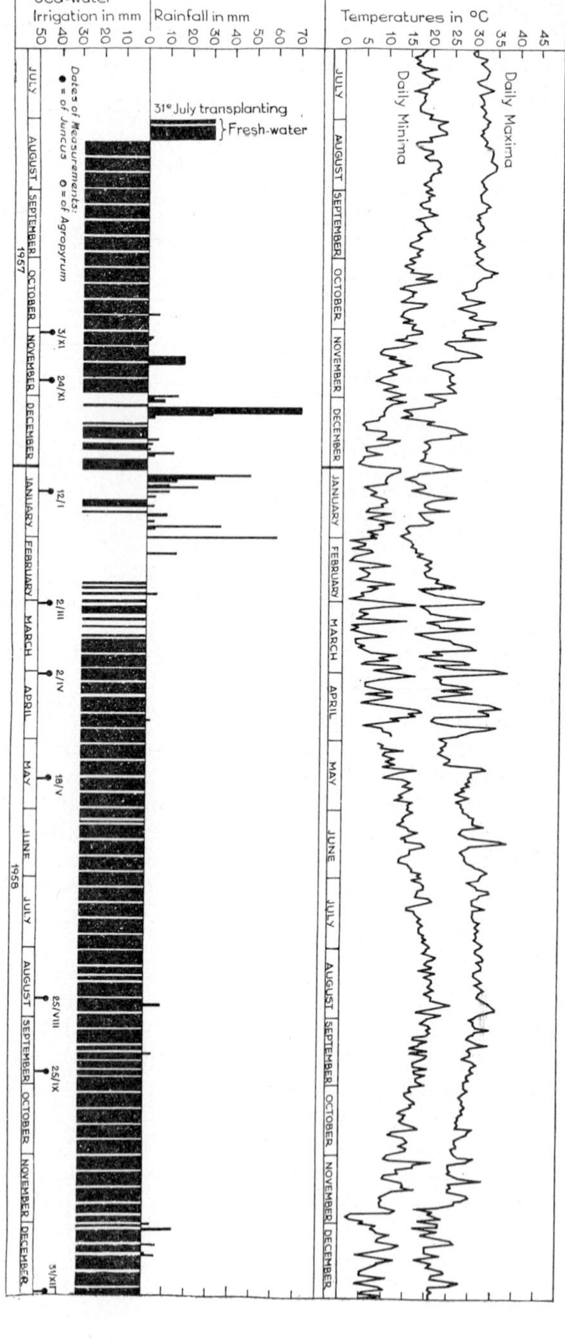


Fig. 2. Sea-water-irrigation, Rainfall, and Temperatures. (Boyko & Boyko, 1959)

Fig. 3. Set-up of lysimeters for the sea-water experiments in Beersheba.

experimental answer to our ecological observations that highly permeable soils, particularly sand and gravel, greatly enhance the salt tolerance of plants in general. After the success with about 180 species, mostly non-halophytes, in the Desert Garden of Eilat, planted in 1949 with saline underground water from the desert, our experiments tried to find the answer to the following questions:

1) Is it possible to grow plants on sand or gravel by irrigation even with sea-water or other extremely saline water? Which are, for various species, the limits under varying conditions?

2) How far is salt accumulation in the root layer a limiting factor, if deep layers of highly permeable soils are used?

3) What is the growth behaviour of those plant species experimented with under these conditions of irrigation?

The questions 1) and 3) will be dealt with later in connection with the plant measurements. In this part we want to bring the data pertinent to question 2).

Fig. 4. Construction of simple lysimeters for the sea-water experiments in Beersheba (first stages): earth work and putting the lower half-barrel without its bottom on a inclined tin plate.
In the background our Technical Assistant LAWRENCE VER TER who made also most of the almost daily chemical analyses of chlorine content in the water before irrigation and after percolation.

In order to obtain answers of general value to the decisive questions of percolation and salt accumulation, we had to compare the relevant data from various places with different climatic conditions, different soil structures, and obtained by irrigation with different concentrations.

This could be achieved, if only with difficulties, for the carrying out of such experiments met with strong administrative opposition. The conviction prevailing here among the official soil and irrigation scientists before that time was that a chlorine content of 400 to 600 mg/l was the maximum permissible in irrigation water without endangering crops and soils. Using water with many thousands mg/l chlorine seemed to them like a crime. A grant from UNESCO and the administrative support of the Negev Institute for Arid Zone Research in Beersheba, however, made it possible to continue the experiments until 1963 in Beersheba, and the great amount of data from the many hundreds of soil- and water analyses obtained during the years 1957—1963 could be compared and evaluated.

The main results from the various places are summarized and presented here for comparison:

Analyses of soil profiles in summer 1958 made it clear then

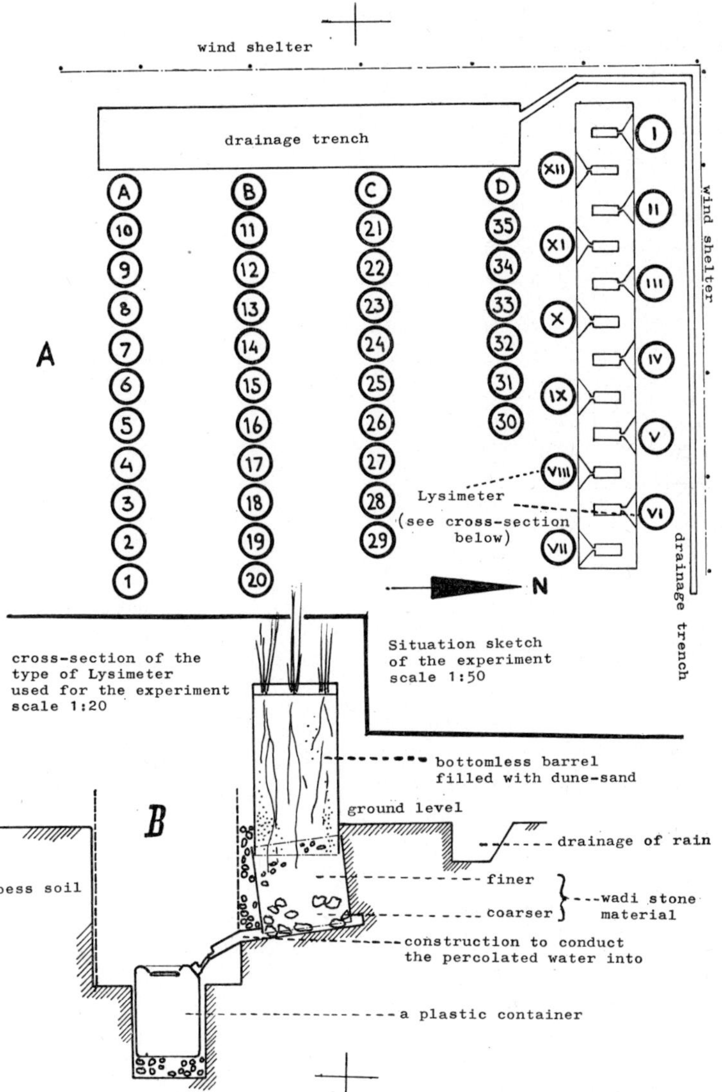


Fig. 5: A) Set-up of the sea-water experiments in Beersheba (altogether 46 units), with 12 lysimeters.
B) Profile through a lysimeter.

already that no danger of salt accumulation in the root layer has to be feared even from irrigation with water of oceanic concentration, particularly not when irrigation is carried out in the open, and a few months later rainfall is to be expected to wash away any possible accumulation of NaCl. The analyses are presented in Table VI.

They show that after 100 times irrigation after the last effective rainfall, no danger of accumulation is indicated.

In his capacity as Chairman of the International Commission on

**Table VI.**

Sea-water experiments: Soil-analyses
(after BOYKO & BOYKO, 1959).
Total Salt and NaCl content in the soil (dune sand) after 100 times irrigation with various sea-water concentrations.
Beginning of irrigation: 31. July 1957.
Date of Analysis: 2. July 1958 (based on dry weight 105° C).

| Series Nr. | Unit barrel No. | Irrigation with: | Depth of soil samples | Moisture % | Cl % | Calculated NaCl % (Coeff. 1.6) | Total sum of salts % |
|---|---|---|---|---|---|---|---|
| Control units | 1 | | 10 cm | 6.7 | 0.004 | 0.0064 | 0.017 |
| | | | 30 cm | 0.5 | 0.003 | 0.0048 | 0.015 |
| | | 0% sea-water | 60 cm | 1.0 | 0.004 | 0.0064 | 0.015 |
| | 1a | (Tap-water) | 10 cm | 6.8 | 0.013 | 0.0208 | 0.032 |
| | | | 30 cm | 1.5 | 0.004 | 0.0064 | 0.015 |
| | | | 60 cm | 1.0 | 0.004 | 0.0064 | 0.013 |
| | 5 | | 10 cm | 11.9 | 0.358 | 0.5728 | 0.675 |
| | | | 30 cm | 14.1 | 0.463 | 0.7408 | 0.950 |
| I | | 100% sea-water | 60 cm | 15.9 | 0.460 | 0.7360 | 0.900 |
| | 5a | NaCl 3.170% | 10 cm | 10.6 | 0.510 | 0.8160 | 0.950 |
| | | Total 4.360% | 30 cm | 13.2 | 0.465 | 0.7440 | 0.900 |
| | | | 60 cm | 15.3 | 0.540 | 0.8640 | 0.950 |
| | 4 | | 10 cm | 1.8 | 0.138 | 0.2208 | 0.280 |
| | | | 30 cm | 1.8 | 0.115 | 0.1840 | 0.235 |
| II | | 75% sea-water | 60 cm | 2.4 | 0.075 | 0.1200 | 0.145 |
| | 4a | NaCl 2.378% | 10 cm | 5.0 | 0.160 | 0.2560 | 0.300 |
| | | Total 3.270% | 30 cm | 2.4 | 0.107 | 0.1712 | 0.220 |
| | | | 60 cm | 2.9 | 0.080 | 0.1280 | 0.255 |
| | 3 | | 10 cm | 1.8 | 0.097 | 0.1552 | 0.190 |
| | | | 30 cm | 2.2 | 0.083 | 0.1328 | 0.180 |
| III | | 50% sea-water | 60 cm | 1.5 | 0.058 | 0.0928 | 0.120 |
| | 3a | NaCl 1.585% | 10 cm | 5.8 | 0.115 | 0.1840 | 0.235 |
| | | Total 2.180% | 30 cm | 4.6 | 0.098 | 0.1568 | 0.190 |
| | | | 60 cm | 3.3 | 0.110 | 0.1760 | 0.210 |
| | 2 | | 10 cm | 4.6 | 0.050 | 0.0800 | 0.100 |
| | | | 30 cm | 4.3 | 0.061 | 0.0976 | 0.120 |
| | | 25% sea-water | 60 cm | 2.2 | 0.044 | 0.0704 | 0.090 |
| IV | 2a | NaCl 0.793% | 10 cm | 8.1 | 0.056 | 0.0896 | 0.115 |
| | | Total 1.090% | 30 cm | 3.8 | 0.061 | 0.0976 | 0.120 |
| | | | 60 cm | 1.9 | 0.046 | 0.0736 | 0.095 |

Applied Ecology of the International Union of Biological Sciences, the first author asked, in his regular Circular Letters, scientists in other countries to verify these experiments or to start similar ones. The first requests for information on saline water irrigation were made in 1950 and for information on sea-water irrigation in 1957, and during the following few years a number of stations carried the work forward with similar results (see chronological table facing page 17 of this volume).

After the place in Bet Dagon had to be abandoned, an analysis of the soil beneath the spot where 100% East Mediterranean sea-water had been used, resulted in the fact that also in 1.30 m depth beneath the root collar of the experimental plants no traces could be discovered of the high amount of chlorides supplied during the two years of sea-water irrigation. The soil there on the base of the sand-filled frame (bottomless asphalt barrels) was a medium, dark grey soil containing a considerable amount of clay particles, but also limestone, gravel and residues from building material. The date of the analysis was the 20th of January 1960 and a few effective rainfalls only preceded the day of taking the soil sample after interruption of the sea-water irrigation in July 1959.

The result is all the more interesting because we could calculate the amount of NaCl poured into this soil during the time July 1957—July 1959 (see Table VII).

**Table VII.**

Amount of salt poured into the bottomless barrel framing an area of 1667 square centimetre (= 1/6 square metre) by irrigation with 100% sea-water during the period from July 31st, 1957 to July 31st, 1959 in Bet Dagon.

| Year | Number of irrig. days | Amount of irrig. water in l | T.D.S. in g/l | NaCl in g/l | Total amount of salts poured in kg | NaCl poured in in kg |
|---|---|---|---|---|---|---|
| 1957 | 110 | 550 | 43.6* | 31.7* | 23.98 | 17.43 |
| 1958 | 262 | 1310 | 43.6* | 31.7* | 57.12 | 41.53 |
| 1959 | 35 | 175 | 43.6* | 31.7* | 7.63 | 5.55 |
| | | | | | 88.73 | 64.51 |

* The analysis of Table VI is taken as basis of these calculations; the average figures are probably higher for the reasons mentioned in the text (evaporation from the tank containing the sea-water for irrigation).

This simple calculation shows that during these two years a total amount of 88.73 kg sea salt, with 64.51 kg NaCl was poured through the bottomless asphalt barrel, serving as a 90 cm high frame and framing an area of 1661 $cm^2$, i.e. 1/6 square metre.

The data of percolation and accumulation collected in Bet Dagon were seen as orientation data only and in order to base final conclusions on a broader basis, we constructed a number of

**Table VIII.**

Example of percolation measurements during the sea-water experiments in Beersheba. Lysimeter Nr. IV. (25% Sea-water = Caspian Sea Type) Plant name: *Juncus punctorius* L.

| Date | Time | Irrigation | | | | | Date | Time |
|---|---|---|---|---|---|---|---|---|
| | | Quantity | | Quality | | | | |
| 1961 | | m. litre | mm | $Cl^-$ | NaCl | T.D.S. | | |
| | | | | mg/litre | | | | |
| — | | | | | | | | |
| 9. July | 14.30 | 10,000 | 60 | 2920 | | 9600 | | |
| 10. July | | | | | | | | |
| 11. July | | | | | | | | |
| 13. July | | | | | | | | |
| 14. July | | | | | | | | |
| 15. July | | | | | | | | |
| 16. July | 14.30 | 10,000 | 60 | 2920 | | 9600 | | |
| 17. July | | | | | | | | |
| 18. July | | | | | | | | |
| 19. July | | | | | | | | |
| 20. July | | | | | | | | |
| 21. July | | | | | | | | |
| 22. July | | | | | | | | |
| 23. July | | | | | | | | |
| 24. July | | | | | | | | |
| 25. July | | 10,000 | 60 | 2920 | | 9600 | | |
| | last irrigation | | | | | | | |
| 8 Sept. | | | | | | | | |

lysimeters in the Negev Institute for Arid Zone Research (see fig. 3, 4 and 5), using the means at our disposal as best as possible.

The percolated water had always a much higher concentration than the water of the preceding irrigation, and was particularly richer in its chlorine content. Table VIII may serve as example for the simple method how the percolation measurements were taken and recorded on stencilled forms during the whole period.

The speed of percolation shows great differences between the various units although they were filled with the same material. Lacking an adequate instrumentation we measured the percolation on the surface and received the following maximal difference: The time of pouring the irrigation water onto the surface was in one case 16 seconds and the last patch of water was percolated beneath the surface after 105 seconds (the 16 seconds included); in the other case the time of the irrigation act (each time by can)

| Rainfall* | | | | Date | Time | Percolated water | | | |
|---|---|---|---|---|---|---|---|---|---|
| Quantity | | Quality | | 1961 | | Quantity | | Quality | |
| ı. litre | mm | Cl⁻ | T.D.S. | | | m. litre | mm | Cl⁻ | T.D.S. |
| | | mg/litre | | | | | | mg/litre | |
| | | | | | | 0 | | | |
| | | | | | | 2,020 | | 5241 | 16,800 |
| | | | | | | 970 | | 4715 | 15,500 |
| | | | | | | 420 | | 4314 | 12,250 |
| | | | | | | 160 | | 4418 | 12,000 |
| | | | | 16. July | 14.00 | 70 | | 5305 | 16,000 |
| | | | | | | 4,000 | | 4775 | 15,000 |
| | | | | | | 600 | | 4988 | 24,000 |
| | | | | | | 14.050 | | 11576 | 17,500 |

* No rainfall occurred during this period but on most of these days heavy dew deposit could be recorded, at sunrise in 100 cm height, and on five days even fog.

was 7 seconds only, and percolation of all water beneath the soil surface was achieved already after 42 seconds.

All analyses of the percolation water and of the soil profiles in the Beersheba experiments confirmed the preliminary results obtained in Bet Dagon: No salt accumulation took place irrespective of the concentration of the sea-water type.

In addition to the very rich material of our own data assembled throughout the years of these experiments we asked Dr. R. MATZ, Head of the well equipped Laboratory for Desalination in the Negev Institute for Arid Zone Research in Beersheba, to investigate this problem, independent of our own experiments, with our sand material and with fresh sea-water. He made a concise study and kindly put the figures at our disposal. After having cleaned the sand he analysed the sea-water before and after percolation and found the following figures (Table IX).

**Table IX.**

Analysis of sea-water before and after a percolation experiment through sand by Dr. R. MATZ (Beersheba):

| | in the sea-water | in the percolated water |
|---|---|---|
| Ca | 407 mg/l | 450 mg/l |
| Mg | 1,492 mg/l | 1,500 mg/l |
| Na (incl. K) | 12,930 mg/l | 12,560 mg/l |
| Cl | 21,968 mg/l | 22,460 mg/l |
| Sulfide | 2,683 mg/l | 2,690 mg/l |
| $H_2CO_3$ | 263 mg/l | 180 mg/l |
| | 39,743 mg/l | 39,840 mg/l |

plus minor elements including trace elements.

**Graphical presentation of "Biological Desalination"**

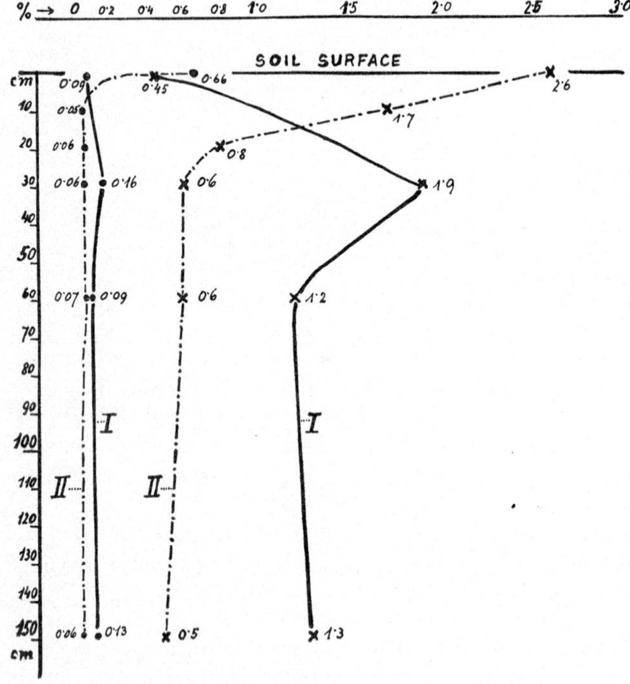


Fig. 6. Biological Desalination in Neot Hakikar.
(8 km South of the Dead Sea) (H. BOYKO, 1965).

Sodium (= ×) and chlorine (= ●) content:

I. in the unirrigated soil profile,

II. in the soil profile, irrigated with saline water through 2½ years and after the second harvest of a dense *Juncus maritimus* crop. (for details see text).

The analyses were kindly made by Dr. (Mrs.) KÖRTESZ, Negev Institute for Arid Zone Research, Beersheba.

This phenomenon of non-accumulation seems to be of particular significance in view of the great amount of salts and particularly of sodium chloride applied to the soil during these irrigation experiments with sea-water.

The respective calculations are based on the analysis of the original sea-water brought regularly in a tank to the spot. The actual amount is considerably higher, because under the hot and arid climate of Beersheba the evaporation process in the tank itself is faster than in the Mediterranean climate of Bet Dagon. Some of the percolation tables show a total amount of diluted solids in the irrigation water (100%) of more than 50,000 mg/l, with 53,000 mg/l as the highest amount. It must be added that the years from 1957—1963 were part of an about 10 years long dry period and rainfall remained in all those years far below the yearly average (about 250 mm in Beersheba).

A field experiment in the Dead Sea region with saline underground water of a T.D.S. of about 300 mg/l with about 1000 mg/l chlorine complemented the total picture. Irrigation applied with this water amounted to 2000—3000 cubic metres per dunam/year, i.e. 2—3 cubic metre per square metre and year. The soil consisted mainly of fine sand (79.2%) of several meters depth with a relatively small percentage of stones (0.6%), coarse sand (8.4%), silt (9.3%) and clay (2.5%) of rather even distribution to 2 metres depth or more. Profiles were taken from irrigated and non-irrigated parts and the results presented in the graph of Fig. 6.

As is to be seen from the graph, neither here could salt accumulation be observed; on the contrary, with the exception of spots with a tiny layer of salt efflorescence on the surface, the soil profile irrigated with saline water for two years contained *less* sodium and chlorine than the profile from the non-irrigated control part of this field, with the same soil structure in both spots down to 2 metres beneath the surface as the deepest layer investigated.

The reason for this is to be seen in the fact that the crop plant of this field, *Juncus maritimus*, accumulates salt in its stalks, and leaves, which had already been harvested twice before the second soil profiles were taken. It seems that this is the first practical test of the new principle of "Biological Desalination" on a field scale (H. Boyko, 1966).

Although irrigated with saline water of a much lower T.D.S. than in the experiments in Bet Dagon and Beersheba, these soil profiles are of particular significance as a field experiment on an area of 10,000 square metres. Further important points to be taken into consideration in comparing these experiments with the former ones are: the much finer soil structure, and consequently the slower percolation, the hotter climate with less rainfall (less than 100 mm yearly average), and finally the nearness of the groundwater. (At the time when the last soil samples were taken the groundwater level was 2.20 m.)

The crop plants grown in this field developed very well and so did other crop plants, grown there by the settlement Neot Hakikar, such as tomatoes, Rhodos grass, sugar melons, date palms and various others. In general, after our experiments in the desert garden of Eilat showed so good results, we distributed seeds and seedlings from 1950 on to the Agricultural Research Station, to the Hebrew University, and to the small groups of pioneers of that time. Settlements along the whole length of Wadi Araba are now using our principles and methods, and the number of crop plants grown there successfully on an economic scale is steadily rising.

Summarizing we can say that the results from all places where these experiments were made justify the conclusion that the danger of salt accumulation is negligible if the general rules formulated in a former article (see p. 145) are followed. The final result of salt accumulation or non-accumulation is, however, not only a function of climate, soil structure, water composition and concentration, but also of the crop plant used for harvesting. It is to be foreseen that mathematical formulas of general validity may emerge from international cooperative work. Such work should be carried out under various conditions along broad climate profiles from the tropics to the subarctics. Only by such a cooperating transnational teamwork, as UNESCO, FAO and the World Academy of Art and Science are aiming at, final results can be expected from research on the actual relation between climate and soil structure on the one hand and permissible salt concentrations and compositions of saline irrigation water on the other. Efforts in this direction are now under way. The Proceedings of the International Symposium on "Irrigation with Highly Saline Water and Sea-water With and Without Desalination" in Rome, 1965, under the auspices of the World Academy of Art and Science and the official Italian authorities and sponsored by UNESCO elucidate these efforts. (WAAS-Series, Vol. IV).

The general rules, however, can already be regarded as confirmed. (see also the article of H. Osvald, Sweden, in this volume and in the WAAS-newsletter, 1966).

## Plant Species and Their Behaviour

*1. Agropyrum junceum* L.

This perennial grass inhabits as a pioneer shifting sand dunes along the Mediterranean coasts and also occurs along the Atlantic coasts of Europe. In Europe its plantgeographical area extends

Fig. 7. Sea-water experiments with *Agropyrum junceum* L. Young plant (half a year old) in the Bet Dagon Series, irrigated with sea-water of North Sea type (20,000—27,000 mg/l T.D.S. with 75% NaCl in it). (see text for details)

from the Atlantic eastwards to the coastal dunes of the North Sea and of the Baltic Sea.

We have, however, to distinguish two distinct ecotypes, differing not only in their geographical area but also in their number of chromosomes: according to the Chromosome-Atlas of DARLINGTON & AMMAL (1945), the boreo-atlantic ecotype has 28 and the mediterranean ecotype 42 chromosomes.

The genus *Agropyrum* is closely related to *Triticum* (wheat) and crossing experiments between certain species of *Agropyrum* with wheat varieties gave successful results. This may be mentioned here because it opens vistas into other lines of research in continuation of our experiments, e.g. in the direction of achieving salt resistant cereals. *Agropyrum junceum* itself is a valuable binder of shifting dunes and has been converted in Israel to a cultivated (irrigated) fodderplant of high nutritive value on very poor and

Fig. 8. Sea-water experiments with *Agropyrum junceum* L.

Young plant (half a year old) in the Bet Dagon Series, irrigated with sea-water of Oceanic type (30,000—40,000 mg/l T.D.S. with 75% NaCl in it).

Its height is considerably shorter (about 40%) than that of the plant in fig. 7; width and other developments are, however, almost the same (this individual is an exception, the others irrigated with this high concentration were much weaker or died).

salty-sandy soils. It has, for instance, successfully been used for the productivization of sterile sandy lands suffering from the influence of salt spray from the sea. Thus a sterile area north of Tel Aviv could be won for production after many trials with other crops had failed. Quantitative plantsociological records carried out by the first author of this paper in summer 1955 revealed the poverty and salinity of these soils.

In spite of heavy irrigation with fresh water in order to desalt the soil, only a very sparse vegetation of halophile weeds could be counted (see Table X).

**Table X.**

Plantsociological square record of a sterile field in Kfar Vitkin (north of Tel Aviv). Recorded area: 400 square metres (20 × 20 m), July 1955.

| Number of of in-dividuals | Species | Abun-dance | Domi-nance | Remark | Indicator of |
|---|---|---|---|---|---|
| 3 | *Cakile maritima* SCOP. | 2 | 1 | succulent leaves | sand and salt spray |
| 1 | *Salsola Kali* L. | 1 | 1 | dead | |
| 17 | *Senecio joppensis* DINSM. | 2 | 1 | succulent | sand and salt spray |
| 1 | *Silene succulenta* FORSK. | 1 | 1 | succulent leaves | sand and salt spray |
| 1 | *Spergularia marginata* (DC) KITTEL | 1 | 1 | succulent leaves | saline soil |
| 2 | *Sporobolus arenarius* (GOUV.) DUV.-JOUV. | 1 | 1 | | sand |
| 6 | *Stellaria media* (L.) VILL. | 2 | 1 | succulent leaves | saline soil |

The specific micro-distribution of these seven species is highly significant and in accordance with the Geo-ecological Law (BOYKO, 1947) its indications are more reliable than a few chemical analyses of the soil. They showed immediately the causes of the failures of the trials with other crops and, as a result, the introduction of *Agropyrum junceum* as crop plant was recommended (BOYKO, BOYKO & TSURIEL, 1957). Soon after the establishment of an *Agropyrum* field the area became a valuable asset to the respective settlement. A chemical analysis taken from a paper by D. TSURIEL (1954) shows the high nutritive value of this plant species.

All these considerations in addition to its natural sand habitats in contrast to the heavy soils which form the natural habitats of *Juncus maritimus* induced us to choose *Agropyrum* as one of the first plant species for our sea-water experiments.

Including the orientation experiments (see BOYKO & BOYKO,

**Table XI.**

Fodder analysis of *Agropyrum junceum*, after TSURIEL (1954).

| | |
|---|---|
| Water | 72.6% |
| Fats | 1.1% |
| Raw cellulose | 8.5% |
| Proteins | 4.4% |
| Ashes | 3.3% |
| Carbohydrates | 10.4% |
| | 100.0% |

1959) the outlay of these experiments was extended in various directions:

a) dense sowing, i.e. about 1 seed per square centimetre and irrigated with sea-water dilution of 11,000 mg/l T.D.S. of which 8,400 mg/l were sodium chloride (February 1958 to January 1960).

b) individual sowing and measuring under controlled irrigation with three types of sea-water and a parallel control experiment with fresh water.

The three sea-water types were:

1) oceanic type with about 30,000—40,000 mg/l T.D.S. with 75% NaCl.

2) North Sea type with about 20,000—27,000 mg/l T.D.S. with 75% NaCl.

3) Caspian Sea type with about 10,000—13,000 mg/l T.D.S. with 75% NaCl.

c) as in b) with the addition of fertilizers.

d) Comparison with experiments carried out by M. P. D. MEIJERING on the dunes of the North Sea island Spiekeroog in Western Germany, at the recommendation of UNESCO, on the basis of our experiments.

We found already in the first experiments that a relatively high concentration can be used to grow *Agropyrum junceum* on sand with great practical and perhaps even economic results. This valuable plant gave a yield of 116.80 g fresh weight and an air dry yield of 75.62 g per square dm. This means a transcalculated theoretical yield of 116 tons fresh weight per hectare and 75 tons air dry per hectare. This surprisingly good economic outlook is weakened by the high amount of irrigation water used in this first orientation experiment. It was, per 1/6 square metre, i.e. the surface of each frame, 30 l a week with a T.D.S. of 11,000-13,000 mg/l.

Cutting was carried out only once for the purpose of this weighing and of determining the osmotic value of the cell sap. According to the experience with grasses in general, it can be assumed that more frequent cutting may still raise the yield of this valuable fodder plant. On the other hand, the amount of saline irrigation water was intentionally kept exceedingly high and further experiments will have to reduce this amount in order to find out for each region separately the right biological and economic balance between the water factor on the one side and the yield on the other.

There were about 1600 single individuals in the circular frame and not one of them died during the whole time. Under these density conditions, however, only four individual plants reached a flowering and fruiting stage in the second year. In contrast to the later experiments with single plants, the whole plant mass consisted of rather thin halms and leaves, but the high total yield was interesting if viewed from the angle of practical applicability as a very nutritive perennial fodder plant under irrigation with highly saline or sea-water on shifting dunes.

Instructive as these experiments with densely grown plants were, they did not permit the observation and measurements of single plant individuals. For this purpose, in the winter 1958/59 the following lay-out was made in Bet Dagon: Ten 90 cm high frames (bottomless asphalt barrels) with a diameter of 46 cm were put on sand and filled with sand, and on Feb. 15, 1959, 10 sets of two seeds were laid in each of them in 2 cm depth. Germination was left to rainfall as in nature and supported by fresh water irrigation. 175 seeds germinated between the 8th and 11th of March 1959. From the 25th of March on the plants received only sea-water in three types, as to be seen from Table XIII and Table XIV, except for the individuals in the control units irrigated with fresh water.

The development of the single individuals was very different. Two plants of particularly good development in spite of the high salinity of the irrigation water are to be seen in fig. 7 and 8, one, fig. 7, irrigated with sea-water of the North Sea Type with 20—27,000 mg/l T.D.S., the other, fig. 8, even with sea-water of oceanic concentration 30,000—40,000 mg/l T.D.S. The latter plant individual was, however, an exception; the others, irrigated with these extremely high concentrations were much weaker or died early. The two plants are both six months old and promised good seed production in the next vegetation period. Unfortunately, the experiments in Bet Dagon had to be abandoned and could be observed only until January 1960.

Here we want only to show by means of these pictures that even irrigation with oceanic concentration promises success under otherwise favourable conditions, if by selection through several years strains are bred from mother plants like that shown in Fig. 7 and 8.

The experiments were, however, continued at the Negev Institute for Arid Zone Research in Beersheba from July 1959 to October 1963.

Since this book is mainly concerned with principles and methods, we shall in the following demonstrate by means of the three Tables XII, XIII and XIV the methods of the lay-out and of our measurements.

Altogether about 220 individual plants of *Agropyrum junceum* were experimented with, with 120 of them subjected to periodic measurements of their development. The subjects of these measurements are to be seen from Table XII, showing as an example the development of one single individual plant (Plant No. 1 in Unit (frame) Nr. 3) under irrigation with sea-water of the North Sea type in the arid climate of Beersheba, from July 1960 to October 1963.

The originally recorded amount of 20,000 mg/l T.D.S. is the possi-

**Table XII.**

Sea-water experiments with *Agropyrum junceum* L. in Beersheba: Development of one individual plant as example (irrigated with 20,000—27,000 mg/l T.D.S. until June 1961)

| Nr. of Unit | Nr. of Plant | Date | number of green halms | number of green leaves | average height in cm | maximum height in cm | basal area in $cm^2$ | Remark |
|---|---|---|---|---|---|---|---|---|
| 3 | 1 | 28 July 1960 | 1 | — | — | 14 | | |
| | | 14 Sept. 1960 | 2 | 5 | 10.5 | | 3 | |
| | | 28 Dec. 1960 | 2 | 15 | 14.3 | | | |
| | | 9 Apr. 1961 | 2 | 13 | 16.0 | | | |
| | | 15 May 1961 | 2 | 9 | 17 | 29 | | |
| | | 16 May 1962 | 11 | 18 | ? | 38 | 72 | spikes flowering, the upper one or two leaves of each halm and the whole halms and spikes are green |
| | | 30 Apr. 1963 | 16 | 22 | ? | 26 | | |
| | | Oct. 1963 | 2 | — | 6 | 8 | | |

ble minimum of this specific dilution, because not only in this but also in all other cases the given concentration fluctuated between the respective minimum and an about 30% higher amount for the following reason:

Sea-water was brought every 2—3 weeks from the shallow East Mediterranean seashore south of Tel-Aviv and filled into a tin-tank with an opening of about 10 square decimetres. Evaporation in the sunbeaten horizontally laying cylindric tank of about one cubic metre volume was considerable and an analysis of this sea-water brought in with a concentration of about 40,000 mg/l T.S.C. after two weeks showed a concentration of 53,000 mg/l T.S.C., i.e. about 30% more. This fluctuation applies, as mentioned before, to all our applications of sea-water and its dilutions for irrigation. Since the main purpose of this article is to stimulate similar experiments in other countries, it seems necessary to draw the attention as much to the weakness of the experiments as to its merits. Such weaknesses can easily be avoided with an appropriate budget and instrumentation.

The natural vegetation cycle of *Agropyrum junceum* on the coastal dunes of Israel is as follows: Germination occurs soon after a few effective rainfalls during September to February, seemingly with a relatively broad amplitude of temperature. Flowering depends on the temperature-climate in spring and usually starts

**Table XIII.**

Sea-water experiments with *Agropyrum junceum* in Beersheba:
Minimum and Maximum data of measurements of plants under different Sea-water concentrations

| Date | Data of | irrigated with fresh water | irr. with 10,000-13,500 mg/l (Caspian Sea type) | irr. with 20,000-27,000 mg/l (North Sea type) | irr. with 30,000-40,000 mg/l (Oceanic concentr.) | Remark |
|---|---|---|---|---|---|---|
| 28. July 1960 | a) number of green shoots per plant | 1-1 | 1-1 | 1-1 | 1-1 | |
| | b. number of green leaves per plant | 0-2 | 0-0 | 0-0 | 0-0 | |
| | c. height per plant | 7-21 cm | 8,5-18 cm | 8-22 cm | 11-19,5 | |
| 14. Sept. 1960 | a } see above | 1-2 | 1-1 | 1-2 | 1-1 | |
| | b } see above | 2-9 | 2-10 | 2-8 | 1-1 | |
| | c } see above | 6.5-24.5 cm | 8-21 cm | 8-23 cm | 7.5-20 cm | |
| 15. Dec. 1960 | a } see above | 1-11 | 1-4 | 1-6 | 1-2 | |
| | b } see above | 6-62 | 4-16 | 2-26 | 1-7 | |
| | c } see above | 9-15 cm | 9-27 cm | 7-17 cm | 5-11 cm | |
| 9. Apr. 1961 | a } see above | 3-21 | 1-5 | 1-7 | 1-2 | |
| | b } see above | 10-77 | 7-29 | 3-36 | 1-9 | |
| | c } see above | 8-19,5 cm | 7-25 cm | 10-20,5 cm | 4-12 cm | |
| 15. May 1961 | a } see above | 3-14 | 1-5 | 1-7 | 1-1 | * the highest ones are already flowering |
| | b } see above | 12-48 | 4-22 | 3-24 | 3-4 | |
| | c } see above | 14-51 cm* | 12-28 cm | 5-29 cm | 11-17 cm | |
| 16. May 1962 | a } see above | 3-21 | 3-12 | 2-10 | seemingly dead | ** the higher ones are already flowering and beginning to fruit |
| | b } see above | 5-54 | 9-27 | 4-25 | | |
| | c } see above | 15-108 cm** | 25-65 cm | 20-56 cm | | |
| | d basal area | 8-192 cm² | 10-120 cm² | 2-105 cm² | — | |
| 30. Apr. 1963 | a } see above | 3-7 | 2-8 | 2-13 | — | |
| | b } see above | 4-11 | 2-21 | 4-22 | | |
| | c } see above | 8-13 cm | 24-35 cm | 11-28 cm | | |

at the end of May, fruiting from June to September. After fruiting all parts above the soil surface are drying up in the following chronological order: First the leaves then the spikelets and finally the halms from the top downwards. In autumn new shoots appear in growing numbers from year to year. This perennial grass may live on its natural habitat through several tens of years, with old plants building tufts up to one metre in diameter or more.

A similar yearly cycle was observed on the plants of this experiment, and the three different vegetation cycles are pointed out in Table XII by horizontal lines.

Table XIII is based on such measurements, but of all plant individuals and compares various criteria of their development according to the different concentrations of irrigation water (types of sea-water). The plants in the control frames, irrigated with fresh water, showed in all criteria higher figures and — this seems to be more important — *quicker development*. One plant, for instance, was already shooting into flowering and fruiting spikes, which means also almost doubling its height, whereas in the 60 plant individuals in the frames No. 1—6, irrigated with sea-water of the three different types indicated in the Table, no sign of such development was yet to be seen at that date (Mid May 1961 and 1962).

*The Dying Experiment*

As from July 1961, irrigation had to be stopped and the plants had to rely solely on dew and the scanty rainfall of the following winters.

The consecutive development during the vegetation period in 1962 can be seen from Table XIV. What we had expected in the earlier experiments in 1957—1959 could now be proved in this "Dying Experiment". In 1959/60 already and still more in 1962 it was strikingly visible that the plants irrigated with sea-water dilutions of Caspian and North Sea type developed more satisfactorily than the plants irrigated with fresh water on the one hand or with oceanic concentration on the other hand. The latter plant individuals were very weak from the start and died in summer or autumn 1962. Those irrigated with fresh water partly survived the very dry summer of 1962 but died during the summer of 1963, whereas those irrigated with sea-water of the Caspian Sea type and of the North Sea type survived both of these extremely dry periods in a much more vigourous condition.

Table XIV shows that the vitality*, presenting itself as drought resistance, rose with the salinity of the irrigation water. The upper limit with *Agropyrum* seems to be near oceanic concentration. In Bet Dagon, with its more humid and therefore more favourable climate for this species, one plant individual reached good average dimensions even under irrigation with sea-water of oceanic concentration. This lets us expect good results of further experiments, if

* I am greatly obliged to Prof. Pierre Dansereau, the American plantecologist, for his many constructive remarks to my manuscripts. He proposes the word "Vigour" instead of "Vitality" as the more appropriate one in this connection. In German, which is my mothertongue, I would have said "Lebenskraft". I am convinced that Dansereau is right, but, not feeling myself competent to judge this point, I leave open for discussion the final decision on this wording (H. Boyko).

**Table XIV.**

*Agropyrum junceum:* "Dying experiment" after having stopped irrigation in July 1961.

| irrigated until June 1961 with | Total percentage of green plants or parts | 16. May 1962 | 18. June | 18. July | 14. Aug. | 14. Oct. | 30. April 1963 | October 1963 living | Remark |
|---|---|---|---|---|---|---|---|---|---|
| Fresh water | a) plants with green leaves | 100 | 0 | 0 | 0 | 0 | 20* | 0 | * only two of the ten plants of 16. May, 1962 revived in winter 1962/63 |
| | b) % of stems | 100 | 100 | 100 | 5 | 0 | 100 | | |
| | c) % of spike axis | 100 | 60 | 55 | 5 | 0 | 100 | | |
| | d) % of spikelets | 100 (fr) | 20 | 0 | 0 | 0 | 100 (fl) | | |
| 10,000—13,000 mg/l T.D.S. (Caspian Sea-Type) | a) see above | 100 | 30 | 0 | 0 | 0 | 71* | | * 5 of 7 revived |
| | b) | 100 | 100 | 80 | 55 | 0 | 100 | | |
| | c) | 100 | 80 | 15 | 10 | 0 | 100 | | |
| | d) | 100 (fr) | 20 | 0 | 0 | 0 | 100 (fl)** | new shoots | ** re-adapted to natural cycle |
| 20,000—27,000 mg/l T.D.S. (North Sea-Type) | a) see above | 100 | 70 | 0 | 0 | 0 | 83* | | * 5 of 6 revived |
| | b) | 100 | 100 | 100 | 80 | 0 | 100 | | |
| | c) | 100 | 100 | 35 | 25 | 0 | 100 | | |
| | d) | 100(bud) | 95 | 0 | 0 | 0 | 100 (fl)** | new shoots | ** re-adapted to natural cycle |
| 30,000—40,000 mg/l T.D.S. (Oceanic to Red Sea concentration | did not revive in 1962 | — | — | — | — | — | — | 0 | |

combined with selective breeding, and indicates a highly promising applicability for wide areas.

*2. Calotropis procera* (WILLD.) R.BR.

This succulent species of the Asclepiadaceae family reaches a height of 2 metres and more, and grows either like a tree with a naked stem and densely foliaged branch-tops, or it divaricates immediately above or beneath the soil-surface. In the latter case it is more shrub-like. The large fleshy leaves are sessile on the stem, with a density counted as 8 to 20 per 10 cm stem length. The number of leaves or the percentage of the naked part of the stem corresponds with the amount of water physiologically available, which depends for a given amount mainly on bio-meteorological or phreatic factors on the one hand and on the salt concentration on the other hand. The plant keeps its general water and salt balance by producing or shedding its leaves.

Its geographical distribution stretches through the saharo-indian region from Algier to Sind (Northwest-India) and southward into Sudan. In Israel its main occurrence is in the parts below sea level around the Dead Sea. It throve, however, well with (saline) or without irrigation, when introduced by the authors into the Mediterranean climate south of Tel Aviv (Bet Dagon). Here it was grown from seeds in the nursery and transplanted when about 15 cm high and densely leaved. Contrary to its habitus with its usually naked stems in its original habitat, the plants developed already after one year a rich foliage covering the whole many-branched bushes. The rich fruiting there and the yield of seeds from these few shrub-like individuals indicate a theoretically possible economic yield of this valuable vegetable silk fibre, far stronger and more beautiful than, for instance, cotton. Lacking the possibility for a field experiment, we measured a few fruits and their seeds and seed hairs. The plant individual from which they were taken was the best developed of the four specimen introduced there in 1957 by the present authors. In 1959, two years old, it had a height of 1.80 m and a width of 1.20 m. Apart from the natural rainfall in winter, it had as irrigation water the East Mediterranean sea-water, filtering through an earth wall of about one metre thickness. The figures taken from four fruits are presented in Table XV.

Although not or not yet introduced as a crop plant on a bigger scale, the plant seems to be of high potential economic interest for desert regions because of its manysided usefulness. Its stem contains a very strong fiber durable under water and therefore useful for manufacturing fishing nets, halters and similar products. Old stems are surrounded by a relatively thick but much interrupted layer of cork. The milky juice is the source of a guttapercha-like product and is also used to remove the hairs from hides. The various

**Table XV.**

Data on fruits of *Calotropis procera* (Willd.) R.Br.
irrigated with saline water

| Fruit Nr.: | 1 | 2 | 3 | 4 |
|---|---|---|---|---|
| Length of fruit | 9.0 cm | 8.5 cm | — | 9.3 cm |
| Total weight, fresh | 9.83 g | — | — | 10.2 g |
| Number of seeds | 413 | — | — | 458 |
| Weight of seeds | 5.92 g | — | — | 6.1 g |
| Weight of seed-hairs | 2.5 g | — | — | 2.6 g |
| Number of seed-hairs | about 500 | about 400 | — | — |
| Length of seed-hairs | 30-50 mm | 41-47 mm | 33-50 mm | 32-52 mm |
| Length average | 42.2 mm | 44.3 mm | 44.4 mm | 45.1 mm |

* The measurements of samples No. 1-3 were made by our coworker Mrs. Adiva Ilan, of No. 4, the biggest fruit of all plants, by H. Boyko.

toxic components of its juice led to numerous medical uses among the native tribes in the large distribution area of this plant species. Another kind of utilization is practized in West Africa, where a native beer, Merissa, is produced from its succulent leaves. Finally, most expensive Kefiyahs (Arab head scarfs) are woven from its snow-white silky seedhairs in Yemen. In the Egyptian Sudan the first industrial plant has been erected for manufacturing textiles from these 4—5 cm long seed-hairs.

In his ecological investigations the first author found this species mainly restricted to localities of a hot climate, where either a high drought resistance or high salt resistance or both is necessary for survival. The stems are mainly naked in these surroundings and physiological investigations are recommended to find out if salt accumulation up to a critical content alone is responsible for leaf-shedding, or rather the lack of sufficient water or both these factors. The plant has a divaricated deep going root system, and since underground water in reachable depth, mostly less than five metres, is indicated on all places where the author found this plant species in Israel as well as in Arab countries and in India, it seems almost certain that critical salt accumulation in the succulent leaves is the main cause of leaf-shedding. Their succulence alone allows already a relatively high potential of salt intake. This too is an additional reason why we should pay more attention to the cultivation of *Calotropis* from the point of view also of the biological desalination possibilities for saline soils and waters.

These numerous and diverse reasons led the authors to recommend, since 1949, agricultural experiments for the introduction of *Calotropis* into the economy of Israel, particularly for the desert parts of the Negev and also for other desert regions in general.

**Table XVI.**

Sea-water Irrigation of *Calotropis procera* R.Br.
Selected measurements on one plant individual during the years 1961-1963 as examples:
Plant Nr. 30, irrigated with sea-water of the Caspian Sea-type (10,000-13,000 mg/l T.D.S.) until July 1961.

| Date of measurements | Total number of living stems | Height of stems in cm | Length of naked stems in cm | Length of leaved stems in cm | Length of flowering top part in cm | Number of leaves | Remarks |
|---|---|---|---|---|---|---|---|
| 1. Aug. 1960 | 1 | 7.5 | — | 7.5 | — | 11 | transplanted seedling |
| 15.May 1961 | 3 | 39 | 15 | 24 | ? | 40 | all stems were cut in March 1961 as safeguard against further damage by frost; last irrigation in July 1961 |
| 16.May 1962 | 5 | a) 55 | 15 | 40 | 26 | 46 | stem c) had 3 fruits on 18. July 1962. The fruits were 6.5-8 cm long and 5-6 cm broad. |
| | | b) 46 | 10 | 36 | 28 | 40 | |
| | | c) 108 | 88 | 20 | 20 | 24 | |
| | | d) 36 | 12 | 24 | 20 | 32 | |
| | | e) 85 | 71 | 14 | 14 | 20 | |
| | | | | | | 162 | |
| 30.Apr. 1963 | 4 | a) 67 | — | — | — | — | stems a), b) and d) were dead at this date. stem c) had 26 buds and flowers. stem e) 22 and stem f) 15 buds and flowers. stem g) is a young shoot with leaves only. |
| | | b) 115 | — | — | — | — | |
| | | c) 108 | 104 | 4 | 2 | 7 | |
| | | d) 72 | — | — | — | — | |
| | | e) 90 | 83 | 7 | 4 | 5 | |
| | | f) 31 | 25 | 6 | 1 | 8 | |
| | | g) 15 | 13 | 2 | — | 4 | |
| | | | | | | 24 | |
| 1.Oct. 1963 | 3 | a) — | — | — | — | — | stems a) and b) not more visible; c), d), e) and f) were dead; stem g) has several clusters of flowers and buds; stem h) and i) are young shoots with leaves only |
| | | b) — | — | — | — | — | |
| | | c) ? | — | — | — | — | |
| | | d) ? | — | — | — | — | |
| | | e) ? | — | — | — | — | |
| | | f) ? | — | — | — | — | |
| | | g) 95 | 85 | 10 | 10 | 12 | |
| | | h) 35 | 27 | 8 | — | 11 | |
| | | i) 40 | 30 | 10 | — | 14 | |
| | | | | | | 37 | |

In our experiments with this plant species, carried out in Beersheba, it was necessary to restrict the number of units (frames) to a minimum. The lay-out of the experiment there stressed therefore the trials with sea-water of the Caspian-Sea type, i.e. with a concentration of 10,000—13,000 mg/l T.D.S., since this is the maximum to be expected for irrigation purposes, in the Negev as well as in most other deserts, for larger areas.

Fig. 9. Sea-water experiments with *Calotropis procera* R. Br.
Four units (bottomless asphalt barrels filled with sand and put on sand) of the Beersheba series of experiments, with young *Calotropis* plants, irrigated with sea-water of the Caspian Sea type (10,000-13,000 mg/l T.D.S.). For details see text. (See also Fig. 11 with a two years old *Calotropis* in the background).

Measurements on plant individuals had to be adapted to the specific morphological features of *Calotropis*.

Six frames (bottomless asphalt barrels) (Units Nr. 30—35) were put on sand and filled with sand before planting one single seedling in each of them. A depth total of 130—140 cm sand was thus achieved above the natural loess layer and it could be assumed that the root system had sufficient space at its disposal. Apart from these six units, five lysimeters (Nr. VII—XI) were also used.

The methods of our measurements are to be seen from Table XVI. The development of one single plant individual can be followed there from the time of planting the seedling (July 1960) until October 1963.

**Table XVII.**

Sea-water experiment with *Calotropis procera* R.Br.: Pattern of development*

| | Plant No. 30 irrig. with 10,000-13,000 mg/litre T.D.S. (Caspian Sea Type) | | | | Plant No. 31 irrig. with 10,000-13,000 mg/litre T.D.S. | | | |
|---|---|---|---|---|---|---|---|---|
| Date of measurements | Number of living shoots (dead ones in brackets) | Total number of leaves | Maxim. height in cm | Phenology | Shoots | Leaves | Maxim. height in cm | Phenology |
| 1. Aug. 1960 | 1 | 11 | 7.5 | | 4 | 38 | | |
| 14. Sept. 1960 | 5 | 37 | | | 2(2) | 42 | | |
| 21. Nov. 1960 | 5(0) | 57 | | | 2(2) | 58 | | |
| 4. Jan. 1961 | 5(1) | 66 | | | 2(2) | 58 | | |
| 22. Jan. 1961 | 5(1) | 61 | | | 2(2) | 57 | | |
| 1. March 1961 | 2(1) | 32 | | | 2(2) | 37 | | |
| All plants regenerated after having been cut back to about 5 cm above the soil-surface in March 1961, as they were visibly suffering from a very cold spell. | | | | | | | | |
| 15. May 1961 | 3 | 40 | | | 2 | 68 | 17.5 | |
| 2. Aug. 1961*** | 3 | 70 | | | 6 | 144 | | |
| 16. May 1962 | 5 | 162 | 108 | fl. | 4 | 55 | 120 | fl. |
| 30. Apr. 1963 | 4(3) | 24 | 110 | bud | 1(3) | 12 | 35 | bud. |
| 1. Oct. 1963 | 3(4) | 37 | 95 | fl. | 6 | 47 | 82 | fl. |

Table XVII gives a similar picture for five individuals, but in a more constricted form.

No far-reaching conclusion could be drawn from the "Dying Experiment" with this species for the simple reason that after the destruction of the shoots irrigated with higher solution than 10,000 mg/l T.D.S., only one plant individual irrigated with fresh water in the lysimeters was at our disposal for comparison. Another negative factor for comparison lies in the fact that the lysimeters were filled with coarse building sand of which all finer material has been sieved away*, and not with the same material of the much

* Most of the sand particles had a diameter of 0.4—0.7 mm.

| Plant No. 32 irr. with 10,000—13,000 mg/l T.D.S. | | | | Plant No. 33 irr. with 10,000—13,000 mg/l T.D.S. | | | | Plant No. 34 irr. with 10,000—13,000 mg/l T.D.S. | | | | Plant No. 35 irr. with 20,000—27,000 mg/l T.D.S. (North Sea Type) | | | |
|---|---|---|---|---|---|---|---|---|---|---|---|---|---|---|---|
| Shoots | Leaves | Max. Hght. in cm | Pheno-logy | Shoots | Leaves | Max. Hght. in cm | Pheno-logy | Shoots | Leaves | Max. Hght. in cm | Phenology | Shoots | Leaves | Max Hght. | Phenology |
| 4(3) | 36 | | | 4(1) | 51 | | | 3 | 36 | | | 4 | 58 | | |
| 3(3) | 42 | | | 3(3) | 55 | | | 4(3) | 61 | | | 16 (2) | 192 | | |
| 3(3) | 61 | | | 3(3) | 84 | | | 4(3) | 88 | | | 12 (3) | 221 | | |
| 3(3) | 64 | | | 3(3) | 75 | | | 4(3) | 84 | | | 8 (6) | 176 | | |
| 3(3) | 65 | | | 3(3) | 75 | | | 4(3) | 83 | | | 8 (6) | 181 | | |
| 3(3) | 45 | | | 3(3) | 38 | | | 4(3) | 54 | | | 8 (5) | 111 | | |
| | | | | | | | | | | | | | | | |
| 2 | 32 | | | 5 | 117 | | fl. | 3 | 64 | | fl. | 3 (4) | 12 | 1 | |
| ) | 93 | | | 17 | 222 | | fl. | 13 | 174 | | fl. | 10 | — | | |
| 3 | 63 | 88 | fl. | 6 | 152 | 132 | fl. | 4 | ? | 124 | fl. | —** | — | | |
| 4(2) | 11 | 58 | bud. fl. | 2(5) | 18 | 90 | bud. fl. | 2 | 23 | 56 | bud. | — | — | | |
| 2 | 22 | 65 | fl. | 4 | 56 | 48 | fl. | 5(1) | 49 | 58 | bud. | — | — | | |

* Condensed Table based on measurements of six units (frames) each containing one plant individual, as shown in the example of Table XVI.
** All shoots were torn out by an unknown trespassing person.
*** Irrigation was discontinued in July 1961.

finer sand (mainly 0.2—0.1 mm) as the other frames. Percolation was therefore much quicker and nutrition possibilities much smaller. This led to complete exsiccation much earlier than in the frames Nr. 30—34. In spite of these extremely unfavourable conditions, the plants lived even in the lysimeters unto the end of the experiment in October 1963, if only with 2 very small shoots of 24 and 20 cm height respectively.

Since on the one hand the scale of our experiment with this plant species is too small for final conclusions and, on the other hand, it seems to be of far-reaching importance, to achieve such conclusions, big scale experiments under favourable administrative conditions are strongly recommended. The present authors are of the opinion that this plant may serve as an important raw material for the industrialization of various developing countries with desert areas.

Fig. 10. Sea-water experiments in Bet Dagon with *Juncus maritimus* LAM. No significant difference is to be seen between plants irrigated with fresh water (left) and plants irrigated with sea-water of Caspian Sea type (10,000-13,000 mg/l T.D.S. with 75% NaCl in it) (right).

*3. Juncus maritimus* LAM.

*Juncus maritimus* is well known as a highly salt tolerant plant species, growing mainly on heavy alluvial soils of salty swamps. In Israel there are at least two distinct ecotypes, one growing in the Mediterranean and northern parts, the other in oases of the southern desert, belonging to the variety *arabicus* ASCH. et BUCH. Since also the morphological features are somewhat different, the latter

Fig. 11. Sea-water experiment in Bet Dagon with *Juncus maritimus* LAM.
left: irrigated with sea-water of North Sea type (20,000-27,000 mg/l T.D.S. with 75% NaCl in it).
middle and right: irrigated with sea-water of Oceanic type (30,000-40,000 mg/l T.D.S. with 75% NaCl in it).
In the middle the plants are recuperating after having been cut three weeks before for measuring the osmotic value of their cellsap.
Technical Assistant (Gardener) ARNOLD DOUWES irrigating just the series irrigated with oceanic type. (Note the careful dispersing of water with the fingers in order to prevent holes in the sand and an exposure of roots).

is frequently described under the separate species name *Juncus arabicus* ASCH. et BUCH.

Our own ecological observations date back to more than thirty years on both ecotypes, and include geographical outposts of the atlantic mediterranean type like that on its eastern-most habitat in a brackish swamp near Danzig, or detailed investigation of old plants growing on small rock-islands (Scogliae) in the northern parts of the Adriatic Sea. There, growing in crevices near the sea level,

**Table XVIII.**

Sea-water experiment with *Juncus maritimus:*
Pattern of development during the irrigation with sea-water in Beersheba (June 1960 June 1961)
and during the "Dying experiment" until 1. October 1963*.

| No. of Unit | No. of Plant individual | Sea-water type | Average height of the 10 highest stems (leaves) in cm | | | Total number of leaves living and dead ones (green and yellow) | | |
|---|---|---|---|---|---|---|---|---|
| | | | | | | a | b | c |
| | | | 30. June 1960 | 30. June 1961 | 1. Oct. 1963 | 30. June 1960 | 30. June 1961 | 1. Oct. 1963 |
| 16 | 1 | Fresh water | seedling 6-10 cm | 54 | 70 | 4 | 107 | 226 |
| 23 | 1 | | " | 66 | 93 | 3 | 37 | 192 |
| 9 | 4 | Caspian Sea type 10,000—13,000 mg/l | " | 28 | 66 | 4 | 147 | 185 |
| 18 | 1 | T.D.S. | " | 38 | 69 | 4 | 33 | 154 |
| 11 | 5 | North Sea type 20,000—27,000 mg/l | " | 26 | 58 | 4 | 64 | 105 |
| 20 | 1 | T.D.S. | " | 24 | 54 | 4 | 11 | 304 |
| 14 | 2 | Oceanic to Red Sea type 30,000—40,000 mg/l | " | 13 | 41 | 4 | 13 | 120 |
| 22 | 2 | T.D.S. | " | 21 | 47 | 4 | 10 | 61 |

* This condensed table shows measurements of one plant individual out of five each from different units (5 plants were planted in each unit), representing all four concentration series, with two plant individuals for each concentration series.

they are frequently flooded by higher waves but soon again washed clean by the next of the all year round rainfalls.

Similarly, our studies included extensive investigations of this species in various oases in the deserts of South West Asia.

The plants we are experimenting with were grown by us from seeds taken from a specific strain, which showed particularly promising economic properties as fibre plant. The technological tests and experiments will be dealt with elsewhere. Here we are particularly concerned with the behaviour of this plant under sea-water irrigation.

After various germination experiments carried out in 1949, the first cultivation of this plant with brackish water was carried out in the desert garden of Eilat in January 1950. One of these individuals has, after 15 years, reached a size of 1.80 m height with a basal

| number of green leaves | | | | number of fruiting stems | new shoots (below 10 cm) | basal area in cm² | basal area in cm² | Remark |
|---|---|---|---|---|---|---|---|---|
| 30. June 1961 | in % of b | 1. Oct. 1963 | in % of c | 1. Oct. 1963 | 1. Oct. 1963 | 15. May 1961 | 1. Oct. 1963 | |
| 106 | 99.1 | 76 | 33.6 | 14 | 0 | 60 | 156 | |
| 36 | 97.3 | 64 | 33.3 | 22 | 2 | 96 | 117 | |
| 147 | 100 | 63 | 34.1 | 8 | 3 | 48 | 140 | |
| 33 | 100 | 66 | 42.3 | 11 | 4 | 36 | 110 | |
| 64 | 100 | 15 | 14.3 | 6 | 1 | 42 | 63 | |
| 11 | 100 | 140 | 46.0 | 14 | 12 | 24 | 224 | |
| 11 | 84.6 | 33 | 27.5 | 4 | 2 | 11 | 64 | |
| 9 | 90.0 | 19 | 31.2 | 2 | 1 | 15 | 49 | |

cover of about 1½ square metres (E. Boyko & H. Boyko, 1966).

Successful field experiments with *Juncus* were made since 1953 on various places apart from Eilat, in Jotvata, S'dom, Holon, and in 1957 we started with our experiment to irrigate this plant with sea-water, first in Bet Dagon and from spring 1960 on in Beersheba.

In Beersheba: in the Negev Institute for Arid Zone Research, we used for *Juncus maritimus* 16 units (bottomless asphalt barrels) with 5 plants in each and, in addition to these frames, we constructed lysimeters as to be seen from Fig. 3, 4, and 5 and used 6 of them for our experiments with this species. Some of these lysimeters were filled with coarse sand, and at the bottom with gravel. We could thus compare the behaviour of the plant-individuals with that of the plants in frames filled with the much finer sand of the other units. Percolation speed in the lysimeters was additionally raised by catching the percolated irrigation water in empty plastic containers standing 60 cm below the bottom of the barrels. Soon this comparison made it clear that too quick a percolation resulted in a slower growth on the one hand and in less resistance against the climatic drought on the other. But even under these unfavourable conditions the same vitality pattern in relation to salt concentration of the irrigation water as in the measurements of all plant-individuals in the ordinary frames was clearly to be observed.

In the condensed Table XVIII some significant data and measurements were chosen out of the great amount of figures filling hundreds of pages in the note books. The Table may, however, suffice to show how the influence of sea-water on the plants was measured and, with the means at disposal, their individual development was followed.

Measurements were taken on each individual plant from June 1960 until 1. Oct. 1963, first every month, later every 2—3 months. The following features were measured:

1) Maximum height (average height of the 10 highest stems or leaves respectively);
2) Total number of leaves;
3) Number of green (living) leaves;
4) Number of fruiting stems;
5) Number of new shoots (height less than 10 cm);
6) Basal area.

The percolation studies and the connected chemical analyses were made, as with all other species, in more or less regular intervals. During the whole time, July 1957—Oct. 1963, daily notes of all meteorological features, taken by ourselves or supplied by the Meteorological Service, accompanied the experiments, from the first day to the last one on each place.

Although the scale of these experiments is relatively small, they allow conclusions of practical economic importance. They are certainly indicative enough to promise important information by the continuation of this new line of research on a broad international scale. It is for instance highly interesting that the most vigorous plant of all (Nr. 20/1) developed from a seed taken from a mother-plant, which itself had already been treated with saline irrigation. (The same could be observed with *Agropyrum junceum.*) Height decreased also here with rising salinity, but again Nr. 20/1 indicates that horizontal expansion or growth is, if at all, not much influenced by it, at least not until a certain degree of salinity is reached. In the case of *Juncus maritimus* and the sand type used, this threshold seems to be above 20,000 mg/l T.D.S. and below 30,000 mg/l T.D.S.

The same result in this respect is to be seen also from fig. 10 taken in the earlier experiments in Bet Dagon. The plants irrigated with sea-water of about 11,000 mg/l T.D.S. (Caspian Sea type) are very similar in size, including their height, to those irrigated with fresh water. The plants irrigated with oceanic concentration (fig. 11) grew much slower and needed about 10 more months to achieve the same height as the former.

For the table we have intentionally chosen the strongest plants of each unit, because, in these first experiments, they alone show the potential of large scale experiments and their application. It may even be more correct to say that they show the *minimum* of the potential since we know how far yields of wild plant species can be raised by converting them into cultivated plants with the use of modern scientific and technical methods.

In this connection it is also necessary to draw the attention to the generally higher vitality (vigour) of the plants irrigated with sea-water than of those irrigated with fresh water. All plants of *Juncus maritimus* survived the second dry period (without any irrigation or drop of rainfall) of about 8 months, but those irrigated with sea-water of the Caspian Sea type and of the North Sea type had at the end of the "Dying Experiment", in October 1963, a significantly higher percentage of green leaves and young shoots than those formerly irrigated with fresh water or with sea-water of oceanic and Red Sea type.

The plants of all four series of irrigation, from fresh water irrigation to that with sea-water of the Red Sea type, revealed the extreme drought resistance of this species. The designation of it as a hygrophyte or a halo-hydrophyte seems therefore highly misleading.

It was also with this plant species that we could further our knowledge with regard to the new principle of "Biological Desalination". The graph of fig. 6 and the text to it explains better than many Tables the desalination process in the soil by cropping salt-accumulating plants like this species taken as the first practical example.

*4. Juncus punctorius* L.

Fruiting plants were collected by the second author for our Herbarium in 1500 m altitude near the monastery of St. Katharina on Mount Sinai in November 1956 in a running watercourse coming from melting snow-fields above.

From this habitat it can be learned that this species is frost resistant as well as heat and drought resistant. The experiments showed that in addition to this it is also highly salt resistant, namely to the same degree as *Juncus maritimus*.

The seeds of the collected plants were used for germination and cultivation experiments and 50 individuals were planted in the same pattern as *Juncus maritimus* in Beersheba: Five individual plants each of the 90 cm high bottomless asphalt barrels used thus

**Table XIX.**

Sea-water experiment with *Juncus punctorius* L.
Pattern of development during the irrigation with sea-water in Beersheba (June 1960-June-1961) and during the "Dying Experiment" until 1. October 1963 with two plant individuals as examples.

| No. of Unit | No. of Plant indi-vidual | Sea-water type | Average height of the 10 highest stems (leaves) in cm | | | Total number of leaves living and dead ones (green and yellow) | | |
|---|---|---|---|---|---|---|---|---|
| | | | | | | a | b | c |
| | | | 30. June 1960 | 30. June 1961 | 1. Oct. 1963 | 30. June 1960 | 30. June 1961 | 1. Oct. 1963 |
| 29 | 1 | Fresh water | seedling 6-10 cm | 50 | 80 | seedling 3-4 cm | 66 | 133 |
| 28 | 4 | Caspian Sea type 10,000—13,000 mg/l T.D.S. | ,, | 30 | 58 | ,, | 72 | 180 |

as frames. These frames were, like the others, put on 30 cm deep furrows filled with sand and gravel, and were filled with medium to fine sand, with particles mainly of 0.1—0.2 mm diameter. Thus good percolation was achieved through 1.20 m sand and gravel. The behaviour of *Juncus punctorius* did not significantly deviate from that of *Juncus maritimus* apart from the fact that the *J. punctorius* individuals were somewhat bigger and sturdier (thicker stalks). It may, however, be that the more continental climate of Beersheba was more favourable to them than to *J. maritimus*.

Treatment followed the same pattern as with *J. maritimus*, except that we irrigated it only with two types of sea-water: Caspian Sea type and North Sea type, i.e. with sea-water of a T.D.S. of 10,000—13,000 mg/l and, in the second case, with sea-water of a T.D.S. of 20,000—27,000 mg/l. The latter irrigation type was, however, used in the lysimeter series only, and the results are therefore not comparable, since all plants in the lysimeter series showed a much weaker development caused by the too quick percolation and therefore lack of sufficient nutrition possibility.

Again 5 seedlings were planted in each of the units and measurements were taken at the same intervals as with *Juncus maritimus*. In Table XIX the two strongest plant individuals show the pattern

| number of green leaves | | | | number of fruiting stems | new shoots (below 10 cm) | basal area 15.May 1961 | basal area 1.Oct. 1963 | Remark |
|---|---|---|---|---|---|---|---|---|
| 30.June 1961 | in % of b | 1.Oct. 1963 | in % of c | 1.Oct. 1963 | 1.Oct. 1963 | 15.May 1961 | 1.Oct. 1963 | |
| 66 | 100 | 27 | 20.3 | 20 | 2 | 48 | 120 | |
| 72 | 100 | 68 | 37.8 | 9 | 3 | 46 | 168 | |

of development during the irrigation period as well as after that, i.e. during the time of the Dying Experiment. *Juncus punctorius* as well as *Juncus maritimus*, described frequently as a hygrophyte, seem to be able easily to survive a dry period of eight to nine months (!) without — at least in our case — any possible connection with underground water.

*5. Rottboellia fasciculata* DESF.

Plants collected by my coworkers at that time, A. ABRAHAM and N. TADMOR, in Savannah Land of Erythrea and seeds from Herbarium material from Herbarium Boyko served as the sources for the experiments with this species. *Rottboellia fasciculata* seems to be extremely drought resistant and this observation together with its potential fodder value for arid regions was the main reason why we included this species in our experiments. The plants easily propagated vegetatively and by seeds in the climate of Beersheba.

Five rooted knots were planted in each unit under the same soil (sand) and irrigation conditions as the other species. Apart from the control plants irrigated with fresh water, two types only of sea-water were used: Caspian Sea type about 10,000—13,000 mg/l T.D.S. and North Sea type 20,000—27,000 mg/l T.D.S., two frames together with 20 plants for each concentration.

In all frames growth was approximately the same and the plants covered most of the surface in a relatively short time (see Table XX). Measuring was directed 1) to the number of leaf fascicles

**Table XX.**

Sea-water experiment with *Rottboellia fasciculata* Desf.
Pattern of development: irrigated June 1960-June 1961.
"Dying-experiment" June 1961-October 1963.

| 1<br>Date of measurement | 2<br>Height in cm (approx.) | 3<br>Cover in % of total Unit area (approx.) | 4<br>Total number of knots |
|---|---|---|---|
| **Unit A:** irrigated with 10,000—13,000 mg/l T.D.S. (Caspian Sea-type) | | | |
| 1. July 1960 | 1.5 | 1 | 5 |
| 15. May 1961 | 3 | 25 | 42 |
| 8. May 1962 | incl. halms 12 | 80 | 95 |
| 18. June 1962 | | | ? |
| 18. July 1962 | | | ? |
| 14. Aug. 1962 | | | ? |
| 14. Oct. 1962 | | | ? |
| 10. Dec. 1962 | | | ? |
| 6. March 1963 | | | 131 |
| 30. Apr. 1962 | 12 | 70 | 137 |
| 1. Oct. 1962 | | 70 | 126 |
| **Unit B:** irrigated with 10,000—13,000 mg/l T.D.S. (Caspian Sea-type) | | | |
| 1. July 1960 | 1.5 | 1 | 5 |
| 15. May 1961 | 3 | 20 | 39 |
| 8. May 1962 | 16 | 60 | 89 |
| 18. June 1962 | | 60 | 92 |
| 18. July 1962 | | 60 | 92 |
| 14. Aug. 1962 | | 60 | ? |
| 14. Oct. 1962 | | 60 | ? |
| 10. Dec. 1962 | | 60 | ? |
| 6. March 1963 | | 70 | 128 |
| 30. Apr. 1963 | 16 | 70 | 136 |
| 1. Oct. 1963 | | 70 | 136 |

on the knots, 2) to the percentage of green leaves in the fascicles, 3) to the projected covering area in relation to the soil surface of the frames, 4) to the length of the internodes, and 5) to height. (The frames, like all others, had a diameter of 46 cm and a soil surface area of 1667 cm².) It may be of interest that the plants irrigated with the highest concentration (20,000—30,000 mg/l

| 5<br>number of knots with green leaves | 6<br>in % of 5 | 7<br>length of internodes in cm | 8<br>Remark |
|---|---|---|---|
| | | | |
| 5 | 100 | none | 5 separated and rooted knots with about 5—10 leaves each were planted |
| 42 | 100 | 3 —6 | |
| 95 | 100 | | |
| 68 | appr. 70 | | Internodes yellow, stems green, fruiting spikelets yellow |
| 61 | appr. 60 | | |
| 48 | appr. 45 | | July and August are months of heavy dew |
| 46 | appr. 50 | | |
| 29 | appr. 33 | | |
| 94 | 72 | | good development in spite of the rather cold winter |
| 101 | 73 | | |
| ? | appr. 33 | 3.5—7 | |
| | | | |
| 5 | 100 | none | as in A |
| 39 | 100 | 3.5—6 | |
| 89 | 100 | | |
| 62 | 67 | | as in A |
| 48 | 52 | | |
| 30 | appr. 33 | | |
| 36 | appr. 33 | 4 —7 | |
| 30 | appr. 25 | | |
| 88 | 69 | | as in A |
| 107 | 78 | | |
| 56 | 41 | 4 —7.5 | |

| 1<br>Date of measurement | 2<br>Height in cm (approx.) | 3<br>Cover in % of total Unit area (approx.) | 4<br>Total number of knots |
|---|---|---|---|
| **Unit C:** irrigated with 20,000—27,000 mg/l T.D.S. (North Sea-type) | | | |
| 1. July 1960 | 1.5 | 1 | 5 |
| 15. May 1961 | 3 | 25 | 32 |
| 8. May 1962 | | 70 | 65 |
| 18. June 1962 | | 70 | 77 |
| 18. July 1962 | | 70 | 78 |
| 14. Aug. 1962 | | 70 | ? |
| 14. Oct. 1962 | | 70 | ? |
| 10. Dec. 1962 | | 70 | ? |
| 6. March 1963 | | 80 | 101 |
| 30. Apr. 1963 | | 80 | 106 |
| 1. Oct. 1963 | | 75 | 106 |
| **Unit D:** irrigated with 20,000—27,000, g/l T.D.S. (North Sea-type) | | | |
| 1. July 1960 | 1.5 | 1 | 5 |
| 15. May 1961 | 3 | 25 | 29 |
| 8 May 1962 | 16 | 80 | 76 |
| 18. June 1962 | | 75 | 79 |
| 18. July 1962 | | 75 | 80 |
| 14. Aug. 1962 | | 75 | 80 |
| 14. Oct. 1962 | | 66 | 80 |
| 10. Dec. 1962 | | 66 | 80 |
| 6. March 1963 | 15 | 75 | 105 |
| 30. Apr. 1963 | | 75 | 108 |
| 1. Oct. 1963 | | 75 | 107 |

T.D.S.) developed less fascicles but much longer internodes (see fig. 12). This behaviour of *Rottboellia* is in contrast to the fact that in general higher salinity leads to a shorter length of plants. On the other hand this plant grows by creeping stems and this horizontal direction of its growth is seemingly the reason for this contrasting behaviour.

Similar observations were made on *Agropyrum junceum* up to an irrigation with 10,000—15,000 mg/l T.D.S., and on *Juncus maritimus* and *J. punctorius*, up to an irrigation with oceanic concentration. No significant length- or time-retarding influence could be observed on the horizontal runners or their rhizomes respectively and, in consequence, the maximal "Basal Area" of all plants, irrespective of the concentration of the irrigation water, showed no significant difference, up to oceanic type. The same phenomena may be the reason that in the studies of KURIAN et al. (1966) the horizontally growing tobacco leaves reached a similar size under highly saline irrigation although the length of the internodes and the height

| 5<br>number of knots<br>with green leaves | 6<br>in % of 5 | 7<br>length of inter-<br>nodes in cm | 8<br>Remark |
|---|---|---|---|
| | | | |
| 5 | 100 | none | as in A |
| 32 | 100 | 3.5—6.5 | |
| 65 | 100 | | |
| 77 | 100 | | as in A |
| 62 | 80 | | |
| 59 | appr. 75 | | |
| 50 | appr. 75 | 6 —9 | |
| 31 | appr. 33 | | |
| 78 | 78 | | as in A |
| 80 | 76 | | |
| 58 | 55 | 7 —9.5 | |
| | | | |
| 5 | 100 | none | as in A |
| 29 | 100 | 3.5—7 | |
| 76 | 100 | | |
| 79 | 100 | | as in A |
| 72 | 90 | | |
| 66 | 82 | | |
| 68 | 85 | 6 —10 | |
| 29 | 36 | | |
| 86 | 80 | | as in A |
| 87 | 80 | | |
| 45 | 42 | 6 —10 | |

of the plants were considerably shortened.

In this experiment only those 10 plant individuals formerly irrigated with 10,000—13,000 mg/l and those 10 plants formerly irrigated with 20,000—27,000 mg/l T.S.C. could be observed and measured. The plants irrigated with fresh water were already dead in 1962 and had not survived the 9 months long dry period of that year.

A few general remarks may be allowed in this connection: The intricate biochemical process in plants under saline irrigation and not the least those processes causing this different behaviour are worth to be studied extensively by biochemists in cooperation with ecologists and physiologists. By following the ecologists' observations in nature we can gain much additional knowledge by the work of physiologists in the laboratory or in phytotrons. On the other hand this work in the laboratory or even in the phytotron is not much convincing as long as it does not go conform with the observations in nature. Too many unrecognizable factors can inhibit there the finding of absolute truth or may be directly mis-

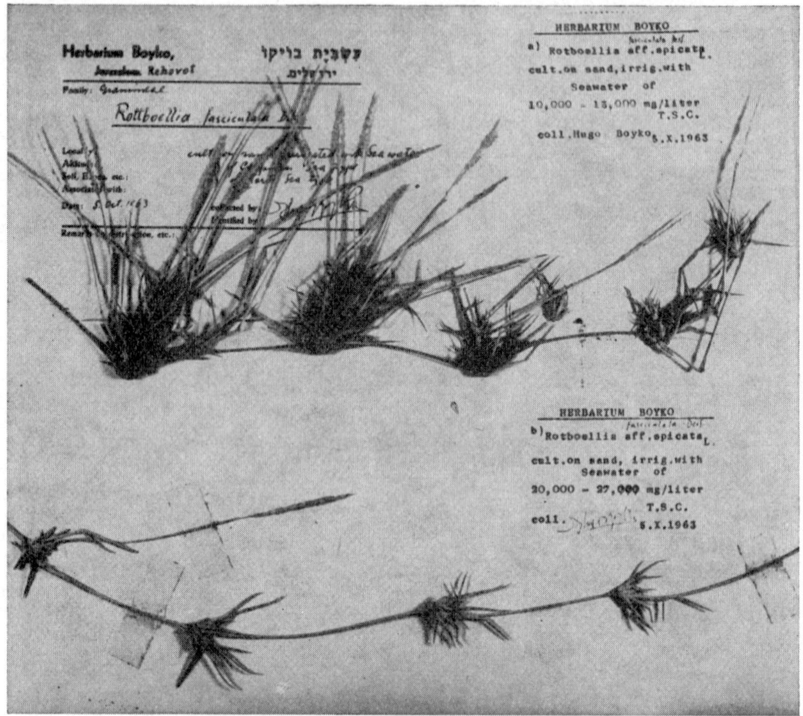


Fig. 12. Sea-water experiments with *Rottboellia fasciculata* Desf.
a) irrigated with sea-water of the Caspian Sea type (10,000-13,000 mg/l T.D.S.)
b) irrigated with seawater of the North Sea type (20,000-27,000 mg/l T.D.S.).
The plants irrigated with the higher concentration showed much longer internodes in the horizontally growing runners (see text for details).

leading. There, as in most other biological problems, biochemistry may come nearest to the basic truth and should, wherever possible, be included in this important new kind of research.

*Orientative Experiments*

The experiments with the following five species can only be considered as orientative experiments, since they extended over one vegetation period only. Originally, it was intended to expand the whole outlay over several years, but it had to be prematurely interrupted.

These species too were chosen for their economic value and particularly with a view for a possibly immediate applicability of positive results in the Negev, where flat sandy soil and dunes as well as highly saline underground water are to be found in great amounts. 600,000 dunams of shifting dunes (i.e. 60,000 hectares or 150,000 acres) alone near Beersheba in the northwestern part of the Negev are waiting for productivization, and huge amounts of saline water are to be expected there by drilling. Similar conditions are, however, spread throughout the semi-arid and arid zones of the globe, and it was felt that such experiments actually carried out — even on the smallest scale — might induce others in other parts of the world to do likewise and so gradually to come to results of high economic and social value for vast areas at present almost empty of population.

The experiments had to be abandoned before they could be finished, but for their general importance it seems to be advisable to mention them here as well.

The five species are:

6) *Hordeum vulgare* L. (barley)
7) *Agave fourcroides* LEMAIRE
8) *Agave sisalana* PERRINE
9) *Ammophila arenaria* LINK
10) *Beta vulgaris* L. (sugar-beet)

*6. Hordeum vulgare* L. (Barley)

Barley and wheat belong doubtlessly to the earliest cultivated cereals, stemming from the various wild forms still abundantly to be found in the irano-turano steppe belt (BOYKO, 1954) of Southwest Asia. Botanists as well as archaeologists agree that here, in Southwest Asia, almost with certainty the cradle of agriculture is to be sought. Both the history of agriculture and observations of salinity are closely connected with this cereal.

Observations on soil salinization can be followed on the basis of historical documents dating back to the fourth Millenium B.C., using the yield of barley and of wheat as yardstick. Irrigation by flooding was used in the earliest times already and the inevitable salt accumulation in these alluvial soils led to a gradual decrease of the yield in the course of time (JACOBSEN & ADAMS, 1958).

The high salt tolerance of barley became manifest in these ancient times already and it endured the growing soil salinity many hundred of years longer than wheat which is to a certain, but lesser degree likewise salt tolerant. This well known salt tolerance was the reason for barley being, also in recent years, subjected to various experiments with high salt concentrations of irrigation water, the highest concentration, however, being as far as we know, far below the concentrations we have used in our experiments, which were carried out up to oceanic concentrations.

Two series of experiments were started, one as orientative experiments only (a), and the second (b) on a more progressed level, interrupted for the reason mentioned above when the spikes were still in a budding stage (see photos of fig. 13 and 14).

Fig. 13. Sea-water experiments with a Bedouin strain of Barley (*Hordeum vulgare* L.) left: irrigated with sea-water of Caspian Sea type (10,000-13,000 mg/l T.D.S. with 75% NaCl in it).
right: irrigated with sea-water of North Sea type (20,000-27,000 mg/l T.D.S. with 75% NaCl in it).
(see text for details).

a) First series:

Seeds were collected during our expedition in the mountainous central parts of the Negev during the years 1941—1952, where we found the well known terrace-cultivations of the Bedouin from 400 up to 1000 m altitude in almost all Wadis (dry valleys flooded occasionally in the rainy winter season) and in artificial catchment areas. The surrounding natural vegetation is that of a semi-desert, mostly represented by an Artemisietum herbae albae (BOYKO, 1949), the density of which is exactly correlated to the IE-factor (Insolation-Exposure). These Wadis are subdivided into terraces by stone walls, many of them originating from the times of the ancient kingdom of Israel and the Nabatean state. The terraces are used by the Bedouin for growing barley and wheat, and the artificial catchment areas for fruit trees, vegetables and tobacco. The present

Fig. 14. Sea-water experiments with a Bedouin strain of Barley (*Hordeum vulgare* L.) irrigated with sea-water of Oceanic type (30,000-40,000 mg/l T.D.S. with 75% NaCl in it).
(Additional fertilizer to this high concentration led to early death of all seedlings in the unit no. 75+ on the left).

average rainfall is from 125 to 200 mm yearly, but according to all known sources, the climate does not seem to be significantly different, compared with that of 2000 years ago.

On these terrace-fields, contrary to the conditions in the alluvial plains, the yearly movement of flood water through all the centuries prevents any accumulation of salt, except as an ephemerical and insignificant efflorescence appearing here and there in small white patches on the soil surface.

The yield on these terraces is sometimes remarkable, considering the most primitive tools and methods used by the Bedouin. The amount depends solely on rainfall, and the height of the barley or wheat fields on these terraces is an excellent biological yardstick for the amount of rainfall according to the catchment area. The first author compared the height throughout several years until 1952 and found that a 40 cm height of barley corresponds to about 200—225 mm rainfall in the area between Rosh Rimon and the newly erected village Mizpeh Rimon. But, as an average, only one in three years is reckoned as worthwhile of harvesting the yield in the central Negev. The strains of barley and wheat used by the Bedouin in this region belong probably to the most drought-resistant ones of all, since they are selected since times immemorial on the basis of drought survival.

It was for this reason that we collected and used these particular seeds for our sea-water experiments.

Since we knew the high salt tolerance, we tried in the first series to add various mineral fertilizers to the sea-water solution, but without significant results.

This may be explained by the fact that the dilutions of sea-water used in this first series contained sufficient nutritive material so as to outbalance the lack of it in the pure sand wherein the plants stood. In any case the scale of this first experiment was too small to allow reliable conclusions.

In this experiment with barley, we had originally in mind to try out the possibility of saving a whole crop in times of catastrophic droughts by applying sea-water or highly saline water after the plants had already reached a certain stage of development.

The seeds were sown on December 5, 1958, and received, for lack of sufficient rain for germination, first fresh water (118 mm) and then 347 mm rainfall. From March 15 on, sea-water served as the only source of water as well as of nutrition (separate units were treated also with the addition of fertilizers, but showed no significant difference). Two units of each series were supplied with sea-water of the Caspian Sea type (11,000 mg/l T.D.S.) and two units with sea-water of the North Sea type (22,000 mg/l T.D.S.). All plants developed well and on May 17, the seeds were collected, counted and weighed (by our co-worker, Mrs. Adiva Ilan) separately for each unit. The unit irrigated with sea-water of the Caspian Sea type yielded, for instance, 40 g. Since the area of one unit is 1/6 square metre, this result indicated a theoretically potential harvest of 2.4 metric tons of barley per hectare.

b) Second Series:

These good results induced us to start immediately a new series of experiments by using these seeds from mother plants already treated with sea-water, and by raising the maximum concentration to the oceanic type, i.e. with 34,000 mg/l T.D.S. with 24,000 mg/l NaCl, or more correct for the reasons repeatedly mentioned, with the fluctuating concentration of 30,000 to 40,000 mg/l T.D.S.

Starting with sowing the seeds on May 17, we eliminated the rainy season completely.

This time we did not sow as in a field, but laid the seeds separately (10 in each unit) in order to allow individual measurements of each plant.

Except for germination and a few days after, only sea-water was used for irrigation.

Fig. 13 and 14 show the achieved budding stage at the time when these experiments had to be interrupted (15. July, 1959).

When the experiments with sea-water were resumed in Beersheba in 1960, barley and the following four species were not included any more, in order to adapt the amount of measurements to the given technical possibilities. For various reasons, however, further experiments in this line with barley as well as with wheat are strongly recommended. It would be best to include them in a somehow coordinated research scheme of a transnational character under various climatic and other conditions, as planned by the World University now in statu nascendi, by the World Academy of Art and Science (see Summary of the Proceedings, III. Plenary Meeting, Rome, September 1965, in WAAS NEWSLETTER, 1966).

The reasons for this recommendation are certain aspects like the following ones:

1) The specific value of barley as staple cereal in subarid regions;

2) the possibility to use one or a few additional saline water or sea-water irrigations in order to save the entire harvest of a region in times of great droughts during the vegetation period. This would apply also to cases of extreme droughts when not a full grains harvest is aimed at but only fodder in order to save the herds of nomads.

3) Investigations in the direction of biological soil-desalination may be of more than only theoretical interest.

4) *Hordeum* or better specific barley strains could be used as a standard plant for various purposes, e.g. as a biological yardstick, if taken from the same motherplants. Such a yardstick could be useful for instance for comparing soil structure, water conditions, climatic factors, etc.

*7. Agave fourcroides* LEMAIRE and
*8. Agave sisalana* PERRINE

These two drought resistant species from Mexico are both of high economic value as fibre plants; the first as the source of a commercial fibre known under the name Henequen or Yukatan Sisal, the second as the source of Sisal Hemp (Bahama Hemp, Yaxci). Both are used mainly for the manufacture of ropes for various purposes and are widely grown, particularly the second one, in the tropics of the Old and New World.

**Table XXI.**

Sea-water experiment with *Agave fourcroides* LEMAIRE
(Af 1—6)

a) Irrigation water: Fresh water (about 500 mg/l T.D.S.) (plus K.P.N. application)

| Date | Plant individual | Number of spread leaves | maximal length of green leaves in cm | Plant individual | Number of spread leaves | maximal length of green leaves in cm |
|---|---|---|---|---|---|---|
| Dec. 5, 1958 | Af 1 | 0 | 6 | Af 2 | 0 | 5 |
| March 22, 1959 | | 2 | 10 | | 4 | 8.5 |
| April 19, 1959 | | 4 | 13.5 | | 5 | 14.5 |
| May 20, 1959 | | 7 | 12 | | 8 | 15 |
| June 24, 1959 | | 12 (two died, 12 new) | 14.5 | | 13 (one died, 10 new) | 14.5 |

b) Irrigation water: about 11,000 mg/l T.D.S. (with 8,000 mg/l NaCl) (plus K.P.N. Application)

| Date | Plant individual | Number of spread leaves | maximal length of green leaves in cm | Plant individual | Number of spread leaves | maximal length of green leaves in cm |
|---|---|---|---|---|---|---|
| Dec. 5, 1958 | Af 3 | 0* | 8.5 | Af 4 | 0* | 6 |
| March 22, 1959 | | 12 | 11.2 | | 7 | 10.8 |
| April 19, 1959 | | 8 | 13 | | 6 | 13 |
| May 20, 1959 | | 8 | 13 | | 7 | 13 |
| June 24, 1959 | | 9 (4 died, 1 new) | 13 | | 7 (one died, 1 new) | 14 |

c) Irrigation water: about 22,000 mg/l T.D.S. (with 16,000 mg/l NaCl) (plus K.P.N. application)

| Date | Plant individual | Number of spread leaves | maximal length of green leaves in cm | Plant individual | Number of spread leaves | maximal length of green leaves in cm |
|---|---|---|---|---|---|---|
| Dec. 5, 1958 | Af 5 | 0 | 6 | Af 6 | 0 | 8.5 |
| March 22, 1959 | | 3 | 8 | | 3 | 9 |
| April 19, 1959 | | 4 | 9.2 | | 2 | 10 |
| May 20, 1959 | | 2 | 7.5 | | 2 | 6 |
| June 24, 1959 | | 2 (3 died, 2 new) | 6 | | 2 (3 died, 2 new) | 6 |

* already spreading.

At the beginning of World War II, in 1939, their drought resistance and potential value as industrial raw material for hot semi-desert and desert areas induced the first of the present authors to recommend these plant species for introduction as economic crop plant into Israel (then Palestine). In 1949 the second author planted them in the desert garden of Eilat with an irrigation water of a fluctuating salt concentration from 2000 to 6000 mg/l T.D.S., mainly sulphates, and in 1958 we started with some orientative trials with sea-water. For this purpose we used sea-water of the Caspian Sea type and of the North Sea type, i.e. a 25% dilution and a 50% dilution of the East Mediterranean sea-water. The concentrations were therefore about 11,000 mg/l T.D.S. (Total Dilution of Salts) with 8,000 mg/l NaCl in the first case and 22,000 T.D.S. with 16,000 mg/l NaCl in the second case.

Eight small off-shoots of *Agave sisalana* (marked in the Tables as As 1—As 8) and six of *Agave fourcroides* (Af 1—Af 6) were at our disposal and we planted them in seven bottomless asphalt barrels (i.e. 90 cm heigh frames of 46 cm diameter) filled with sand and put on sand. Two plants of the same species were planted in each of these seven units. The outlay of the experiment is to be seen from Table XXI and Table XXII.

Thus we had, out of the 14 plant individuals, two plants of the same species and with the same treatment in each of the seven units. With the exception of the two plants of *Agave sisalana* (As 7 and As 8) all plants were supplied with K, P, N in an amount as one would apply to a leafy field crop like cabbage.

These four plants (As 5 and As 6 in one unit and As 7 and As 8 in another unit) were all irrigated with the same high concentration, i.e. sea-water of the North Sea Type, but only those two supplied with fertilizers survived until the end of this experiment in July 1959, whereas the other two plants, As 7 and As 8, were already dead in mid April. Without seeing the possibility to draw conclusions from this behaviour, it seems noteworthy and worthwhile to draw the attention to it for further experiments with this species.

The amount of irrigation given each time was 2½ l per plant from 22. March to 24. June on 25 days. Rainfall during these three months was insignificant (the last and strongest of 1 mm on April 5, 1959). End of June the experiment had to be abandoned. The development of the leaves as the main visible reaction of the plants to the very high salinity of the irrigation water is presented in the Tables XXI and XXII.

This development shows that both species react rather similarly. *Agave fourcroides* may be a little more salt resistant than *Agave sisalana* according to the meagre figures of this small experiment, but for both sea-water of the concentration of 22,000 mg/l T.D.S. without the additional application of any fertilizer was already a lethal concentration in the course of a few weeks. The others died more slowly, and showed after three months still a certain regeneration capacity, but here too at the time of the last measurement almost all the older radially spreading leaves were already wholly or to their greater part dead. This longer survival is probably to

**Table XXII.**

Sea-water experiments with *Agave sisalana* PERRINE
(As 1—8)

| Date | Plant individual | number of spread leaves | maximal length of green leaves in cm | Plant individual | number of spread leaves | maximal length of green leaves in cm |
|---|---|---|---|---|---|---|
| a) Irrigation water: Fresh water (about 500 mg/l T.D.S.) (plus K.P.N. application) | | | | | | |
| Dec. 5, 1958 | As 1 | 0 | 8 | As 2 | 0 | 2.5 |
| March 22, 1959 | | 2 | 11 | | 2 | 2.5 |
| Apr. 19, 1959 | | 4 | 14.5 | | 3 | 4 |
| May 20, 1959 | | 8 | 15.5 | | 6 | 8 |
| June 24, 1959 | | 12 (one died, 11 new) | 15 | | 11 (none died, 9 new) | 12 |
| b) Irrigation water: about 11,000 mg/l T.D.S. (with 8,000 mg/l NaCl) (plus K.P.N. application) | | | | | | |
| Dec. 5, 1958 | As 3 | 0* | 10 | As 4 | 0* | 6 |
| March 22, 1959 | | 5 | 12 | | 7 | 10.8 |
| Apr. 19, 1959 | | 4 | 12 | | 5 | 11.7 |
| May 20, 1959 | | 4 | 11.5 | | 5 | 13.5 |
| June 24, 1959 | | 5 (two died, 2 new) | 11 | | 7 (two died, 2 new) | 14 |
| c) Irrigation water: about 22,000 mg/l T.D.S. (with 16,000 mg/l NaCl) (plus K.P.N. application) | | | | | | |
| Dec. 5, 1958 | As 5 | 0 | 6 | As 6 | 0 | 7.5 |
| March 22, 1959 | | 3 | 8.5 | | 3 | 10.7 |
| Apr. 19, 1959 | | 3 | 7 | | 5 | 11.2 |
| May 20, 1959 | | all dying | | | 4 | 11.5 |
| June 24, 1959 | | all dead (three died, none new) | | | 3 (two died, 2 new) | 11.5 |
| d) Irrigation water: about 22,000 mg/l T.D.S. (with 16,000 mg/l NaCl) (without any additional application) | | | | | | |
| Dec. 5, 1958<br>March 22, 1959<br>Apr. 19, 1959 | As 7 | 1<br>dead | 5<br>5.5 | As 8 | 0<br>4<br>all dead | 8<br>12 |

* already opened for spreading

be ascribed to the potash content of the fertilizer as a counter-balancing factor to the high sodium content in the ionic environment.

More prolonged and statistically more significant experiments would be necessary to find out if the lethal limit was actually reached or if even with this high salinity of the irrigation water a recuperation could perhaps be achieved after a certain *minimum period of adaptation* (see below). This could be achieved by the plants, for instance, by means of a higher succulence and/or a higher amount of critical salt accumulation or both. In consequence, the shedding of leaves would be slower than sprouting and the speed of growing new leaves would surpass the speed of shedding the old ones.

This occurred in all individuals (As 3, As 4, Af 3, Af 4) of both species irrigated with sea-water of a concentration of 11,000 mg/l. Here such a "Period of Adaptation" is clearly indicated. In both species and in all four individuals we found a decline in the number of leaves after four weeks of irrigation with this Caspian sea-water type. But soon, one to two months later, such a recuperation could be observed and the rate of dying was smaller than the rate of growing! In three of the four individuals (As 3, As 4, Af 4) the original number of leaves reached during the short rainy season, mainly January and February, had already been re-achieved.

It would be of high interest, from the scientific as well as from the practical point of view, to find out if this trend of development could lead to economically feasible results. It can reasonably be expected that the usually much lower concentration of saline desert underground water combined with the good percolation and root aeration of sandy areas may offer economic propositions, if adequate experiments will prepare such possibilities of productivizing and industrializing certain parts of hot deserts.

The serious drawbacks for more advanced conclusions from our experiments are not only the small number of plants and the too short period of experimenting, but also the uneven age and size of the off-shoots experimented with. In following experiments this factor has also to be considered.

Although our experiments with these two Agaves are inconclusive, they justify the recommendation that experiments with these economically valuable species should be continued on a much broader scale.

Their survival and even recuperation with a sea-water concentration of 11,000 mg/l with a content of 8,000 mg/l NaCl indicate more promising results of experiments in many deserts and semi-deserts with locally available saline waters. Although these waters usually do not represent such a balanced ionic environment as sea-water and its dilutions, it offers no difficulties to ameliorate this factor by applying the principles of H. HEIMANN (1966).

9. *Ammophila arenaria* LINK

*Ammophila arenaria* is one of the most important sand binders used as such on all continents, particularly on coastal dunes. An orientative experiment was made as follows:

One of the frames (bottomless barrels) filled with dune sand was densely sown with *Ammophila* seeds, several weeks long irrigated with fresh water and then during the whole dry season in 1958, with a sea-water dilution of 11,000 mg/l T.D.S. The plants stood

Fig. 15. Sea-water experiments with *Beta vulgaris* L. (Sugar-beet): from right to left (the biggest and the smallest beet of each group):

a) irrigated with fresh water,
b) irrigated with sea-water of Caspian type (10,000—13,000 mg/l T.D.S. with 75% NaCl in it),
c) irrigated with sea-water of North Sea type (20,000—27,000 mg/l T.D.S. with 75% NaCl in it).

this well, but for lack of time and sufficient help we had to eliminate the continuation of this part and of taking measurements of this species.

Prolonged and scientifically better based experiments seem to be advisable, for positive results may in many difficult cases lead to methods of speeding up or even securing the fixing of dangerously shifting dunes.

*10. Beta vulgaris L.* var. *esculenta* (SALISB.) FIOR (Sugar-beet)

This species is known since long as very salt resistant and numerous experiments with highly saline water have been made with positive results. They were usually made with artificial solutions of sodium chloride. This fact and the economic importance of sugar-beet induced us to start with a small orientative experiment with two types of sea-water. We made our experiments on the same pattern as those with the two species of *Agave*, namely with sea-

**Table XXIII.**

Sea-water experiment with *Beta vulgaris* L. (sugar-beet)
(Measurements on one single plant individual, as example.)

Irrigation water:
11,000 mg/l T.D.S. with 8,000 mg/l NaCl (plus K.P.N. application)

| Plant-Individual | Date 1959 | number of leaves | diameter of leave rosette in cm | maximal width of leaves in cm | maximal length of leaves in cm | Remark |
|---|---|---|---|---|---|---|
| Nr. 6 of Unit Nr. 31 | March 3 | 8 | 21 | ? | ? | steady, healthy development throughout the experiment |
| | March 22 | 19 | 25.5 | 7 | 12.5 | |
| | April 20 | 23 | 37 | 9 | 18 | |
| | May 20 | 31 | 47 | 10.5 | 22 | |
| | June 22 | 36 | 52 | 11.5 | 25 | |

beet (root) measurement

| | Date 1959 | length of the whole plant in cm (incl. root) | length of the beet in cm | circumference of the beet in cm | weight of the beet in g | Remark |
|---|---|---|---|---|---|---|
| | June, 30 | 107 | 11 | 25.5 | 421 | |

**Table XXIV.**

Sea-water experiment with *Beta vulgaris* L. (sugar-beet)
Condensed table of the measurements on all plant-individuals (Nr. 1—12) at the end of the experiment 15. July, 1959).

| Nr. of Units | Nr. of Plant Indi-viduals | Salt con-centration of irrigation water | Number of leaves | Diameter of rosette in cm | Maximal width of leaves in cm | Maximal length of leaves in cm |
|---|---|---|---|---|---|---|
| 29 & 30 | 1— 4 | Fresh water | 32—46 | 58—74 | 10.5—12 | 32—36 |
| 32 & 32 | 5— 8 | 11,000 mg/l | 34—48 | 50—62 | 9.5—11.5 | 22—29 |
| 33 & 34 | 9—12 | 22,000 mg/l | 21—36 | 33—54 | 7—11.5 | 17—28 |

beet (root) measurements

| Nr. of Units | Nr. of Plant Indi-viduals | Salt con-centration of irrigation water | Length of the whole plants in cm | Length of the beets in cm | Circum-ference of beets in cm | Weight of beets in g |
|---|---|---|---|---|---|---|
| 29 & 30 | 1— 4 | Fresh water | 115—125 | 15 —16 | 27—34 | 528—1025 |
| 31 & 32 | 5— 8 | 11,000 mg/l | 90—107 | 9 —14 | 23—28 | 411— 493 |
| 33 & 34 | 9—12 | 22,000 mg/l | 65—110 | 5.5—10 | 16—24.5 | 176— 416 |

water of the Caspian Sea type and that of the North Sea type, apart, as a matter of course, from the control experiments with fresh water.

For reasons mentioned above the smallest possible number of units and plants had to be taken for that series: Six units of the same type as for all others in our sea-water experiments, namely bottomless asphalt barrels filled with sand and put on sand, were used. Two seedlings were planted in each and two units taken for each sea-water type and also for the control units irrigated with fresh water, altogether 12 single plants in the 6 units.

The young seedlings (sown on 1. December 1958) were planted early in March 1959 and irrigated with sea-water from 15. March on. Measurements started on 22. March. Apart from insignificant and ineffective rainfall from then on (the strongest and last of 1 mm on April 5) the sea-water irrigation was the only source of water until the end of June when the beets were taken out and weighed.

The total supply of sea-water of the two concentration types given during this time was 1025 mm. In addition K. P. N. was applied in the same amount as in the experiments with the Agaves. Some details of the results are to be seen from Table XXIII and XXIV.

It is not without interest that in spite of the very great increase in salinity from one concentration to the other, the decline in the speed of vegetative growth is such that the maximum of almost all figures taken with the series of one concentration reaches or even surpasses the minimum of the next lower concentration (see also fig. 11).

Unfortunately no investigation of the sugar content could be made at the end of this experiment. However, since the experiment had only an orientative character, the lack of this chemical test is of no decisive importance.

The results of this experiment are certainly promising and, in view of the decisive importance of sugar in the nutrition of man, they demand an extensive expansion in various directions.

These recommendations are strengthened by the fact that various experiments have shown that the sugar content in sugar-beets is rising up to a certain degree with the salinity of the soil-water environment.

The choice of specific varieties adapted to local conditions and a well based successive selection of seeds from motherplants gradually adapted to higher salinity will have to play a major role for expanded experiments.

## Final Remarks

In describing these experiments we have repeatedly drawn attention to their small scale and to the fact that large scale experiments for each species and in each country separately are necessary before any practical application, such as an economic enterprise, can be undertaken. The main value of our experiments, from the first with brackish water in Eilat, in 1949, to the last with sea-water of the oceanic and even Red Sea type is to be seen in the principle that under certain conditions plants can not only survive but even be grown under irrigation with water of such high concentrations.

In order to prove this, one single growing or even only surviving plant individual would have been sufficient. Factually, since then, many plant individuals and of numerous, very different species, most of them of high economic value for food, fodder or industry, have successfully been grown. Furthermore, it has been proved in various countries along a wide climatic profile from India to the North Sea and the Baltic Sea.

Our experiments have already found a wide response. Even our first experiments in Eilat (H. Boyko, 1952) with desert underground water of a Total Dilution of Salts of 2500—6000 mg/l were soon followed in Sweden by excellent experiments on natural pastures irrigated directly with the water of the Baltic Sea (T.D.S. 6,000 mg/l). The result was a 60% higher productivity of these plots, compared with the non irrigated pastures (HELLGREN, 1959). In Spain, several dozens of vegetables and flowers could be grown under irrigation with water taken directly from the ocean, and so on (see the complementary volume to this book, WAAS, Vol. IV).

All experiments indicate that this new line of research, Saline and Marine Agriculture, is most promising and requires concerted action on the scientific side and highest attention from the appropriate international and national agencies*.

---

Remark:

* Sponsored by UNESCO and organized by the World Academy of Art and Science in cooperation with the National Research Council of Italy and the Italian Academy of Agricultural Sciences, a first step has been made in this direction and an International Symposium has been convened in Rome in September 1965, with the participation also of the Food and Agricultural Organization of the United Nations Organization.

Although a separate volume is dealing with the respective proceedings (WAAS, Vol. IV) it may be appropriate to repeat here the relevant resolution, carried by the 104 participants from 23 countries (Argentine, Australia, Canada, Federal Republic of Germany, France, India, Iran, Israel, Italy, Kuweit, Malta, Morocco, Netherlands, Netherlands Antilles, New Zealand, Poland, Spain, Sudan, Sweden, Switzerland, Tunis, UAR and USA):

Resolution: Rome, 8 September 1965

The participants in the *International Symposium on Irrigation with Saline Water and Sea-water with and without desalination*, organized by the World Academy of Art and Science, the Accademia Nazionale di Agricoltura, and the Consiglio Nazionale delle Ricerche, and sponsored by UNESCO, firmly believe that results already achieved by irrigation with highly saline water indicate clearly that those arid areas at least where such a water supply is available, can be rendered capable of crop production, and they therefore strongly recommend to international and national organizations concerned with human welfare that financial provision continue to be made for the necessary expansion of research and field trials.

The following fields are seen as of particular importance:

1) Ecological, physiological, genetic, biochemical and agrotechnical studies on promising economic plants under direct irrigation with highly saline and sea-water under various climatic conditions;

2) Research in the laboratories and in the field on *biological* desalination of soil and water;

3) Microbiological studies connected with these lines of research.

In consequence of this resolution and in view of the global impact on human welfare of these new lines of pure and applied research, the immediately following Third Plenary Meeting of the World Academy of Art and Science (Rome, Sept. 9-14, 1965) and the newly elected Council for the World University decided to prepare a transnational cooperative research project for this new line of research, covering a wide climatic profile on both sides of the equator, for submission to the competent financing authorities, and to initiate large scale field trials on the basis of its results.

## REFERENCES

BOTTINI, O., 1961. Tradition et recherche en Italie dans l'emploi des eaux saumâtres pour l'irrigation. UNESCO, Arid Zone Research, Vol. **XIV,** Salinity Problems in the Arid Zones, *251—257,* Paris.

BOTTINI, O. & B. LISANTI, 1955. Ricerche e considerazioni sull'irrigazione con aque salmastre praticata lungo il litorale pugliese. *Ann. Sper. Agr.*, N.S. **IX,** *401—436.*

BOYKO, ELISABETH, 1952. The Building of a Desert Garden. *J. Roy. hort. Soc.*, **76,** 4.

BOYKO, ELISABETH & HUGO BOYKO, 1966. The Desert Garden of Eilat since its foundation, 1949, until 1965. Proceedings of the Internat. Symposium on irrigation with highly saline water and sea-water with and without desalination, in WAAS-Series, Vol. IV (in preparation).

BOYKO, HUGO, 1934. Die Vegetationsverhältnisse im Seewinkel, Versuch einer pflanzensoziologischen Monographie des Sand- und Salzsteppengebietes östlich vom Neusiedlersee: II. A) Allgemeines, B) die Gesellschaften der Sandsuccession. *Beihefte, Bot. Zbl.* **51,** II, *600—747.*

BOYKO, HUGO, 1947. On the role of plants as quantitative climate indicators and the "Geo-Ecological Law of Distribution". *J. Ecol.*, **35,** *138—157.*

BOYKO, HUGO, 1949. On the Climax-Vegetation of the Negev, with special reference to arid pasture-problems. *Palest. J. Bot., Rehovot. Ser.*, **7,** *17—35.*

BOYKO, HUGO, 1954. A new plant-geographical subdivision of Israel — as an example for Southwest Asia. *Vegetatio, The Hague,* **5/6,** *309—318.*

BOYKO, HUGO, 1957—1958. Circular Letters of the International Commission on Applied Ecology of IUBS; Rehovot.

BOYKO, HUGO, 1965. Lecture on "Principles of direct irrigation with sea-water" International Symposium on irrigation with highly saline water and sea-water, with and without desalination, Rome, Sept. 5—9, 1965, p. 1—20.

BOYKO, HUGO, 1966. Basic ecological principles of plantgrowing by irrigation with highly saline or sea-water, in: "Salinity and Aridity — New Approaches to Old Problems", *Mon. Biol.* **XVI,** *131—200.*

BOYKO, HUGO & ELISABETH BOYKO, 1959. Seawater Irrigation — a new line of research on a bioclimatic Plant-Soil-Complex. *Int. J. Bioclimat. Biometeorol.* **III,** part II., Sec.B 1, *1—24.*

BOYKO, HUGO & ELISABETH BOYKO, 1964. Principles and Experiments regarding direct irrigation with Highly Saline and Seawater without Desalination. *Trans. N.Y. Acad. Sci.* Ser. II, **26,** suppl. to No. 8, *1087—1102.*

BOYKO, HUGO, BOYKO, ELISABETH & TSURIEL, DAVID, 1957. Ecology of Sand Dunes. Final report to Ford Foundation, sub-project C-le, pp. *353—393,* Jerusalem.

BRYSSINE, G., 1961. Essais sur l'Irrigation à l'Eau Saumâtre Réalisés au Maroc. UNESCO Arid Zone Research, Salinity Problems in the Arid Zones, p. *245—250.*

CHAPMAN, V. J., 1962. Salt Marshes and Salt Deserts of the World. Leonhard Hill, London.

CHAPMAN, V. J., 1966. Vegetation and Salinity, in: "Salinity and Aridity — New Approaches to Old Problems", *Mon. Biol.* **XVI,** *23—42.*

DARLINGTON, C. D. & E. K. JANAKI AMMAL, 1945. Chromosome Atlas of Cultivated Plants. London, pp. 397.

EVENARI, M., 1962. Plant Physiology and Arid Zone Research. UNESCO Arid Zone Research, The Problems of the Arid Zones, *175—195.*

HEIMANN, H., 1966. Plant Growth under Saline Conditions and the Balance of the Ionic Environment in Salinity and Aridity — New Approaches to Old Problems, *Mon. Biol.* **XVI,** *201—213.*

HELLGREN, GUNNAR, 1959. Irrigation with Seawater in Sweden. Report of the Conf. on Supplementary Irrigation, Commission VI, Int. Soc. of Soil Science.

JACOBSEN, T. & R. M. ADAMS, 1958. Salt and Silt in Ancient Mesopotamian Agriculture. *Science, Washington*, **128,** 3334, *1251—1258.*

KURIAN, T., E. R. R. IYENGAR, M. K. NAYARANA & D. S. DATAR, 1966. Effect of Seawater dilutions and its amendments on Tobacco, in: Salinity and Aridity — New Approaches to Old Problems, *Mon. Biol.*, **XVI,** *323—330.*

OSVALD, HUGO, 1966. Salinity Problems, in *WAAS NEWSLETTER*, January 1966, *41—48.*

PANTANELLI, E., 1957. Irrigazione con aque salmastre. *Mem. Staz. Agr. Sper. Bari*, no. 26.

STOCKER, OTTO, 1954. Die Trockenresistenz der Pflanzen (La résistance des végétaux à la séchesse). *VIII. Congres Int. Bot.*, Rapports et Communication, Sections 11 et 12, *223—232*, Paris.

STOCKER, OTTO, 1960. Physiological and morphological changes in plants due to water deficiency, in: Plant-water relationships in arid and semi-arid conditions. Review of Research, Paris, UNESCO, *63—104* (Arid Zone Research XV).

TADMORE, N. H., D. KOLLER & E. RAWITZ, 1958. Experiments in the propagation of Juncus maritimus Lam. *Ktavim*, **9,** 1—2, *77—205.*

TSURIEL, DAVID, 1954. Agropyrum junceum as Fodder and Pasture Plant. *Hassadeh, Tel Aviv*, **9,** *613—614*, (in Hebrew).

UNESCO, Arid Zone Research, 1956. Vol. IV: "Utilization of Saline Water". Reviews of Research, pp. 102, 2nd edition, Paris.

UNESCO, Arid Zone Research, 1961, Vol. XIV: "Salinity Problems in the Arid Zones" Proceedings of the Teheran Symposium, pp. 395.

UNESCO, Arid Zone Research, 1962. Vol. XVIII: "The Problems of the Arid Zone" Proceedings of the Paris Symposium, pp. 481.

World Academy of Art and Science: WAAS NEWSLETTER, January 1966, pp. 48, Rehovot.

World Academy of Art and Science: WAAS Series, Volume IV: "Productivizing of Deserts by irrigation with highly saline water and seawater with and without desalination. Proceedings of the Rome Symposium, September 1965 (in print, Den Haag, 1966).

WALTER, HEINRICH, 1951. Einführung in the Phytologie, III. Grundlagen der Pflanzenverbreitung, I. Teil: Standortslehre. Eugen Ulmer, Stuttgart.

WALTER, HEINRICH, 1961. The adaptation of plants to saline soils in UNESCO, Arid Zone Research XIV "Salinity Problems in the Arid Zones", p. *129—134.*

WERBER, A., 1936. Is Irrigation with Saline Water possible? *Hadar, Tel Aviv*, **IX,** 9. *201—203* (in Hebrew).

# EXPERIMENTS ON THE BASIS OF THE PRINCIPLE OF THE "BALANCE OF IONIC ENVIRONMENT"

(Field Experiments with Groundnuts and Cow Peas)*

BY

H. HEIMANN AND R. RATNER

## Introduction

The author previously developed a new approach to the problem of plant growth under saline conditions (HEIMANN, 1959). Soil reaction, salts, and mineral nutrition are all put together under a common heading, i.e. that of the ionic environment. Healthy plants and good crops can only be obtained if the ionic environment is physiologically well balanced. When soil or water are saline, the preponderance of the sodium cation is the chief feature. This excess of sodium has to be counteracted by other cations. Since long, the view was held that calcium will best perform this function. Though this may be right in respect to the physical properties of the soil, it is not so in respect to what is going on within the living plant. Starting from field observations, facts incidentally reported in literature, and considerations based on a general cellular physiology, potassium was claimed to be the natural antagonist of sodium. Replacing the plant nutrition concept by that of the balanced ionic environment, a new role was assigned to the potassium ion and its bearers, viz. the protection of plants against excessive concentrations of sodium. Based on earlier work by HEIMANN & RATNER (1961), it was assumed that relatively low concentrations of potassium may keep in check much higher concentrations of sodium provided both ions appear together. The aim of the work presented herewith was to verify by field experiments the usefulness of this new approach.

## Experiments with Groundnuts

### Design of Experiment

*Plant*

Groundnuts (*Arachis hypogaea*) were chosen as test plant considering their economic value in irrigated agriculture on semi-arid lands. The variety used was "Virginia-4". Though no indication is given in the literature in respect of their salt tolerance, groundnuts as Leguminosae should be rather sensitive to salinity.

* The research work presented in this paper was carried out at the Israel Institute of Technology in Haifa and supported by UNESCO.

The experiments were carried out on the fields of the experimental farm at Acre. The soil was of the medium-heavy calcareous type of neutral to slightly alkaline reaction, with a total exchange capacity of 14—20 milli-equivalent per 100 g.

*Soil*

The potassium status in a composite sample from the total plot of 2 dunams (= 0.5 acre) is given in the following table:

**Table I.**

K in m.e. per 100 g.

| Depth in cm | total | exchangeable |
|---|---|---|
| 0— 30 | 8.0 | 0.42 |
| 30— 60 | 7.7 | 0.35 |
| 60— 90 | 9.5 | 0.30 |
| 90—120 | 8.0 | 0.26 |

These figures indicate a fairly good supply of available potassium and an ample stock of total potassium.

The complete composition of the ion exchange complex in the different parts of the experimental field is given in Table II.

**Table II.**

Composition of the ion exchange complex in the different parts of the experimental field (2 dunams = 0.5 acre).

| Soil layer | Depth in cm | exchangeable cations in m.e. per 100 g | | | | |
|---|---|---|---|---|---|---|
| | | total | Ca | Mg | K | Na |
| I | 0— 30 | 20.2 | 15.1 | 4.44 | 0.42 | 0.23 |
| | 30— 60 | 18.5 | 13.7 | 3.99 | 0.35 | 0.46 |
| | 60— 90 | 15.2 | 11.8 | 2.66 | 0.30 | 0.42 |
| | 90—120 | 14.4 | 10.8 | 2.82 | 0.26 | 0.50 |
| II | 0— 30 | 19.4 | 14.4 | 4.40 | 0.43 | 0.16 |
| | 30— 60 | 18.8 | 13.5 | 4.64 | 0.41 | 0.21 |
| | 60— 90 | 15.0 | 11.2 | 3.18 | 0.30 | 0.28 |
| | 90—120 | 14.3 | 10.4 | 3.24 | 0.25 | 0.43 |
| III | 0— 30 | 19.7 | 14.8 | 4.30 | 0.41 | 0.17 |
| | 30— 60 | 17.6 | 13.1 | 3.78 | 0.33 | 0.35 |
| | 60— 90 | 14.8 | 10.6 | 3.54 | 0.29 | 0.36 |
| | 90—120 | 14.6 | 10.6 | 3.36 | 0.24 | 0.41 |
| IV | 0— 30 | 20.2 | 14.6 | 4.97 | 0.43 | 0.21 |
| | 30— 60 | 18.9 | 13.2 | 4.90 | 0.36 | 0.44 |
| | 60— 90 | 16.5 | 10.8 | 4.91 | 0.30 | 0.46 |
| | 90—120 | 15.4 | 10.5 | 4.15 | 0.26 | 0.45 |

pH was found to be in the upper layers generally between 7.7 and 7.8 and in the lower layers between 8.1 and 8.2.

The figures prove a uniform composition of the soil on all parts of the field. The exchangeable potassium is highest in the top layer — about 0.42 m.e. per 100 g and decreases gradually — to about 0.25 m.e. per 100 g — in 1 metre depth. The higher figures in the top layer are probably caused by cultivation and fertilization of the field for long times. The exchangeable sodium is rather high and equals in order of magnitude the potassium content.

As a salty groundwater table does not occur in this place, the sodium is probably derived from irrigation with slightly brackish water for tens of years. The sodium migrates slowly downwards under the influence of the calcium ion created from within the soil by the dissolving action of rain and irrigation water. The exchangeable calcium accordingly is highest in the top layer and decreases with the depth.

The size of the plots was 4 by 10 metres.

*Water*

The tap water used had the following compositions:

| | | ppm. | m.e./l |
|---|---|---|---|
| Cations: | Na | 51.5 | 2.24 |
| | K | 3.0 | 0.08 |
| | Ca | 100.0 | 5.00 |
| | Mg | 36.6 | 5.05 |
| Anions: | Cl | 103.0 | 2.90 |
| | $SO_4$ | 8.6 | 0.18 |
| | $HCO_2$ | 415.0 | 6.80 |

As the figures show, the water is slightly brackish, rich in calcium and magnesium. Its sodium adsorption ratio (SAR) amounts to 1.1, a rather low value, indicating a low sodium hazard.

*Fertilizers*

All the plots got a uniform rate of fertilizers ploughed in before sowing at the following quantities:

Phosphorus: 90 kg superphosphate (of 16% $P_2O_5$) per dunam (or 900 kg per ha),

Nitrogen: 30 kg ammonium sulphate per dunam (or 300 kg per ha). An additional 20 kg were given as top dressing four weeks after sowing.

*Treatments*

The figures for six different treatments are recorded in Table VIII. They show three grades of salinity, with or without potassium added on top of the sodium chloride. The different kinds of water were produced by preparing concentrated solutions of the additives and feeding the concentrates into the stream of irrigation water. The device used for this purpose was an aluminum container of milk-churn type, delivering the concentrate at a pre-set rate into the stream. It is activated by the dynamic pressure of the flowing water. The proprietary name of the apparatus is "Diluter"*). The rate of dilution and final concentration of the salts in the irrigation water was found to be reliable within the range of $\pm$ 5%. Each treatment was in four replications, distributed at random.

*Irrigation*

Irrigation was carried out by overhead sprinkling, a technique preferentially used in Israel. The sprinkler heads were of a static type permitting the covering of relatively small rectangular plots. Though the uniformity of water distribution was far from being satisfactory, they were used for lack of better devices. Water was given at intervals of 7 to 12 days according to weather conditions. A total quantity of about 400 cubic metre per dunam (equal to 1.3 acre-foot) was applied during the whole growing season. In overhead sprinkling, the water and the salts in it act on the plants through the roots as well as through the foliage.

*Time Table*

| | |
|---|---|
| Preparing the field and ploughing: | April 8—10, 1959 |
| Sowing | May 14—20, 1959 |
| Appearance of the plants | June 14—25, 1959 |
| Cropping | Sept. 24, 1959 |

*Remarks:* Sowing of groundnuts usually takes place in the middle of April. The delay was caused by technical reasons. The cropping was well in season. This means that the total growth period was shortened by 3 to 4 weeks as compared with the customary practice.

*Observations*

The growth of the plots irrigated with water of the highest salinity, without potash added (treatment No. 5) was retarded from the beginning. At the end of July, chlorosis and marginal scorch of the leaves was observed. Much less so on the plots irrigated with water of the same salinity, but with potash added (Treatment No. 6).

Attack by diseases, especially by *Cercospora arachidicola* was definitely more serious in No. 5 than in No. 6.

On July, 23rd, fully developed leaves from close to the top of the plants were sampled from the check plot (Treatment No. 1) and from the saline treatments Nos. 5 and 6. The leaves were

---

* Supplied by the Cameron Irrigation Company Ltd., London.

weighed, and measured by drawing the contours of thirty leaves from each sample on paper, cutting out and weighing the paper replicas. The following results were obtained:

**Table III.**

| Treatment No. | Average weight per leaf in g | Average surface per leaf in square cm (one side) |
|---|---|---|
| 1 | 0.29 | 16.06 |
| 5 | 0.21 | 11.99 |
| 6 | 0.24 | 14.21 |

As the figures show the growth retarding influence of the high sodium salt concentration was partially checked by the addition of potassium salt.

*Sampling of the Crop*

Due to the very unequal distribution of the irrigation water, the appearance of the plots was not at all uniform. After much deliberation, it was decided to choose from every plot the eight plants best developed, so that each formula in its four replications was represented by 32 plants.

*Results*

The results of weighing the plants, green matter as well as pods and nuts, and chemical analysis of the different parts are reported in Tables IV-VII.

**Table IV.**

| Treat- No. | Cl in water ppm. | Pods | | | | Air dried plants | |
|---|---|---|---|---|---|---|---|
| | | Number of pods 32 plants | total weight of pods g | weight of pods p. plant g | average single weight g | weight of 32 plants without pods, g | average weight of single plant, g |
| 1 | 100 | 262 | 346 | 10.5 | 1.32 | 7910 | 247 |
| 2 | 170 | 587 | 1052 | 33.0 | 1.78 | 7410 | 231 |
| 3 | 1010 | 282 | 485 | 15.2 | 1.72 | 7640 | 239 |
| 4 | 1080 | 570 | 1012 | 31.6 | 1.78 | 6260 | 195 |
| 5 | 1610 | 285 | 451 | 14.1 | 1.40 | 5910 | 185 |
| 6 | 1730 | 347 | 626 | 19.5 | 1.80 | 5230 | 163 |

*Discussion*

Table IV may be interpreted as follows:

The vegetative development was not in parallel with that of the

**Table V.**

Composition of shelled nuts.

| Treatment | | per cent | | | m.e. per 100 g | | | ratio |
|---|---|---|---|---|---|---|---|---|
| No. | Added ppm | Ashes | Fat | Crude Protein | Ca + Mg | K | Na | K/Na |
| 1 | 0 | 3.02 | 37.8 | 28.2 | 26.0 | 22.3 | 1.4 | 15.9 |
| 2 | 150 KCl | 2.91 | 42.6 | 26.2 | 22.1 | 22.7 | 1.2 | 18.9 |
| 3 | 1500 NaCl | 2.98 | 41.0 | 26.4 | 19.5 | 20.6 | 9.8 | 2.1 |
| 4 | 1500 NaCl + 150 KCl | 2.99 | 42.5 | 25.1 | 20.9 | 20.1 | 9.8 | 2.0 |
| 5 | 2500 NaCl | 3.15 | 41.5 | 26.2 | 19.0 | 20.4 | 13.2 | 1.5 |
| 6 | 2500 NaCl + 250 KCl | 2.93 | 38.3 | 26.2 | 17.9 | 22.0 | 10.4 | 2.1 |

pods. Green mass was highest in the check (Treatment No. 1), followed by treatment No. 3. The crop of pods, however, was highest in Treatment No. 2, closely followed by No. 4. The crop of pods was remarkably low in the check (No. 1) actually the lowest of all treatments, in spite of the rich vegetative development. We ascribe this fact to the delay in sowing and the very short period of growth. The addition of salts to the irrigation water exerted a general stimulative influence and caused an earlier blooming and some shifting in the growth process from the formation of green matter to that of fruits. Both these influences have also been observed by others as reported in a number of publications from Italy, North Africa, and other places (see references). The average weight per pod was also lowest in the check (1.32 g).

The addition of potassium chloride to the non-salinized irrigation water caused an increase in pods by 200%, while green matter decreased by about 6%.

**Table VI.**

Composition of dried shells.

| Treatment | | % | m.e. per 100 g | | | | ratio |
|---|---|---|---|---|---|---|---|
| No. | Added ppm. | Ashes | Ca | Mg | K | Na | K/Na |
| 1 | 0 | 5.57 | 26.7 | 15.3 | 43.5 | 2.9 | 15.3 |
| 2 | 150 KCl | 5.12 | 26.0 | 18.0 | 37.0 | 1.6 | 23.1 |
| 3 | 1500 NaCl | 5.87 | 25.3 | 16.8 | 38.3 | 24.0 | 1.6 |
| 4 | 1500 NaCl + 150 KCl | 6.45 | 29.0 | 20.6 | 39.0 | 22.0 | 1.95 |
| 5 | 2500 NaCl | 8.12 | 32.5 | 20.0 | 38.0 | 48.5 | 0.78 |
| 6 | 2500 NaCl + 250 KCl | 7.90 | 30.5 | 18.0 | 44.3 | 38.5 | 1.15 |

**Table VII.**

Mineral composition in the different parts of the mature plants.

| Treatment | | Ashes in % | | | Milli-equivalents per 100 g dry matter | | | | | | | | | | | |
|---|---|---|---|---|---|---|---|---|---|---|---|---|---|---|---|---|
| | | | | | Ca | | | Mg | | | K | | | Na | | |
| No. | Added ppm. | r | p | l | r | p | l | r | p | l | r | p | l | r | p | l |
| 1 | 0 | 7.23 | 6.86 | 12.10 | 44.3 | 52.0 | 153.5 | 34.0 | 49.4 | 84.5 | 24.0 | 34.0 | 18.0 | 6.1 | 2.5 | 1.4 |
| 2 | 150 KCl | — | 8.33 | 14.07 | 55.4 | 60.0 | 170.0 | 43.6 | 66.4 | 92.0 | 27.0 | 30.5 | 12.0 | 5.7 | 4.3 | 1.2 |
| 5 | 2500 NaCl | 10.85 | 10.35 | 13.33 | 50.5 | 82.6 | 182.5 | 40.5 | 71.8 | 73.5 | 14.0 | 32.0 | 13.0 | 63.0 | 25.0 | 6.9 |
| 6 | 2500 NaCl + 250 KCl | 10.61 | 12.23 | 13.18 | 46.5 | 98.0 | 170.0 | 38.5 | 78.0 | 70.0 | 19.0 | 39.5 | 14.0 | 54.0 | 15.0 | 3.9 |

*Remarks*: r = roots,; p = petioles; l = leaves.

Salinization by the addition of 1500 g/l sodium chloride (Treatment No. 3) did not significantly reduce green matter, and the crop of pods was somewhat higher than in the check. The same salinity but potassium chloride on top of it (Treatment No. 4) amounting to 10% of the sodium salt, more than doubled the crop of pods and brought it close to that obtained in Treatment No. 2 while the green matter declined.

Salinization by the addition of 2,500 g/l of sodium chloride (Treatment No. 5) reduced still further the production of green matter and more so the crop of pods. The same salinity but potassium chloride on top of it (Treatment No. 6) again at a ratio of 10% of the sodium chloride considerably improved the crop of pods (by close to 40%) while reducing green matter production to its lowest figure.

The average size of the pods was lowest in the check (1.32 g) and highest in the high-salinity — plus — potash treatment (1.80 g). The three no-potash treatments taken together resulted in an average weight of 1.48 g.

Table V gives the composition of the groundnuts by their main organic and mineral constituents. No clear correlation is discernible in respect to the organic constituents. In the minerals, the sodium is the only element fluctuating within very wide limits, i.e. from 1.2 to 13.2 m.e. per 100 g, whereas potassium remains rather constant.

Table VI reports the mineral composition of the empty shells. Again the fluctuations in sodium are considerable reaching a peak of 48.5 m.e. per 100 g corresponding to 1.1% by weight. The presence of potassium in the irrigation water apparently depressed the uptake of sodium.

Table VII contains the mineral composition in the different parts of the mature plants, roots, petioles and leaves, separately as influenced by four different treatments.

The total minerals (ashes) were highest in the leaves, the place of the assimilatory activity. The treatments do not seem to have exerted a major influence. Calcium was mainly concentrated in the leaves, which generally contained three to four times more than the roots. Magnesium was less highly enriched in the leaves when it surpassed its concentration in the roots by the factor of two only. The concentration of potassium was still more evenly distributed though smallest in the leaves. The low concentration of the potassium in the leaves is certainly due to the mature state of the analyzed plants. Samples of leaves taken during the peak of the growth period proved to contain from 32 to 54 m.e.K per 100 g. Sodium shows a state known from other cases:

It is mainly concentrated in the roots where its concentration reaches a value ten and more times higher than that in the leaves. Apparently there exists a blocking mechanism in the transport of this element from the roots into the shoot. The presence of extra

**Table VIII.**

Growth Experiments with Cow Peas for Green Fodder

Sown (After Groundnuts): 26th September 1959
Cropped: 22nd December 1959.

| No. | Added to tap water | Cl in water p.p.m. | Na(+K) m.e./l | S.A.R.* | Cropped in kg | Crop in relative figures |
|---|---|---|---|---|---|---|
| 1 | 0 (check) | 103 | 2.2 | 1.1 | 88.0 | 100 |
| 2 | 150 ppm KCl | 174 | 4.2 | 2.1 | 89.2 | 101 |
| 3 | 1500 ppm NaCl | 1013 | 27.9 | 14.0 | 45.1 | 51.3 |
| 4 | 1500 ppm NaCl + 150 ppm KCl | 1084 | 29.9 | 15.0 | 66.6 | 75.7 |
| 5 | 2500 ppm NaCl | 1623 | 45.1 | 22.6 | 16.8 | 19.1 |
| 6 | 2500 ppm NaCl + 250 ppm KCl | 1694 | 48.4 | 24.2 | 45.9 | 45.9 |

*Remarks:*
* S.A.R. = Sodium Adsorption Ration. In computing the S.A.R., K was included with Na.

The quality rating of treatments No. 3 and 4 in the Riverside diagram would be C 5—S 2, and No. 5 and 6 would be C 5—S 3.

potassium in the growth medium reduces in this case too the uptake of sodium by the different parts of the plant.

## Experiments with Cow Peas

After removing the crop of groundnuts, the plots were sown with cow peas, on September 26, and irrigated with the same water treatments as received by the groundnuts. The peas were cropped as green fodder on December 22, after a growth period of 87 days.

The results are recorded in Table VIII.

*Discussion*

The addition of potassium to the tap water had no significant influence on the crop. In the conventional language of "plant nutrition" that means there was no lack in available potassium. But when salinized water (with 1500 ppm sodium chloride added) was used, the crop decreased by 50%, and recovered to 75% of the check on the addition of 150 ppm potassium chloride on top of the salt. Salinization with 2500 ppm sodium chloride caused a drop to about 20% of the check, and the further addition of 250 ppm potassium chloride a recovery to 52% of the check.

It may be concluded that the very salt-sensitive pea plant responded most distinctly to the adjustment of the sodium-potassium ratio in the water coming into contact with the foliage and also percolating into the rootzone of the plants.

## Summary

A new concept for growing crops under saline conditions stresses the importance of a balanced ionic environment with special attention to the sodium-potassium relationships.

In order to verify this new approach, field experiments with groundnuts as test plant were carried out, using irrigation water artificially salinized by addition of 1500 and 2500 g of sodium chloride to one cubic metre of water.

In parallel water was used salinized to the same degree but receiving on top of the sodium chloride 150 or 250 g of potassium chloride per cubic metre.

The irrigation was by overhead sprinkling so that besides the roots the foliage was exposed to the salinized water also.

A stimulative effect of salinity and especially of potassium salt on blooming and fruiting was observed and the total period from sowing to cropping seems to be shortened by the effect of salts.

Salinity at both degrees had a depressing effect on yields. But the addition of potassium chloride at a ratio of 10% of the sodium chloride present in the water completely restored the yield at medium salinity and partially at the highest degree of salinity.

The addition of potassium salt to the salinized water reduced the susceptibility of the groundnut plants to attack by fungal diseases such as *Cercospora arachidicola*.

After removal of the groundnuts, cow peas were grown on the experimental plots which continued to be irrigated with salinized water of the same treatment as before. The peas were cropped as green fodder after a growth period of 87 days. The plants responded significantly to the adjustment of the sodium-potassium ratio in the irrigation water.

## REFERENCES

BERNSTEIN, L. & HAYWARD, H. E., 1958. Physiology of salt tolerance. *Ann. Rev. Plant Phys.* **9**: *43*.

CERIGHELLI, R., & DURAND, V., 1954. Influence du chlorure de sodium sur la germination et développement du riz. *J. Riz* **4**: *34—45* (1954).

HEIMANN, H., 1959. Irrigation with saline water and the ionic environment. Potassium Symposium 1958: *173—220* (Berne).

HEIMANN, H. & RATNER, R., 1961. The influence of potassium on the uptake of sodium by plants under saline conditions. *Bull. Res. Counc. Israel* **10** A: *55—62*.

LIPMAN, C. B., DAVIS, A. R., & WEST, S. S., 1926. The tolerance of plant for NaCl. *Soil Sci.* **22**: *303—322*.

NOVIKOFF, V., 1946. Note sur l'utilisation des eaux salées. *Ann. Serv. Bot. Agron. Tunisie* **19**: *138—162*.

PANTANELLI, E., 1937. Irrigazione con acqua salmastre. *Staz. Agrar. Sper. Bari*, No. 26, 80 pp.

PURVIS, E. R. & BRILL, G. D., 1957. Can use brackish water to irrigate some crops. *New Jersey Agric.* **1957** (Jan-Febr.): *11—13.*

YANKOVITCH, L., 1946: Récherches d'une méthode d'étude de la résistance des plantes aux chlorures. *Ann. Serv. Bot. Agron. Tunisie* **19**: *165—177.*

# IRRIGATION WITH SALINE WATER IN PUGLIA*

BY

GIACOMO LOPEZ

*Agricultural Experimental Station — Bari*
*(Dir. Prof. Vincenzo Carrante)*

## Introduction

World population, which today comprises some three milliard people is expected to double within a few decades. This makes it essential to produce more food by the utilization of vast areas of land which, up to now, have not been cultivated because they are unsuitable for normal agriculture. Furthermore, it necessitates the searching out and putting to use of new water resources, such as waters which are more or less brackish.

This is even more essential in those regions where sweet waters, whether surface or underground, are scarce, and where atmospheric precipitations are equally insufficient. But in order to utilize brackish water practically, we must know not only its chemical composition and concentration, but also the complex inter-relationship between water, soil and plants, even if only a few varieties of crops may be grown.

As is known, the salinity of water may have different origins and compositions, mostly due to the nature and structure of the soils through which the water has passed.

In the case of the Puglia underground waters PANTANELLI (1937) has found that the two principle factors of salinity are:

1. Permeable strata along the sea-shore, allowing sea-water infiltration where the sea-bound underground flow is weak.
2. The existence of impermeable concave strata, for instance of clay or tufo (tufa), which hinder the renewal of the underground water layer, thereby retaining salinity, ancient or recent.

According to PANTANELLI (1937) the first of the above factors, which has a permanent stratigraphic cause, is more serious and important than the second. This second factor is, in fact, the relict of old lagoons which have disappeared because of the gradual rising of the land, and is, therefore, liable to improve as the causes, which produced the accumulation of salts in prehistoric times, no longer exist. Furthermore, this condition causing salinity may be the result of an accidental contamination and, as such, may even more easily be eliminated.

The abundance of such brackish waters raises complex problems

* Translated from the manuscript written in Italian.

that must be solved before planning extensive irrigation programs, as the different salts may affect the plants as well as the soil. We must, therefore, not only consider the total or partial salt resistance of the plants, and select only the most resistant varieties, but must also take into consideration the interactions between water and soil salts.

These interactions, in fact, may dangerously alter the physico-chemical properties of the soil, such as the structure, permeability, state of exchangeable cations on the colloidal complex, etc.

## Water Characteristics

In the province of Bari, as in the longest part of the Salentino coastline, the presence of a rocky shore, cracked, permeable and calcareous, permitted sea-water infiltration which resulted in a more or less brackish layer of underground water; salinity varies between 2 and 10‰ along the Bari sea-shore, and between 2 and 22‰ along the Salento (Lecce) coast.

However, salinity does not penetrate more than 1 to 2 km near Bari and 4 to 5 km near Lecce and — according to PANTANELLI — gradually diminishes as we get farther from the sea. BOTTINI & LISANTI (1955), in a study of the brackish wells along the Bari sea-shore, classified the wells into two groups: those not further than 600 m, and those between 1,000 and 3,000 m from the sea.

In the first group, although there were recurrent variations, they found salinity ranging mainly between 4 and 6‰; seldom was it above 6 or as low as 1 to 2‰. For the interior group, although there was even greater variability, the salinity remained mostly below 4‰.

LOPEZ (unpublished data), analyzing the waters of five wells in 1952 in the Egnazia (Brindisi) zone, found the following values of particular interest, even if limited to few samples: (see Table I).

These waters contain mainly sodium chloride, which amounts to 60-80% of the total residue. Aside from the effects caused by their high salinity and composition, the waters of the Puglia coast are also rich in plant nutritional elements, foremost among them, the sulphuric and the calcium ions.

After sodium, calcium and magnesium are the most frequent cations. This is very important as, both in the soil and in the plants, they partly or totally neutralize the sodium damage.

However, the different proportions of the single salty constituents are much more important from the viewpoint of modern theories concerning the utilization of brackish waters, and the more or less harmful effects of such irrigation on the plants or on the irrigated soil. In fact, the most important characteristics of water for irrigation, aside from the total salt concentration, are: a) the proportion of sodium in respect to the other cations; b) the presence of some

**Table I.**

Chemical composition of irrigation water from five wells, used in the Egnazia (Brindisi) zone.

| Origin of sample (well) | Dist. from the sea (m) | Residue 100-105°C g/l | Cl g/l | $SO_4$ g/l | $HCO_3$ g/l | Na g/l | K g/l | Ca g/l | Mg g/l | pH |
|---|---|---|---|---|---|---|---|---|---|---|
| 1. Fonte Caretta | 1500 | 9.160 | 4.390 | 0.664 | 0.050 | 2.670 | 0.029 | 0.173 | 0.192 | 7.4 |
| 2. Trappeto del Re | 3000 | 2.640 | 1.150 | 0.168 | 0.010 | 0.663 | 0.011 | 0.099 | 0.150 | 7.3 |
| 3. L'Abbate | 1500 | 6.240 | 2.920 | 0.408 | 0.213 | 1.208 | 0.013 | 0.152 | 0.188 | 7.5 |
| 4. De Bellis | 1500 | 3.600 | 1.660 | 0.238 | 0.053 | 0.839 | 0.013 | 0.114 | 0.106 | 7.3 |
| 5. Contrada Coccaro | 2000 | 1.880 | 0.760 | 0.101 | 0.122 | 0.702 | 0.007 | 0.094 | 0.082 | 7.4 |

toxic ions, such as boron and, in the case of some crops, chlorine, sodium and bicarbonates.

Irrigation water may be satisfactorily evaluated by determining its electrical conductivity; this has become a standard, reliable method for measuring the salt concentration by now. According to electrical conductivity values, the U.S. Salinity Laboratory (RICHARDS et al., 1954) has adopted the following four classes of conductivity values: the first, below 250 micro-mhos/cm; the second, between 250 and 750 micro-mhos/cm; the third, between 750 and 2,250 micro-mhos/cm; and the fourth, between 2,500 and 5,000 micro-mhos/cm.

According to the above classification, the first group, corresponding to a total salinity of about 0.150‰, may be used without any harmful effects; the second group, corresponding to a total salinity between 0.150 and 0.450‰, may be considered practically safe under all conditions; the third group, corresponding to a salinity between 0.450 and 1.36‰, may be used only on permeable soils with moderate liscivation (alcalination) while the waters of the fourth group are totally unusable.

Recently, the same U.S. Salinity Laboratory has modified the above classification, proposing the following:

(1) from 0 to 250 micro-mhos/cm: *waters of low salinity*, usable for most crops and most soils, with a low probability of increasing soil salinity.

(2) from 250 to 750 micro-mhos/cm: *slightly brackish waters*, usable also for salt-sensitive plants under the condition that the soil be of medium or high permeability. On soils of low permeability, it is necessary, even with normal irrigation practices, to ensure a sufficient leaching of the soil, and, in some cases to raise only crops of low salt sensitivity.

(3) from 750 to 2,500 micro-mhos/cm: *medium brackish waters*, usable only on soils of good or moderate permeability. In such cases, regular drainage must be ensured in order to prevent increased salt concentration, and crops of moderate or high salt resistance must be selected.

(4) from 2,500 to 4,000 micro-mhos/cm: *highly brackish waters*, usable only on permeable soils where sufficient drainage is ensured in order to eliminate excess salt, and only for salt resistant crops.

(5) from 4,000 to 6,000 micro-mhos/cm: *very highly brackish waters*, generally unusable except for very permeable soils with frequent leaching and for very salt resistant crops.

(6) above 6,000 micro-mhos/cm: *exceedingly brackish waters;* unusable in any case.

As a result of the analysis of many irrigation waters, the Bari Agricultural Experimental Station has, for some time, adopted the following classification in six groups:

(I) less than 0.5‰, *sweet waters,* drinkable if bacteriologically pure, and with a total hardness of under 32 French degrees.

(II) from 0.5 to 1.5‰, *slightly mineralized waters,* inadvisable for human consumption even if bacteriologically pure; generally suitable for animal consumption and irrigation.

(III) from 1.5 to 3‰, *slightly brackish waters,* unfit for human consumption, hardly fit for animal consumption, but usable on permeable soils for almost all crops, except for those which are exceedingly salt sensitive.

(IV) from 3 to 6‰, *brackish waters,* generally unfit for irrigation, only usable with caution for salt resistant crops.

(V) from 6 to 10‰, *very brackish water,* usable only with great caution on very permeable soils for extremely salt resistant crops.

(VI) above 10‰, *exceedingly brackish waters,* unusable in all cases.

## Seasonal Variations in Salt Concentration of Irrigation Waters

The water salinity is in many cases variable. In contrast to wells of constant salinity, also after prolonged exploitation, there are others whose salinity varies seasonally in accordance with atmospheric precipitation and with exploitation. As an example, we list in Table II data concerning a well in the experimental field for irrigation with brackish waters, established by the Ente Irrigazione di Puglia e Lucania, for the period from June 1954—December 1957 (Ficco, 1960).

During the period November—February, which is the period of maximum rainfall (averaging 70% of the total annual rainfall), total salinity may drop to a minimum of 1—2‰ and $Cl^-$ to 0.5—0.7‰, as a result of increased water flow from the hinterland. On the contrary, during the hotter season, coinciding with the need for irrigation, as a result of the diminished hinterland flow and of increased exploitation, the salinity may reach a maximum of approximately 8‰.

Such a high concentration just at the time when irrigation is required (June-August) makes these waters usable only on very permeable soils and for very salt resistant crops.

However, as will be illustrated further on, the use of these waters is justified by some favourable factors and by the application of particular agronomical practices, which allow the removal of the salts brought by those waters in a relatively short time.

Another important factor in respect to the variation of salt concentration in underground waters is the thickness of the bottom-water layer over the sea-water layer and the way in which this salty layer affects the sweet-water layer.

It is known that the Puglia region south of the region of the Ofanto River may be considered as constituted of a foundation

**Table II.**

Seasonal fluctuations of salinity and chlorine content of the water from a well in Egnazia (Brindisi).

| Date | | Total Salt Content ‰ | Chlorine ‰ |
|---|---|---|---|
| June 16 | 1954 | 7.252 | 3.368 |
| July 22 | | 7.865 | 3.936 |
| August 14 | | 8.746 | 4.149 |
| September 13 | | 8.200 | 3.901 |
| October 19 | | 4.726 | 2.145 |
| November 20 | | 1.218 | 0.571 |
| December 10 | | 1.694 | 0.794 |
| January 15 | 1955 | 1.892 | 0.858 |
| February 12 | | 2.384 | 1.085 |
| March 15 | | 2.230 | 1.014 |
| April 20 | | 3.100 | 1.439 |
| May 5 | | 7.364 | 3.561 |
| June 16 | | 7.200 | 3.840 |
| July 10 | | 7.560 | 3.680 |
| August 26 | | 7.700 | 3.760 |
| September 20 | | 6.600 | 3.480 |
| October 20 | | 2.280 | 1.140 |
| November 21 | | 5.830 | 2.840 |
| December 20 | | 4.750 | 2.310 |
| January 20 | 1956 | 5.22 | 2.52 |
| February 20 | | 1.13 | 0.48 |
| March 20 | | 1.65 | 0.72 |
| April 9 | | 1.98 | 0.89 |
| May 20 | | 6.56 | 3.35 |
| June 20 | | 7.13 | 3.70 |
| July 10 | | 7.27 | 3.65 |
| August 10 | | 7.24 | 3.75 |
| September 10 | | 7.45 | 3.73 |
| October 20 | | 7.30 | 3.62 |
| November 15 | | 6.12 | 3.22 |
| December 15 | | 4.17 | 2.13 |
| January 15 | 1957 | 4.28 | 2.15 |
| February 15 | | 4.19 | 2.10 |
| March 15 | | 2.54 | 1.13 |
| April 15 | | 4.15 | 2.08 |
| May 10 | | 4.72 | 2.14 |
| June 10 | | 7.87 | 4.11 |
| July 10 | | 7.62 | 3.76 |
| August 10 | | 7.81 | 3.98 |
| September 10 | | 6.26 | 3.33 |
| October 15 | | 4.68 | 2.34 |
| November 15 | | 1.91 | 0.89 |
| December 15 | | 3.80 | 2.13 |

layer of stratified cretaceous lime-stone of considerable thickness, covered in the coastal regions by chronologically younger soils of pleistocene origin (Carrante et al., 1956; Pantanelli, 1937; Zozzi & Reina, 1954).

The above-mentioned calcareous foundation, in itself impermeable, is today remarkably fissurated and broken as a result of the geological changes of the Region. Through those fissures and splits, increased by the solvent activity of dissolved bicarbonic acid, the water is rapidly absorbed through a complicated system of (natural) ducts, channels and tunnels, thus originating the well-known "carstic circulation".

In a general way, as a result of gravity, atmospheric water tends to drop as much as possible, stopping only when it meets the salty sea-water which, in a similar way, through lateral permeation invaded the calcareous strata. Nevertheless, as there exists a difference in density between the two types of water (percolated and sea-water) and as a result of the relatively slow movement of the percolated one, the two waters do not mix, thus sweet water stratifies above the salt water to form a layer of increasing thickness as the distance from the sea increases.

The level of such a sweet layer tends to converge with the sea-level, but stays always some metres above it as a result of the difference in salinity.

The thickness of the sweet layer may be theoretically calculated by the use of Herzberg's theory when the difference in level between the surface of the layer and that of the sea is known, by means of the following equation:

$$H = h\frac{d}{D - d}$$

in which h is the difference in level between the surface of the layer and that of the sea, H is the thickness of the layer, D is the density of sea-water, and d the density of sweet water.

As a matter of fact, however, the contact surfaces between sweet and salt water are not quite delineated, as the determining conditions are not stationary; there is always a mixed zone caused by pumping, by tide movements and by sea-currents or Adriatic winds.

## Sodium Percentage

The cations absorbed on the colloidal fraction of the soil are in equilibrium with those present in the solutions circulating in the same soil. It is therefore evident that the equilibrium is broken, when the proportion of some of these cations in the irrigation waters increases, causing cationic exchange phenomena between the two systems.

The cations normally found in irrigation waters are in decreasing proportion: sodium, calcium and magnesium, and in a small quantity, potassium. The cationic exchange most important to us is the sodium one, i.e. the ease with which that cation may be absorbed in the clay, according to the following reaction.

$$2\,Na^{+} + Ca\,/\underline{clay}/ \leftrightarrows \overset{Na}{\underset{Na}{}}\,/\overline{clay}/ + Ca^{++}$$

Such a phenomenon, called by the U.S. Salinity Laboratory "risk of soil alkalinization" is caused by the absolute and relative cation concentrations. The risk increases with the proportion of sodium, while it is low or nil if calcium and magnesium predominate.

The various mineral constituents, especially cations, are therefore essential in characterizing such waters and in modifying the soils irrigated by them. When planning irrigation, the problem of sodium on the basis of the relationship $Na/Ca + Mg$, with the concentrations in mg. equ., must be considered. In the past it was suggested to consider the sodium percentage according to the following formula:

$$\text{Na percentage} = \frac{Na \times 100}{Ca + Mg + K + Na}$$

However, in order to determine the limits of the risk of sodium alkalinization, for a soil irrigated with brackish water, the following formula suggested in 1953 by the U.S. Salinity Laboratory seems to be more correct:

$$\text{Sodium absorption relationship} = \frac{Na^{+}}{\sqrt{(Ca^{++} + Mg^{++})/2}}$$

The above formula is founded on cationic exchange equations of the type bassa actions, which are theoretically more strictly correlated with the percentage of sodium absorbed in the soil-absorbing complex. We shall indicate the said sodium percentage by Na.

On the basis of said relationship and of the sodium percentage, the U.S. Salinity Laboratory has suggested the following classification of irrigation waters; such classification being mainly related to the effects of the exchangeable sodium on the physical and structural characteristics of the soil.

1) *Low sodium content water* (Na $<10$), may be used on almost any soil with little danger of the accumulation of small quantities of exchangeable sodium;

2) *Medium sodium content water* (Na between 10 and 18), may cause alkalinization in soils rich in clay and poor in organic matter, especially when the drainage is scanty, unless there are appreciable quantities of sulphates (gypsum) in the soil. Such water, however, may be used on loose and permeable soils.

3) *High sodium content water* (Na between 18 and 26) may cause

noxious accumulation in almost all soils not containing gypsum; may require special treatment, such as excellent drainage, high liscivation and the addition of organic matter to improve the physical conditions;

4) *Excessive sodium content water* (Na $>$26), as a rule, unsuitable for irrigation.

Another important factor in water is the content of carbonates and bicarbonates. EATON (1950) maintains that the carbonates and bicarbonates may indirectly affect the quality of water by means of calcium and magnesium precipitation according to the following process:

$$Ca^{++} + N^{+} + 3\ HCO_3 \rightarrow CaCO_3\ \text{(precipitate)} + Na^{+} + HCO_3 + CO_2 + H_2O$$

Such a process would decrease the total salinity as a result of calcium and magnesium precipitation, but increase the proportion of sodium against the other cations, with more risk of sodium contamination as explained above.

This phenomenon is more evident when in the irrigation waters carbonic and bicarbonic ions exceed calcium and magnesium ions, and it is increased when soil containing solutions rich in such ions gets dry as a result of plant absorption or of evaporation. In such an event, after calcium and magnesium have precipitated as carbonates, the excess carbonic ions combine with sodium, thus causing primary soil alkalinity, called "residual sodium carbonate" by EATON.

The increase of such carbonates raises the soil pH, and accentuates, as will be shown further on, the characteristic symptoms of the black alkaline soils.

Analyses by several authors of papers on irrigation waters in the Puglia region show that the relationship indicating the sodium absorption risk rises to high values (above 10) only in the case of medium or high saline waters, which are very dangerous for irrigation.

So far, however, these waters, even if used for a very long time on the same fields, have not caused alkalinization; and this is due to favourable natural conditions such as: High soil porosity and permeability permitting sufficient drainage; and also concentrated winter pluviosity, causing intense washing out of the excess salts that have accumulated during the irrigation period.

Another favourable factor is the local practice of alternating irrigated horticultural crops every second on third year with dry cereal crops.

## The Soil, with Respect to Irrigation and Salinity

Exaggerated irrigation by brackish water may, in the process of time, so badly damage the soil as to no longer permit rational agriculture and, as a final result, cause more or less serious alkalinization.

The first stage in soil salinization and alkalinization is the increase of soluble salts in the circulating solutions; this phenomenon is more marked in those zones where rains are scarce, especially when drainage is insufficient.

The said increased salinity is also due to evaporation and to water absorption by the plants. This is followed by a higher proportion of sodium (either soluble or exchangeable), precipitation of calcium and magnesium carbonates, and, later in a second stage, by precipitation of calcium sulphates, the formation of sodium carbonate and the increase of the reaction.

Such a series of phenomena, through their negative influence on the characteristics of clay type and structural colloids and on the utilization of some nutritive elements, may cause the reduction or, in extreme cases, the total destruction of soil fertility.

On the experimental field of Savelletri (Brindisi) during 1955—57, Ficco (1960) found at the end of his tests on average salinity, a decrease of 28% from the initial test and a decrease of 81.3% for the chlorine.

In such experiments, the salt content of the soil, although reaching 2—3‰ during and after irrigation, dropped again in the following winter season to a normal 0.2—0.3‰. During all these tests, soil salinity was measured every year in April, before the sowing, and in October after harvest.

As a matter of fact, salt accumulation was never found in a quantity sufficient to damage the soil through alkalinization.

Such happy results are due to a number of favourable conditions existing all along the coastal area under consideration. These conditions are: remarkable soil porosity, good natural drainage, concentrated winter rains and reconstitution of the deep water reserves.

Table III shows analytical data concerning physical and hydrological characteristics of ten soil samples; N. 1—5 were taken along the coastal zone between Bari and Mola and are typical red soil on calcareous rock of the cretaceous layer*.

The other samples were taken around Lakes Alimini and Fontanelle (Lecce) and are of soils on pleistocene "tufi". They are characterized by the presence of coarse materials, sometimes in large proportions and of high calcareous content.

In lysimetric tests at the Bari Agricultural Experimental Station, it was found that winter rains wash away soluble salts from the

* The clay fraction of red soil, according to Cecconi, is mainly composed by "illite", which is much less expansable, plastic and dispersible than other clay minerals, particularly montmorillonite, typical of soils that swell easily and are dispersible.

**Table III.**

Physical and hydrological characteristics of some soils in the Apulian coastal Region (Lopez)

| No. of sample | skeleton % | soil % | coarse sand % | fine sand % | silt % | clay % | $CaCO_3$ % | organic substance % | permea-bility cm/h | porosity % | moisture equi-valent % | water capacity | hygrosco-pical co-efficient | pH |
|---|---|---|---|---|---|---|---|---|---|---|---|---|---|---|
| 1 | 1.7 | 98.3 | 2.0 | 24.4 | 20.2 | 51.5 | Tr. | 1.84 | 6.76 | 56.1 | 27.5 | 54.4 | 3.56 | 7.75 |
| 2 | 1.2 | 98.8 | 6.1 | 22.9 | 18.8 | 50.3 | Tr. | 1.78 | 11.40 | 56.5 | 30.9 | 52.1 | 3.66 | 7.65 |
| 3 | 5.9 | 94.1 | 8.4 | 21.4 | 17.4 | 50.1 | 0.5 | 2.15 | 1.87 | 57.1 | 36.4 | 52.5 | 3.36 | 8.50 |
| 4 | 1.2 | 98.2 | 15.5 | 25.9 | 9.4 | 46.8 | 0.7 | 1.65 | 9.56 | 53.9 | 27.9 | 55.1 | 3.70 | 8.50 |
| 5 | 17.8 | 82.2 | 10.1 | 15.9 | 23.7 | 36.1 | 11.2 | 2.96 | 8.53 | 54.9 | 24.4 | 52.3 | 2.64 | 8.50 |
| 6 | 1.5 | 98.5 | 10.9 | 32.1 | 29.8 | 27.1 | 3.0 | 2.26 | 5.5 | 53.4 | 24.8 | 51.9 | 5.16 | 7.40 |
| 7 | Tr. | 100.0 | 14.1 | 25.8 | 29.4 | 30.7 | 0.5 | 3.01 | 7.1 | 53.0 | 27.8 | 53.7 | 5.77 | 8.00 |
| 8 | 22.0 | 78.8 | 41.3 | 44.8 | 8.1 | 5.7 | 51.0 | 1.32 | 24.6 | 49.6 | 13.9 | 35.6 | 0.68 | 8.40 |
| 9 | 24.5 | 75.5 | 36.8 | 35.7 | 15.7 | 11.8 | 40.0 | 1.84 | 20.0 | 52.5 | 17.7 | 45.2 | 1.75 | 8.40 |
| 10 | 0.5 | 99.5 | 7.2 | 38.7 | 35.1 | 17.9 | 3.0 | 2.42 | 8.7 | 54.2 | 47.5 | 47.5 | 4.11 | 8.30 |

cultivable layers, thus allowing winter crops on soils into which, during the preceding summer, irrigation had brought large quantities of salts.

The winter washing carries away, as a rule, chlorine first and then sulphates, calcium and alkalis. In soils fertilized by sulphates, these are washed away much faster than the chlorine.

In calcareous soils, the alkalis, particularly sodium, are washed away to a higher extent than calcium, and their delavation (leaching) is inversely proportionate to the absorbing power of the soil.

These soils are loose and permeable, which is important in view of the fact that the easy drainage of gravitational waters renders excessive soil impregnation impossible and, in the case of brackish water, prevents the rise of salinity in the circulating solutions when these are concentrated by evaporation or transpiration.

According to BOTTINI & LISANTI (1955), additional fundamental characteristics are connected with such looseness of the soil, i.e. capillary movements of only moderate intensity and amplitude. Therefore, it can be assumed that should gravitational waters accumulate underground, the emergence to the surface of the water layer newly reformed would be only slight and the danger of capillary return to the cultivated soil layer of the salt accumulated in the percolated waters would be very slight. Such an occurrence is even further prevented by another important factor, namely, the particular location and thickness of the soils irrigated by brackish waters along the whole "Puglia" coastal zone.

In most cases, the thickness is limited to an average of a few tens of centimetres, so that the cultivated layer comprises it almost in full. A real and true subsoil is therefore missing, and in its place there are banks of compact lime-stone, more or less crystallized, or of thick pleistocene "tufo". The lime-stone is — as already mentioned — stratified and cracked, while the "tufi" are very permeable, being composed of calcareous sands more or less cemented by calcareous cement, mixed with fossil debris.

Thus, the excess irrigation water, after percolating through the cultivated layer, is dispersed into deeper strata, and can no longer return to the surface through capillarity. Nothwithstanding such favourable conditions, there is, at the end of each irrigation season, a remarkable rise in the salinity of the solutions circulating in the cultivated layer, due to the high quantities of accumulated salts, which — according to FICCO (1960) — may in a single season reach and surpass 25,000 kg/ha. But after the heavy winter rains (approx. 700 mm from October to February) the salinity drops again to normal levels.

However, it cannot be ruled out that the contact of salt rich irrigation water with the various soil components — and particularly with its clayey colloidal fraction — may disturb the balance between the cations in the solutions circulating in the soil and those absorbed in the absorbing compound, modify the proportions of the absorbed metallic cations with an increase of exchangeable sodium, and cause unfavourable phenomena, such as dispersion and collapse of the clay, increase of salinity, decrease of permeability, loss of acration, etc.

The absorption of sodium ions on clay surfaces depends on the salt concentration and the sodium proportion in respect of the other cations present in the irrigation water and, when this comes into contact with the soil, in the circulating solutions.

According to CHANG (1961), waters of low salinity and poor sodium, little affect sodium absorption (soil alkalinization); on the other hand, highly saline waters with high Na/K relationship greatly increase the proportion of absorbed Na, up to 50% of exchange capacity.

As shown before when discussing the qualities of irrigation waters, formation of residual sodium carbonate and the resulting rise of soil reaction are favoured by the concentration of circulating solution and by the precipitation of calcium and magnesium carbonate, particularly when the sum of carbonic and bicarbonic ions exceeds the sum of the above mentioned bivalent ions recorded.

HAUSENBUILLER, DAQUE & WAHAB (1960) studied the effects of irrigation on the levels of exchangeable sodium and of $CaCO_3$ in the soil; they found that for water with a relationship Na:Ca equal to 1:1, the exchangeable sodium accumulated in the subsoil more than in the surface.

WILCOX, BLACK & BOWER (1954) found a remarkable rise in the exchangeable sodium also when the concentration in water was 5 meq/l or more and soluble sodium percentage was 75%. The rise of the reaction beyond alkalinity limit ($pH > 9$) was strictly connected with the exchangeable sodium when waters rich in carbonates and bicarbonates were used, while no significant relationship between those two variables existed when no bicarbonated waters were used.

Thus the excess irrigation water, after percolating through the cultivated layer, is dispersed into deeper strata, and can no more by capillarity return to the surface.

In the Puglia coastal zone discussed here, no remarkable cases of alkalinization or sodiumization have been found as yet, thanks to the intrinsic characteristics of soils and subsoils, although irrigation by brackish waters has been practised as far back as the Graeco-Roman colonization.

We have but little data concerning the condition of exchangeable sodium in irrigated soils; but these few (LISANTI, 1958; LOPEZ, 1958) authorize us to consider the danger of soil-degradation as non existent or, at least, as very remote.

Actually the exchange capacity of these soils, amounting to

30—32 meq per 100 g of soil, is mainly saturated by calcium which in some cases may reach 80—90% of the total of all absorbed cations. If the other two cations (K and Mg) are also considered, we can easily infer that sodium must exist in a minimum measure, or at least only to an extent which renders soil alkalinization very unlikely.

A favourable factor, in this connection, can be the presence of more or less subdivided $CaCO_3$, particularly in soils laying over pliocene tufi.

This component, according to BOTTINI & LISANTI (1955), tends to transfer new quantities of calcium ions into the circulating solutions, continuously; thus in the same solutions the relationship sodium/calcium drops further, to the point that the action of sodium ions becomes almost nil.

To the natural factors of the edaphic set up structure of the Puglia coastal zone where brackish waters are used for irrigation, human activity must be added, namely special agronomical technique and organic fertilization.

The agronomical technique consists in alternating irrigated horticultural crops and dry grain crops every second or third year. Organic fertilizers consist of seaweed gathered along the sea-shore and previously washed by rain-water.

These two methods make the removal of soluble salts easier and quicker, even where the natural conditions of the soil are not so favourable.

In addition, according to DABELL (1955), no less important is the natural conservation of hydro-equilibrium between atmospheric waters percolated into the deep layers and useful for irrigation purposes without harmful mixing with unfiltered sea-water. Such an equilibrium is actually perfectly kept, thanks to the use of hand or animal-driven pumps.

Such equilibrium-maintaining conditions prevailed up to some ten years ago. But there is a danger of serious deterioration if — as is already happening in many other countries — the irrigated areas are extended and modern mechanical pumps used, resulting in disturbing the deep water layers, the emerging of underlying brackish waters up to the point of salinization of wells which were formerly quite suitable for irrigation.

## Effects on Plants by Irrigation with Brackish Waters

Various and complex is the behaviour of various plants as a result of high salt concentration in the soil, be it natural or brought about by irrigation waters.

Without entering into the physiological details of the problem, which would exceed our limits, it seems from many experiments on sand or on nutritive solutions, that the growth inhibiting effect of the salts in the circulating solutions is more due to the higher osmotic pressure than to the concentration itself, as less water is absorbed in the soil.

This means that plant tolerance or resistance to higher salt concentrations in the circulating solutions depends only on three factors:

a) the ability to raise the osmotic pressure in the cell sap, so as to balance or compensate the higher external osmotic pressure;

b) the ability to regulate the absorption of the ions from the circulating solutions in a way to increase the internal osmotic pressure without excessive ion accumulation inside the vegetal cells,

c) the ability of the protoplasm to resist higher ion accumulation.

In addition to the reduced growth caused by the increased osmotic pressure, no less important effects are due to the toxicity of the single salts or ions present in the solutions.

Such toxicity may manifest itself not only by direct action of the salt or the ion on root membranes or on vegetal tissues, but also indirectly by affecting the absorption and the metabolism of essential nutritive elements.

On the other hand, some salts or ions may stimulate growth if they do not appear in too high concentration. For instance, sodium in small doses may improve the quantity and quality of tomatoes.

Plant salt resistance may be valued in different ways:

1) The simple survival against high salt concentration without any economic production. This valuation is more ecological than agronomical, and utilizes in effect the edaphic surrounding, as plants thus surviving constitute the natural spontaneous flora of brackish soils.

2) A more agronomical way is to valuate salt tolerance through the productive ability of different species or varieties at a certain salinity level, considering as salt resistant those that in par conditions gave the best production.

3) To compare the product of the same variety or species cultivated in saline surroundings with the same in normal surroundings.

This last criterion seems to be the best, as it makes it easier to understand the resistance of the different species to the various salt doses.

Pantanelli (l.c.) sets the value of 2.25‰ as upper limit between sweet waters and the usable brackish waters. This value is very approximate, since the effect of brackish water on plants and soils depends not only on the total concentration but also on the single components.

In nutritive solutions above 2 g/l many plants show the symptoms characteristic of saline medium; the more resistant ones even thrive better such as turnip (*Brassica* sp.), mustard *(Sinapis alba)*, tomato *(Solanum lycopersicum)*, eggplant *(Solanum melongena)*, pepper *(Solanum capsicum)*, rice *(Oryza sativa)*, beetroot *(Beta vulgaris)*, cotton (*Gossypium* sp.) which resist up to 10‰.

In the soil these limits increase little if the soil is sandy or poor in lime, iron and free aluminium; they rise very much if it is clayey or rich in said elements.

According to VAN DEN BERG's (1950) list, the following plants are most resistant (above 10‰):

Date palm *(Phoenix dactylifera)*, some varieties of cabbage (*Brassica* sp.), barley *(Hordeum vulgare)*, sugar-beet, turnip, cotton, etc.

Medium resistant:

Olive *(Olea europea)*, vine *(Vitis vinifera)*, fig *(Ficus carica)*, tomato, cauliflower, lettuce *(Lactuca sativa)*, potato *(Solanum tuberosum)*, carrot *(Daucus carota)*, cucumber *(Citrullus* sp.), pea *(Pisum sativum)*, onion *(Allium cepa)*, some varieties of clover *(Trifolium* sp.), alfalfa *(Medicago sativa)*, wheat *(Triticum* sp.), oat *(Avena sativa)*, rice, sorghum *(Sorghum* sp.), maize *(Zea mays)*, flax *(Linum usitatissimum)*, sunflower *(Helianthus annuus)*.

Non resistant:

Pear *(Pyrus communis)*, apple *(Pyrus malus)*, orange *(Citrus aurantium)*, lemon *(Citrus limonium)*, almond *(Prunus amygdalus)*, peach *(Prunus persica)*, strawberry *(Fragaria vesca)*, bean *(Vicia faba)*, celery *(Apium graveolens)*.

Experiments at the Bari Agriculture Experimental Station proved that irrigation with brackish water lifted from the phreatic layer and having a total approximate salinity of 3 g/l improves the quantity and quality of some vegetables, such as tomato, eggplant, pepper, turnip, cabbage, fennel *(Foeniculum vulgare)*, asparagus *(Asparagus officinalis)*, artichoke *(Cynara scolymus)*.

On the other hand the following plants proved to be very sensitive also to low salinity:

Cucumber *(Cucumis sativus)*, melon *(Cucumis melo)*, pea and lentil. At the same Bari Experimental Station, on medium-heavy soil, PANTANELLI obtained 14,900 kg/ha tomatoes by irrigating with water of total salinity between 3.19 and 3.30 g/l and Cl-ion between 1.35 and 1.45 g/l, as against using Sele river water with 0.23 g/l total salinity and 0.014 g/l Cl. Water salinity also enormously improved the quality of the product. Freshwater tomato was more liable to *Fusarium* disease and ripened later; the fruit was less firm, more flaccid, less durable, sugar-poor and less tasty.

Similar results were obtained with eggplant, pepper and other vegetables that grew bigger, tastier, etc.

At the Fasano Experimental Station, FICCO (1960) studied with several vegetables, besides the effects of salt concentration, those of manuring and of irrigation treatment (quantities and turns frequent)). The soil here was sandy lime, 15—40 cm thick on calcareous tufo, cracked and very permeable. Water salinity increased from 4‰ in spring to 9‰ in August, i.e. in the full irrigation season. Varieties were: Tomatoes, fodder-maize and grain-maize, sorghum gentilé, sunflower, sinensis vine, cabbage, soybean, eggplant, pepper, alfalfa and celery.

*Tomato:* Best quantity from short irrigation turns (8 days with 14 turns). Each turn 254 $m^3$/ha; season total 3500 $m^3$/ha — Irrigation method: furrow infiltration.

*Fodder and grain maize:* Similar results.

After 15—20 days after germination, plants more sturdy and green where treated with mixed fertilizer. Same with *fodder beet, sorghum,* etc. More sugar content in *sugar-beet* irrigated with brackish water.

*Cabbage:* Turn of 8 days (instead of 4) produced stunted growth, less production and tendency to early blossoming with economic loss.

*Alfalfa:* Brackish water gave results which were not bad, but inferior to those of freshwater irrigation.

Poor results were obtained from eggplant, pepper and celery which, contrary to former observations by PANTANELLI, gave products not quite saleable. FICCO attributed this to the higher salinity of Fasano waters as against those of Bari.

We have so far considered the effects of the total water salinity without examining those of the single ions, in particular chlorine and sodium; these effects are not less important as these two ions by themselves are much more harmful than a much more complex group of ions, as is found in irrigation waters.

Chlorine and sodium are the most frequent ions, as they amount, on an average to about 2/3 of the total ions.

Under the form of sodium chloride, they stimulate all plants up to 3 g/l, and wheat and oats even up to 6 g/l. According to PANTANELLI, it is inadvisable to feed them for a long time to those plants that cannot resist sodium chloride in a concentration exceeding 1.5 g/l.

It must also be considered that plants resist much better to sodium chloride in water used for soil irrigation than in water cultivation (hydroponics) and that the resistance increases with the age of the plant.

In Bari province (PERNIOLA & PORCELLI, unpublished) the following plants proved sodium chloride resistant above 5‰: tomato, eggplant, cabbage, asparagus, artichoke, and lupin. Cucumber, melon, pea, bean, lentil and groundnut proved to be medium resistant, and least resistant were meadow fodder grasses (Gramineae), meadow clover *(Trifolium pratense)* and incarnate clover *(Trifolium incarnatum)*.

As already remarked by PANTANELLI, these grades of resistance to sodium chloride are relative, as they change with the nature of the soil and climatic conditions.

Sulphuric ion is usually better tolerated than chlor ion. As sodium sulphate it is resisted up to 7—8‰ by almost all the plants cultivated in the region, while oats resist even up to 15‰.

But no sodium sulphate exists in the irrigation waters of the region considered, as sodium is completely saturated first by chlorine and secondly by carbonic acid; therefore nothing can be stated concerning its effects on horticultural plants.

Only in some pliocene and miocene clayish zones of Lucania and Calabria, are there waters rich in sulphates, especially in calcium sulphate. But, considering (a) that this salt is little soluble, (b) that it is able to compensate the damage of some other ions, (c) that the sulphur ion has some nutritive value, we may rule out that these waters, even if chalky, may harm the vegetation.

Between the components of brackish waters, the most damaging is sodium, especially when, in the presence of an excess of carbonic ions in respect of calcium and magnesium ions, it produces sodium carbonate. Under this form, as already mentioned, besides being toxic to the plants, it damages the structure and permeability of the soil.

On no calcareous soils, even a 1‰ concentration would be harmful to cotton, sorghum, barley, asparagus, alfalfa and wheat.

In the Puglia regions, no irrigation was tried with water rich in sodium carbonate, as these waters have no carbonates and little bicarbonate, always less than 500 p.p.m., always saturated by calcium and secondarily by magnesium. It might be that some waters from clayish-siliceous soils in the appenine zone of Lucania or Capitanata be rich in sodium carbonate, but in these cases as well no danger is to be feared, as this component is always accompanied and neutralized by sulphates.

In conclusion, we may say that considering the particular nature of the soils and of the brackish waters, if these are not too concentrated, their use for irrigation is useful as it reduces transpiration, it increases drought resistance, it renders the tissues more fleshy and it improves the quality of the products by means of accumulation of aromatic principles, of sugars and of soluble compounds of nitrogen and phosphorus.

## Final Considerations

The removal from the soil of excess salts brought by irrigation with brackish water or originally existing presents various and complex problems, and a solution must be reached if we want to be able to use brackish waters for irrigation.

On the other hand, the use of moderately brackish waters is, on these soils, useful as it improves the quality of a number of products especially horticultural vegetables. For these waters to be useful, such conditions must exist which favour the "liscivation" of the soil without damaging its structural characteristics and without it becoming alkaline, and as such totally unproductive and sterile.

From this emerges the importance of the systems of soil cultivation, of irrigation and of drainage for the removal of excess waters.

The irrigation system, the water volumes and the turns are especially important when brackish waters are used.

In the zone under consideration, the irrigation system most used is by lateral infiltration, with high water volume and short turns; this practice has proved the best for most horticultural products, especially tomatoes.

High volumes and short turns on the one hand prevent the concentration of the water of a single shift as a result of evaporation and transpiration; on the other hand they cause a vast movement of percolation reaching below the rocky layer under the cultivated soil. These movements effect a first washing out of the soil and the subsequent — at least partial — removal of the excess salts which, once brought below the rocky layer, can no more ascend by capillarity.

Fortunately, in the Puglia littoral zone, these practices are favoured by the nature of the rocky bank layer and by the fact that the soils are rather loose. Even where they are clayish (as in some red earth on cretaceous chalks), thanks to the mineralogical structure of their clay fraction they are always permeable enough to prevent water stagnation and salt concentration.

Such practices are not always sufficient to completely prevent the danger of salt accumulation, especially of those brought with the last seasonal irrigations. But these salts are removed later by the winter precipitations themselves.

Furthermore, local farmers usually interrupt brackish water irrigation for a whole year or at least from October to the following July in the fields that have become too salty, raising on them beans or cereals.

The results of several researchers and the practical local experience have proved that irrigation with brackish water may be successful when a number of rules are observed such as appropriate culture planning, irrigation techniques (small fields, large water quantities and short turns), letting the soil rest at least every two years in order not to worsen its structural characteristics, abundant organic fertilizers, etc.

These norms, however, are valid only for loose soils whose chalk content and permeability ensure the natural washing out of excess salts; for heavier soils and particularly for the clayish soils, the best irrigation technique must be selected individually for each case.

## REFERENCES

Berg, C. Van den, 1950. The influence of salt in the soil on the yield of agricultural crops. *Trans. Fourth int. Congr. Soil Sci.* **1**: *411—413.*

Bottini, O., & Lisanti, E., 1955. Ricerche e considerazioni sull'irrigazione con acque salmastre praticata lungo il litorale pugliese. *Ann. Sper. Agr. n.s.* Roma **IX**: *401—436.*

Carrante, V., Della Gatta, L., Perniola, M. & Lopez, G., 1956. I Terreni agrari della provincia di Taranto. *Ann. Sper. Agr., Roma, n.s.* **XI**: Suppl. al n.2: LXXIX-CXLV.

Chang, C. W., 1961. Effects of saline irrigation water and exchangeable sodium on soil properties and growth of alfalfa. *Soil Sci.* **91**, *29—37.*

Dabell, J. P., 1955. Some observations on the use of brackish water for irrigation along the Bari-Brindisi coastal plain. FAO, Roma.

Eaton, F. M., 1950. Significance of carbonates in irrigation waters. *Soil Sci.* **69**: *123—133.*

Ficco, N., 1960. Un triennio di sperimentazione irrigua con acque salmastre. Ente per lo Svil. dell'Irrig. e la Trasfor. Fond. in Puglia e Lucania, Bari.

Hausenbuiller, R., L., Daque, M. A., & Wahab, D., 1960. Some effects of irrigation waters of differing quality on soil properties. *Soil Sci.,* **90**: *357—364.*

Lisanti, L. E., 1958. The red hearts and the colloids of the red hearts. *Ann. Fac. Econ. Commerc., Univ. Bari,* **XLI.**

Lopez, G., 1958. Stato del potassio nei terreni agrari di Puglia e Lucania, Conv. per la concimazione potassica per l'Italia Meridionale e insulare, Portici.

Pantanelli, E., 1937. Studio chimico-agrario dei terreni della Provincia di Bari. *Ann. Sper. Agr. Roma,* **XXII.**

Pantanelli, E., 1937. Irrigazione con acque salmastre. *Staz. Agr. Sper. Bari, Publ.* **136.**

Perniola, M. & Porcelli, A., — Ricerche sulla influenza del sodio nella nutrizione minerale delle piante. Unpublished.

Richards, L. A., et al., 1954. Diagnosis and improvement of saline and alkali soils. *U.S. Dept. Agric.,* n.**60.**

Wilcox, L. V., Black, G. Y. & Bower, C. A., 1954. Effect of bicarbonate on suitability of water for irrigation. *Soil Sci.* **77**: *259—266.*

Zoczi, L. & Reina, C., 1954. Le acque sotteranee in terra d'Otranto. Cassa per il Mezzogiorno Roma, documento n.**1.**

# EFFECT OF SEA-WATER AND ITS DILUTIONS ON SOME SOIL CHARACTERISTICS*

BY

M. R. NARAYANA, V. C. MEHTA & D. S. DATAR
*Central Salt & Marine Chemicals Research Institute, Bhavnagar, India**

(with 2 figs.)

Many countries have taken recourse to investigate the possibilities of bringing into cultivation the vast stretches of dune sands of coastal regions, under sea-water irrigation and the results obtained in Israel, Spain and Sweden have been reviewed elsewhere (IYENGAR & NARAYANA, 1964). Recently, similar experiments were tried on some crops in India and encouraging results have been obtained with tobacco and wheat (unpublished data).

In the use of highly saline water and sea-water for irrigation, the main danger involved is the salinity of the soil and BERNSTEIN (1962) rightly pointed out that a water of good quality may cause trouble when improperly used or when applied to a very impermeable soil, while a water of relatively poor quality may produce good results when used on permeable, well-drained soils. Investigations in Israel (BOYKO & BOYKO, 1959) showed that no salt accumulation takes place in the root zone even when the soils are irrigated with sea-water, if the soils are deep and highly permeable. The present communication describes the permeability characteristics of a sandy soil and a silty clay soil for sea-water and its various dilutions, to come to a better understanding of the suitability of sandy soils for such irrigations.

## Materials and Methods

Sea-water for the investigation was collected from Bhavnagar coast and the sandy and silty clay soils from the coasts of Mahuva and Bhavnagar respectively. Only surface samples up to a depth of 8 cm for silty clay soils and 30 cm for sandy soils are studied, since the soils of the respective profiles compared well with their surface soils.

The soils were air-dried and the portion passing through 2 mm sieve was used for all the determinations. Soil pH was determined on 1:2.5 soil water suspension using a Beckman electronic pH meter with glass electrodes. Soils were equilibrated with various dilutions

* Lecture delivered at the Symposium on the problems of the "Indian Arid Zone" held at Jodhpur, India, November, 1964.

of sea-water by continuous leaching for a minimum period of about 5 hours. The salt content in the soils was gravimetrically determined in 1:2 soil water extract. The dispersion coefficient was measured by determining the percentage of clay by the pipette method after leaving the soil in contact with water for 24 hours (referred to as the dispersion factor) and expressing it as percentage of the total clay content of the soil obtainable on complete dispersion (PURI, 1949). For the determination of the dispersion factor, 1% and 10% soil water suspensions were used respectively for the silty clay and sandy soils. The set-up of the constant head permeameter as described by CHRISTIANSON (1944) was used for the permeability determinations. A glass tube of 28 cm length and 4 cm diameter was used as the permeameter. Glass wool was used as the base in the tube, over which 20 g of quartz, 20 g of soil and 15 g of quartz were put in this sequence. Compaction was effected by dropping the tube 20 times on a soft wood from a height of one inch. Viscosity corrections were applied to the permeability results.

Silt was separated from sea-water by filtration and the total salt content in sea-water was gravimetrically determined. The conductivity was recorded using a Serfass conductivity bridge. The analysis for Ca and Mg was carried out by the versenate titration method, for Na and K by the flame photometric method and for $CO_3$, $HCO_3$, Cl and $SO_4$ by the conventional methods.

## Results and Discussion

Of the various classifications proposed from time to time to categorize water for irrigation use, the one developed by the United States Salinity Laboratory is widely followed, wherein the waters are classified into salinity grades in terms of specific conductance and alkalinity grades in terms of sodium adsorption ratio (SAR). The conductivity, composition and SAR values of sea-water (Table I) reflect in the first instance, that it is absolutely unsuitable for irrigation purposes. The values for sea-water dilutions have also been presented in the table since the trend of research as at present is to try various dilutions of sea-water either to find the maximum sea-water tolerance of the species and varieties of plants or to utilise sea-water during the season when its salinity is low. The successful use of sea-water for irrigation experiments in Israel and Spain is explained on the theory of partial root contact connected with the formation of subterranean dew in soils (BOYKO & BOYKO, 1964).

The effect of sea-water and its dilutions on some of the soil characteristics such as pH, permeability and dispersion coefficient for sandy and silty clay soils are discussed in the following.

The mechanical analysis of the soils is presented in Table II. The pH of the soils (Table III) tends to fall with increasing salinity

**Table I.**

Conductivity, composition and SAR values of sea-water and its dilutions

| Sea-water & its dilutions | Total salts me/l | Conductivity micromhos/cm at 25° C | Ionic composition me/l | | | | | | | | SAR values |
|---|---|---|---|---|---|---|---|---|---|---|---|
| | | | Na | K | Ca | Mg | Cl | $SO_4$ | $CO_3$ | $HCO_3$ | |
| Sea-water | | | | | | | | | | | |
| (36,322 ppm) | 1253.2 | 56,849 | 504.4 | 7.3 | 19.5 | 102.8 | 560.1 | 56.1 | 0.4 | 2.6 | 64.7 |
| 30,000 ppm | 1035.6 | 47,760 | 416.7 | 6.1 | 16.1 | 84.9 | 462.9 | 46.4 | 0.3 | 2.1 | 58.6 |
| 20,000 ppm | 690.4 | 32,890 | 277.8 | 4.1 | 10.8 | 56.6 | 308.6 | 30.9 | 0.2 | 1.4 | 47.9 |
| 10,000 ppm | 345.1 | 17,222 | 138.9 | 2.1 | 5.4 | 28.3 | 154.3 | 15.5 | 0.1 | 0.7 | 33.9 |
| 5,000 ppm | 172.6 | 8,371 | 69.5 | 1.0 | 2.7 | 14.2 | 77.2 | 7.7 | 0.05 | 0.35 | 23.9 |

Distilled water used in preparing the dilutions was of pH 5.8.

**Table II.**

Mechanical composition of the soils.
(Expressed as percent on oven dry basis).

| Soil No. | Moisture* | Total soluble salts | $CaCO_3$ | Coarse sand | Fine sand | Silt | Clay | Textural class |
|---|---|---|---|---|---|---|---|---|
| 1 | 1.72 | 0.0172 | 18.03 | 1.55 | 71.66 | 3.68 | 3.52 | sandy |
| 2 | 11.58 | 2.0089 | 4.13 | — | 13.61 | 24.77 | 53.02 | silty clay |

* Per cent on air dry basis.

of the water used in preparing the suspension. The fall is well marked in the initial levels of salinity. The practical importance of this observation would be that the pH of soil as determined in the laboratory with distilled water would not be the same as it exists in the field, particularly under saline water irrigation, and the pH of the soil would vary in a profile depending on the addition of salts and the rate of evaporation of water, apart from such other factors as carbon-dioxide concentration of the soil air etc.

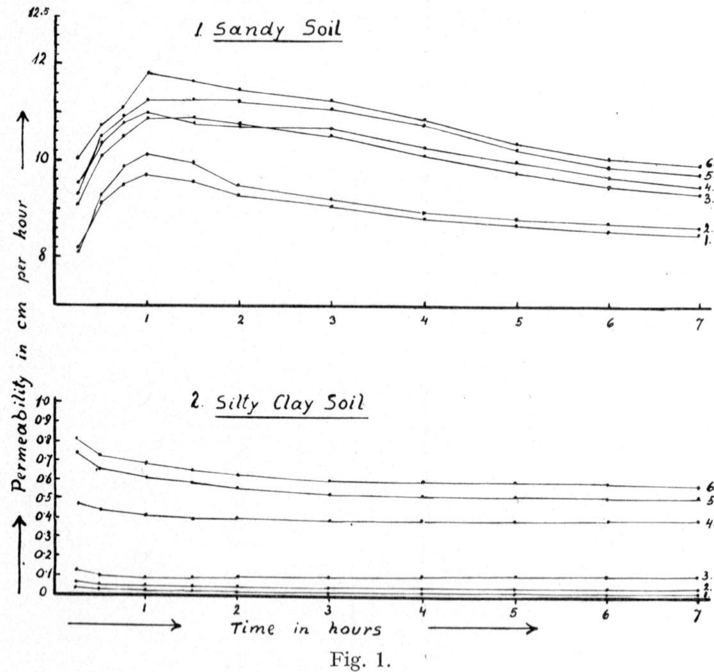


Fig. 1.

Permeability curves (Fig. 1) for the two soil types indicate that the permeability increases with increase in the salinity of the water and the permeability of the sandy soil for distilled water is hundreds of times more than that for the silty clay soil. This indicates that any salt left out by irrigation with saline water will be easily washed down into the deeper layers during the subsequent irrigations. Further, the water-holding capacity of the sandy and silty clay soils was found to be 33 and 48% respectively, which indicates that the salt content of the soils should accordingly approximate to about one third and half the salt concentrations of the irrigation water. This is confirmed by the values of salt content (Table III) in the soils equilibrated with sea-water of various dilutions. Boyko & Boyko (1959) observed that the salt concentration in the soils was

**Table III.**

pH and salt content of soils with sea-water dilutions.

| Sea-water and its dilutions | pH of water | pH of soils in sea-water and its dilutions | | Salt content of soils equilibrated with sea-water and its dilutions | |
|---|---|---|---|---|---|
| | | No. 1 | No. 2 | No. 1 | No. 2 |
| Sea-water | | | | | |
| (36,322 ppm) | 8.00 | 7.90 | 8.30 | 11,228 ppm | 15,350 ppm |
| 30,000 ppm | 7.95 | 7.90 | 8.30 | 8,492 ppm | 14,550 ppm |
| 20,000 ppm | 7.85 | 7.90 | 8.30 | 6,068 ppm | 9,180 ppm |
| 10,000 ppm | 7.50 | 7.90 | 8.40 | 3,040 ppm | 4,785 ppm |
| 5,000 ppm | 7.30 | 7.95 | 8.50 | 1,608 ppm | 3,000 ppm |
| Distilled Water | 5.80 | 8.10 | 8.60 | — | — |

only 0.2% even after 100 times irrigation with sea-water containing 3.4% salts. The soils used by them contained more of stones and coarse sand and much less of silt and clay. The high permeability and very low water holding capacity of the soils might account for the low salt content left in the soils.

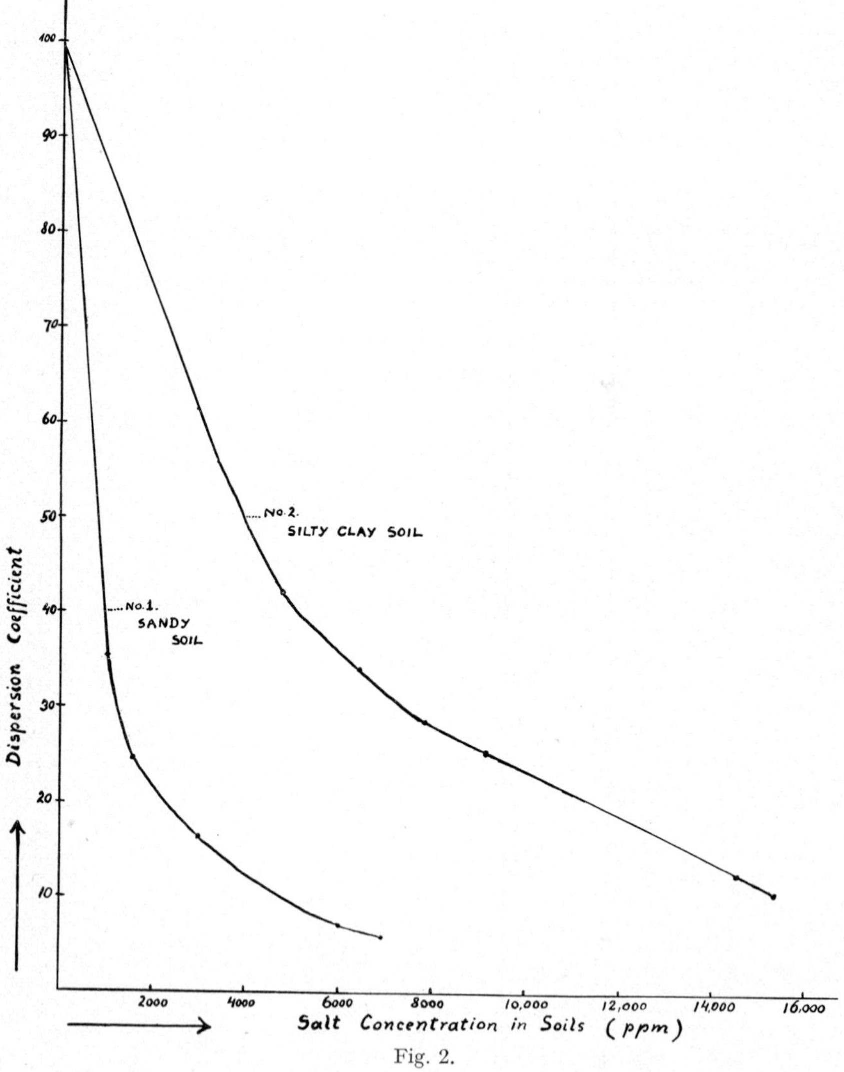


Fig. 2.

The predominance of sodium in sea-water is reflected in its high SAR value and a soil irrigated with such water would be left with a high exchangeable sodium percentage (ESP). But, the extent of deflocculation of clay that can be brought about by the high ESP is controlled by the salt content of the soil-water system and the effect of deflocculation on soil tilth would obviously be determined by the amount of clay present in the soil.

The dispersion coefficient is a measure of the percentage of total clay that can pass into the suspensoid state by simple contact with water, and its value varies from 0 to 100%, depending on whether the soil is completely flocculated or completely dispersed. The dispersion coefficient for the soils (Fig. 2) equilibrated with waters of different salinity levels and dispersed, after drying, in distilled water, shows that as the salt content in the soil decreases, the dispersion coefficient increases. These results indicate that, in nature, if a heavy textured soil is either irrigated or flooded with sea-water, it comes into dispersed state as soon as the excess salts are removed by rain or freshwater irrigation. The soil looses its structure and remains in puddled condition and becomes very hard on drying. Such observations on clayey soils flooded with sea-water have been recorded by DYMOND (quoted by RUSSEL, 1953). On the contrary, a sandy soil would maintain good tilth, since the low amount of dispersed clay will be insufficient to bring the soil to any unfavourable physical condition.

## Conclusion

The permeability of the soils increases with increase in the salinity of the water. Low clay content, high permeability and low water-holding capacity of sandy soils explain their suitability for highly saline or sea-water irrigation. In such soils no salt accumulation can take place, since any salts left out during one irrigation can be washed down into deeper layers during subsequent irrigation. Unlike a heavy textured soil, a sandy soil will not come into a puddled state even when saline water irrigation is interrupted by freshwater irrigation or by rain.

## REFERENCES

BERNSTEIN, L., 1961. Salt affected soils and plants. Proc. Symp. on the problems of Arid Zone. UNESCO, Arid Zone Res. Ser. XVIII, p. 139—174; Paris, 1962.

BOYKO, H. & BOYKO, E., 1959. Sea water irrigation — A new line of research on a Bioclimatological plant soil complex. *Int. J. Bioclimatol. Biometeorol.* **3** (II, B), *1—24.*

BOYKO, H. & BOYKO, E., 1964. Principles and experiments regarding direct irrigation with highly saline and sea water without desalination. *Trans. New York Acad. Sci.;* Supplement to No. **8**, Ser. II., **26,** *1087—1102.*

CHRISTIANSON, J. E., 1944. Effect of entrapped air upon the permeability of soils. *Soil Sci.* **58,** *355—365.*

IYENGAR, E. R. R. & NARAYANA, M. R., 1964. Sea Water Agriculture. Paper presented at the Symposium on "Science and Nation during the Third Five Year Plan". ASWI, New Delhi.

PURI, A. N., 1949. Soils: Their Physics and Chemistry. Reinhold Publishing Co., New York.

RUSSEL, E. W., 1953. Soil conditions and plant growth, Longmans, Green & Co., New York.

# EFFECT OF SEA-WATER DILUTIONS AND ITS AMENDMENTS ON TOBACCO*

BY

T. KURIAN, E. R. R. IYENGAR,
M. R. NARAYANA & D. S. DATAR

*Central Salt & Marine Chemicals Research Institute, Bhavnagar, India*

Plant growth under saline conditions, confronting the investigators, has resurrected many a research to reveal the degree and kind of plants resistant to soil salinization (HAYWARD & BERNSTEIN, 1958). Gradually more and more information is available on the use of saline water for irrigation in arid and semi-arid regions and its quality, its influence on crop production and the effects on soil characteristics (USDA 1954, UNESCO 1961).

Recently, investigations on the utilization of sea-water for plant growth has emanated from the findings of BOYKO & BOYKO (1959), who explained that salt accumulation in the root region does not occur, if the soils are sandy and extend deeper than the root zone. Irrigation with sea-water in humid regions on sandy soils in order to supplement the rainfall is discussed in relation to the salinity effect on growth stages of plants and the cation composition of plants and soils (BATCHELDER et al., 1963).

HEIMANN (1958) explaining the concept of "Balanced ionic environment" under saline conditions stressed the need for evaluating the sodium-potassium relationship of irrigation waters. The antagonistic effect of ions on the uptake of the solutes by barley roots indicates that potassium and phosphate impede the accumulation of sodium and chloride, but the intake of nitrate is not influenced by the presence of chlorides and sulphates (REIFENBERG & ROSOVSKY, 1947).

The proper dose of potassium salts in overcoming the effect of salinity under irrigation with highly saline waters on tobacco is yet obscure. In the present study an attempt is made to determine the effect of sea-water dilutions and its amendments with potassium salts on the growth and chemical composition of Tobacco, Keliu-20.

## The Experiment

Tobacco K-20 was raised in the nursery in 1963. Sixty days old seedlings were transplanted in earthen pots, 15″ × 25″ containing sandy soil (coarse sand 68%, fine sand 26%, silt and clay 3%, carbonates 3%). The plants were irrigated with tap-water for

* Lecture delivered at the Symposium on the problems of the "Indian Arid Zone", held at Jodhpur, India, November, 1964.

complete establishment. One month after transplantation the seedlings received the following treatments:

| Sea-water dilutions | Total K meq/l | Total $H_2PO_4$ meq/l |
|---|---|---|
| 10,000 ppm | 2.1 | — |
| 10,000 ppm amended | 88.8 | 77.3 |
| 15,000 ppm | 3.1 | — |
| 15,000 ppm amended | 133.2 | 115.9 |
| 20,000 ppm | 4.2 | — |
| 20,000 ppm amended | 177.6 | 154.6 |
| *Control* | | |
| Tap-water | — | — |
| Tap-water amended | 108.4 | 96.6 |

Sea-water: 34.10 g/l (TSC); Na 10.6; Cl 19.0; K 0.30 g/l.

Six replications were maintained for each of the treatments. The amendments of the media were made by using $KNO_3$ and $KHPO_4$ of C.P. quality. Depending on the need for irrigation the above solutions were given at an interval of 7—11 days. The crops received 8 irrigations and per plant 8 litres of the media were supplied in each of the above irrigations.

Measurements of plant height, number of leaves, area of leaves, and length of the root systems were recorded and the results were statistically analysed. Chlorophyll and nicotine estimations of the mature leaves were made by adopting the methods of COMAR (1942), and AVENS & PEARCE (1939) respectively. The chlorophyll analysis was made by using a Hilger and Watts spectrophotometer H 700.

Soil salinization at different depths was measured by sampling the soils at 10 cm, 30 cm and 60 cm depths and determining the electrical conductivity of soil water extract of 1:1 using the Serfass conductivity bridge RCM 15B1. Sodium chloride in the extract was calculated by estimating the sodium content with the Flame-photometer. The pH determination of these soil samples was carried out in a soil water suspension of 1:2.5 and by employing the Beckman electronic pH meter with glass electrodes.

## Results

Irrigation with sea-water dilutions depresses the growth of tobacco plants but their amendments remarkably improves the growth (Table I). There was a heavy mortality in plants treated with 20,000 ppm and its amendment; therefore, the study was confined to the other treatments listed in the Tables I—III.

**Table I.**

Growth behaviour of Tobacco K-20 irrigated with sea-water dilutions.
(mean of six replications)

| Treatments | Height of the plant cm | Root length cm | Number of leaves | Leaf area in $cm^2$ | Fresh weight of shoot g | Dry weight of shoot g | Fresh weight of root g | Dry weight of root g |
|---|---|---|---|---|---|---|---|---|
| *Control* | | | | | | | | |
| Tap-water | 14.62 | 34.74 | 9.40 | 38.21 | 14.9270 | 4.2994 | 1.6613 | 0.6653 |
| Tap-water amended | 34.10 | 44.04 | 14.60 | 125.26 | 63.6475 | 13.4408 | 5.8611 | 3.5744 |
| *Sea-water dilutions* | | | | | | | | |
| 10,000 ppm | 10.90 | 20.14 | 6.60 | 28.78 | 10.6578 | 1.4222 | 0.6224 | 0.2726 |
| 10,000 ppm amended | 16.88 | 24.78 | 9.60 | 54.08 | 24.0340 | 3.9118 | 1.3522 | 0.3398 |
| 15,000 ppm | 11.10 | 14.52 | 5.40 | 24.66 | 8.0565 | 1.0305 | 0.1849 | 0.1132 |
| 15,000 ppm amended | 12.10 | 22.80 | 6.60 | 29.14 | 8.2904 | 1.2512 | 0.3204 | 0.1703 |
| LSD 5% | 6.13 | 11.50 | 1.85 | 19.82 | 12.6828 | 2.7743 | 1.7689 | 1.5861 |
| LSD 1% | 8.36 | 15.73 | 2.53 | 27.03 | 17.2976 | 3.7833 | 2.4136 | 2.1622 |

### Growth Behaviour

The control amended series show significant increase in all the characters recorded and hence, the comparison of growth is made between sea-water treated sets and tap-water control.

The heights of the plants recorded exhibit no significant effect of the treatments. But the root length is affected by the sea-water dilutions except in the 10,000 ppm. amended series. The number and area of the leaves indicate a similar tendency, with the interesting exception of a greater leaf area in plants receiving 10,000 ppm. amended medium. Sea-water dilutions do not prove to have much decreasing effect on the fresh weight of shoot. The dry weight of the shoot, on the other hand, is practically affected with less significant values in plants treated with 10,000 ppm. amended solution. The differences of fresh weight and dry weight of the roots are much less significant to the treatments (Table I) than those of the shoots.

### Soil Study

The soil samples collected at 10 cm, 30 cm and 60 cm do not tend to have any accumulation of salts at the root zone (Table II). Salt accumulation at 60 cm depth is always higher in all the cases (due probably to the insufficient depth of the pots). The pH of the soil solutions has no marked differences either in different depth or at the different salinity levels, while the pH of the sea-water amended media was towards the acidic side.

### Chemical Composition

The chlorophyll analysis shows a decrease in total chlorophyll, α chlorophyll and β chlorophyll with increase in salinity of the media from 10,000 ppm upwards. The amendments overcome such effects. The ratio of chlorophyll α to β increases slightly with the increase in sea-water concentrations.

It is interesting to note that a higher nicotine percentage is obtained in all the series and particularly in the 10,000 ppm amended group when compared with the control (Table III).

## Discussion

An effective approach in growing plants under saline conditions warrants the better understanding of the physiological basis of salt tolerance of plants. The salinity tolerance is specific to each of the species and varieties (Bernstein & Hayward, 1958). Some times, the plants grow in linear proportion to the concentration of salts present in the irrigation water (Lunin et al., 1961). In tobacco the antagonistic effect of potassium on the uptake of sodium is influenced when the potassium concentration increases in the external media (Parikh et al., 1957). Similarly, in the experiments described

**Table II.**

Degree of soil salinization and pH.
(Average of duplications).

| Treatments | Soil Water extract of 1:1 | | | | | | Soil-water suspension of 1:2.5 pH values | | |
|---|---|---|---|---|---|---|---|---|---|
| | Ece mmhos/cm soil depth in cm | | | calculated NaCl per cent soil depth in cm | | | soil depth in cm | | |
| | 10 | 30 | 60 | 10 | 30 | 60 | 10 | 30 | 60 |
| *Control* | | | | | | | | | |
| Tap-water | 0.3738 | 0.1242 | 0.1177 | 0.0075 | 0.0031 | 0.0029 | 8.5 | 8.1 | 8.5 |
| Tap-water amended | 1.1775 | 0.8831 | 1.1120 | 0.0058 | 0.0028 | 0.0033 | 7.9 | 8.3 | 7.5 |
| *Sea-water dilutions* | | | | | | | | | |
| 10,000 ppm | 1.1447 | 1.7681 | 3.1418 | 0.0305 | 0.0485 | 0.0740 | 8.8 | 8.8 | 8.6 |
| 10,000 ppm amended | 1.4391 | 1.9625 | 3.7941 | 0.0365 | 0.0312 | 0.0637 | 8.2 | 8.2 | 7.8 |
| 15,000 ppm | 1.5045 | 1.7008 | 3.7941 | 0.0725 | 0.0775 | 0.1250 | 8.2 | 8.4 | 8.4 |
| 15,000 ppm amended | 2.1587 | 4.2520 | 5.2333 | 0.0630 | 0.1400 | 0.2300 | 8.2 | 7.8 | 8.0 |

**Table III.**

Chlorophyll and nicotine content of Tobacco K-20 irrigated with sea-water dilutions.

| Treatments | Chlorophyll mg/100 g of the fresh weight of leaves | | | | Nicotine percentage of leaves on oven dry basis |
|---|---|---|---|---|---|
| | Total chlorophyll | Chlorophyll α | Chlorophyll β | Ratio α : β | |
| *Control* | | | | | |
| Tap-water | 16.82 | 7.51 | 9.31 | 0.81 | 0.268 |
| Tap-water amended | 36.04 | 15.77 | 20.30 | 0.78 | 0.548 |
| *Sea-water dilutions* | | | | | |
| 10,000 ppm | 20.24 | 9.06 | 11.19 | 0.81 | 0.748 |
| 10,000 ppm amended | 33.31 | 14.48 | 18.85 | 0.77 | 1.048 |
| 15,000 ppm | 15.68 | 7.14 | 8.53 | 0.84 | 0.458 |
| 15,000 ppm amended | 25.66 | 11.38 | 14.30 | 0.79 | 0.609 |

here, the vegetative growth of tobacco plants decreased with the increase of the salinity but improved considerably, by a marked increase in the leaf area, on amendment with potassium salts.

BOYKO & BOYKO (1959) state that accumulation of salts at the root zone need less be feared even under oceanic water irrigation, if the soils are highly permeable and sufficiently deeper. The soil salinity in this experiment shows an increase in the salt concentration at 60 cm depth. This can be attributed to the prevention of further leaching of salts in the pots. The pH of the soils is not markedly varying at different depths of the various treatments despite the low pH values of the amended media; probably due to attaining equilibrium with the calcium carbonate of the soil.

Salinity affects the chemical composition of plants grown on salt affected soils or irrigated with saline solutions. The level of salt concentration, rather than the other factors of the soil, is the determining factor for the reduction in chlorophyll content of the plants (KIM, 1958). It is noteworthy that with the increase in salinity of sea-water dilutions above 10,000 ppm the chlorophyll content decreased, but on the contrary the ratio of $\alpha/\beta$ chlorophyll showed a tendency to increase with the increase in salinity. Application of sodium chloride in higher doses impares the quality of tobacco by reducing nicotine, sugars, and nitrogen (PEELE et al., 1960; GOPALACHARI, 1961; PAL et al., 1963). HUTCHINSON et al. (1959) reported a reduction in nicotine content in high concentration of sodium chloride, but noted an increase in nicotine on the supply of potassium. The sea-water dilutions and their amendments exhibit an increase in nicotine percentage of the leaves.

## Conclusion

Plant response to sea-water dilutions is governed by the level of salinity. Addition of potassium salts in proper ratios may create a balanced ionic condition thereby preventing the toxic effects of sea-water ions on the metabolic activity of the plant.

Nevertheless, the soils should be highly permeable and extend deeper that the root zone to prevent any salt hazard on the plant growing in them.

More exploratory investigations may prove useful for achieving a successful crop production even under irrigation with highly saline waters of inland and coastal regions.

ACKNOWLEDGMENT

The authors wish to express their thanks to Shri V. H. VAIDYA, for the help in recording the observations on various instruments.

## REFERENCES

AVENS, A. W. & PEARCE, G. W., 1939. Silicotungstic acid determination of nicotine. *Ind. Eng. Chem. Anal.* Ed. **38,** *505—508.*

BATCHELDER, A. R., LUNIN, J. & GALLATIN, M. H., 1963. Saline irrigation of several vegetable crops at various growth stages. II. Effect on cation composition of crops and soils. *Agron. J.* **55,** *107—110.*

BERNSTEIN, L. & HAYWARD, H. E., 1958. Physiology of salt tolerance. *Ann. Rev. Plant Physiol.* **9,** *25—46.*

BOYKO, H. & BOYKO, E., 1959. Sea water irrigation — a new line of research on bioclimatological plant soil complex. *Int. J. Bioclimatol. Biometeorol.* **3** (II, B1), *1—24.*

COMAR, C. L., 1942. Analysis of plant extract for chlorophyll α & β. *Ind. Eng. Chem. Anal.* Ed. **14,** *877—879.*

COPALACHARI, N. C., 1961. Studies in the physiology of Spangle formation in "bidi" tobacco (*Nicotiana tabacum* L.). *Indian J. Plant Physiol.* **IV,** *47—53.*

HAYWARD, H. E. & BERNSTEIN, L., 1958. Plant growth relationships on salt affected soils. *Bot.Rev.* **24,** (8/10), *584—635.*

HEIMANN, H., 1958. Irrigation with saline water and the ionic environment. Potassium symposium, *173—220.* Madrid.

HUTCHINSON, T. B., WOLTZ, W. G. & MACALLB, S. B., 1959. Potassium-sodium Inter-relationships: I. Effect of various rates and combinations of K and Na on yield value and physical and chemical properties of flue-cured tobacco grown in field. *Soil Sci.* **87,** *28—35.*

KIN, C. M., 1959. Effect of saline and alkaline salts on the growth and internal components of selected vegetable plants. *Physiol. Plant.* **11,** *441—450.*

LUNIN, J., GALLATIN, M. H. & BATCHELDER, A. R., 1961. Effect of stage of growth at time of salination on growth and chemical composition of beans: I. Total salinization accomplished in one irrigation. *Soil Sci.* **91** (3), *194—202.*

PAL, N. L., BANGARAYYA, M. & NARASIMHAM, P., 1963. Influence of various nutritional doses of chloride on the growth of tobacco plant and burning capacity. *Soil Sci.* **95,** *144—148.*

PARIKH, N. M., DANGARWALA, R. T. & MEHTA, B. V. ,1957. Effect of saline water on yield and chemical composition of "Bidi"-tobacco. Conference tobacco research worker's Sum. Proc. (1), *61—62.*

PEELE, T. C., WEBB, H. J. & BENOCK, J. F., 1960. Chemical composition of irrigation water in South Carolina coastal plain and effect of chlorides in irrigation water on the quality of Flue-cured Tobacco. *Agron. J.* **52,** *464—467.*

REIFENBERG, A. & ROSOVSKY, R., 1947. Saline water irrigation and its effects on the intake of ions by barley seedlings. *Palestine J. Bot.* **IV** (1), *1—13.*

UNESCO, 1961. Salinity problems in the arid zones. Proceedings of the Teheran Symposium.

UNITED STATES SALINITY LABORATORY STAFF., 1954. Diagnosis and improvement of saline and alkali soils. USDA. Handbook, No. **60.**

# DILUTED SEA-WATER AS A SUITABLE MEDIUM FOR ANIMALS AND PLANTS, WITH SPECIAL REFERENCE TO CLADOCERA AND AGROPYRON JUNCEUM BOREO-ATLANTICUM SIMON. & GUIN.

BY

MEERTINUS P. D. MEIJERING

*Hermann Lietz-Schule, Spiekeroog, Western Germany*

(with 8 figs.)*

## Introduction

The term "diluted sea-water" is not identical with the term "brackish water". Brackish water is a mixture of fresh water and sea-water (REDEKE, 1933). This definition holds for sea-water, which is diluted by various fresh waters. These may have flowed in as rivers from other areas. Thus the total salt content of sea-water can be lowered; and, within certain limits, it is also possible for the qualitative composition of sea-water to be altered. On the other hand it is often the case that sea-water pours into fresh waters so that the salt content of such a water is raised to a certain degree. Large variations in the composition of brackish waters can be observed when inland waters flow through stone layers containing various soluble salts. Brackish waters from this origin differ very much from diluted sea-water when analysed qualitatively. For this reason REMANE (1958) characterises all brackish waters as "waters having a mean salt content".

Qualitatively, diluted sea-water is identical with sea-water. It differs only quantitatively from the latter. Following the "Venice System for the Classification of Marine Waters according to Salinity" we can classify diluted sea-water among the "mixohaline waters".

In nature mixohaline waters are never of constant salinity. Evaporation and rainfall bring about changes in the salinity of a water. Life in mixohaline water has to deal with this fact. An organism can settle in mixohaline water, when it can dispose of a certain degree of resistance adequate to meet the fluctuations in the salt content of the medium. The concept of "resistance", then, is of central interest in the biology of brackish waters.

Under natural conditions mixohaline waters are subject to the inconstancy of some other factors, which are active physiologically. These may be temperature, $O_2$-content etc. Since fluctuations of different factors usually influence each other in their physiological effects, living-conditions vary continuously. In physiological ecology these circumstances are taken into consideration.

* I am grateful to Mr. J. B. REDFERN for correcting the English manuscript.

In the following cases diluted sea-water is understood as sea-water diluted with aqua destillata. This solution does not exist in nature. It could be compared with mixohaline waters diluted by rainfall. This solution is suitable for research work in the laboratory, insofar as all those factors, which vary under natural conditions, can be kept under control, i.e. can be kept constant. The main factors which are active physiologically are salt content, temperature, light, feeding, space per individual, $O_2$-, $H_2S$- or $CO_2$-content, and the acidity of the medium. It is possible to present combinations of factors which remain constant throughout an organism's life. Optimal living-conditions can be looked upon as an optimal combination of all factors which influence the duration of life. From such an optimum we can take measures to calculate the degree of retardations occurring under natural conditions, where organisms are subject to various fluctuations. In mixohaline waters above all, plants and animals are forced continuously to adjust themselves to variations in one factor or another.

HAYWARD & WADLEIGH (1949) discussed different ways, in which salt tolerance can be evaluated. According to them the ecological criterion is largely the organism's power of survival under conditions where salinity steadily increases. Another criterion, more useful to the agronomist, would be, as indicated by the example given by these authors, "the relative performance of a crop on a given level of soil salinity as compared to its performance on a comparable non-saline soil". This latter criterion is similar to that used in our experiments, which will be described here.

The words "tolerance" and "resistance" have a somewhat negative character when they are used in a general sense. This made people and nations having to deal with sea-water as highly saline water give it a bad name. This is not astonishing, since floods of sea-water have often ruined agricultural land. The author had even to rescue parts of this manuscript during the flooding of areas along the North Sea coast in February 1962, and some experiments, which will be mentioned here, had to be stopped, as they were reached by the floods.

Yet in spite of these facts one can hardly believe that such an equilibrated solution as sea-water, which covers large parts of the world, would damage creatures of terrestrial or limnic habitats on account of the qualitative composition of its salts. The damage seems to be caused only by the quantitatively high salinity.

Since it is generally stated that life of limnic or terrestrial regions descends from marine life, we can accept the view that marine animals for instance would be suited to fresh waters, which differ markedly from sea-water with respect to their various salt compositions. There has been no evidence until now that this also means that animals lose their capacity to live in qualitative sea-water as they lose their ability to settle in quantitative sea-water. In other words, diluted sea-water of the one or other degree of salinity could be an optimal medium for non-marine organisms just as sea-water is optimal for marine creatures.

The author started his experiments with the collaboration of his teacher, the late Prof. Dr. R. H. FRITSCH (Gießen and New Delhi).

## Physiological Adaptations of Cladocera to Salinity

In order to evaluate diluted sea-water as a suitable medium for freshwater animals it seems desirable to keep an organism under observation throughout its life, or even to follow a population throughout its life-cycle. A first consideration, then, is to find a short-lived animal for investigations in this direction. A second consideration is to select an animal which is well-known, since such a complicated solution as sea-water demands several measures for comparison.

*Cladocera* do well in these respects. These widely distributed Phyllopoda are suited to a considerable range of different fresh-water habitats mainly because of their reduced size, their ability to reproduce parthenogenetically, and because they are short-lived (BANTA, 1939). They can even settle in small seasonal ponds. Some *Cladocera* have returned to the sea or to mixohaline waters near the sea-shores (BERG, 1933; REMANE, 1950; GIBITZ, 1922; LAGERSPETZ, 1955; MEIJERING, 1961) or live in inland brackish waters (MEGYERI, 1955).

The fact that freshwater *Cladocera* may return to mixohaline waters has stimulated physiological investigations dealing with the adaptation of *Daphnia magna* STRAUS to media of increasing salt content. FRITSCHE (1917) found that in this animal the osmotic pressure of the body liquid increases in proportion to the increasing salinity of the surrounding water so that the osmotic pressure inside the animal will always be a little higher than that of the medium. The rate of this regulation depends on the intensity of metabolism, i.e. on the animal's age, state of feeding and, to a certain degree, on temperature. FRITSCHE used various dilutions of Mediterranean sea-water for his experiments; this is of course of special interest for our purposes.

We have now to consider metabolism with respect to regulation and resistance. There is a lot of literature dealing with cladoceran metabolism as indicated by the heart rate. The results of many investigators are, however, scarcely comparable, since their methods of working and the media in which the animals were kept differ greatly from each other (SCHWARTZKOPFF, 1955). Accordingly we have analysed the heart rate of male and female *Daphnia magna* throughout life when reared in diluted sea-water and kept under constant conditions of the environment (FRITSCH, 1958a; MEIJERING, 1958; FRITSCH & MEIJERING, 1958).

I would like to single out the time just after moulting, when we found a marked fall in the heart-beat curve, followed by the laying of parthenogenetic eggs in the female, after which the curve rose to the normal level throughout the instar. This fall of the curve after the moulting of the animal is caused by changes in osmotic pressure, by several ions, and probably by hormones (MEIJERING 1960, 1962a). On using artificial salt solutions according to FLÜCKIGER & FLÜCK (1949) or PASSOWICZ (1935) we found that the fall of the heart rate was much more violent so that in many cases the animals were not able to regain the normal heart rate and died. The fact that *Cladocera* mainly die at the time of ecdysis was already observed by BANTA. The diluted sea-water which we used in our experiments contained about 2.3‰ sea-salts. Following the "Venice System" we can classify this solution under the β-oligohaline type of water. The media of PASSOWICZ and FLÜCKIGER & FLÜCK contain 5.1‰ and 4.5‰ salts. Both solutions differ very much from sea-water qualitatively, since they contain $CaCl_2$ (PASSOWICZ) and $Na_2CO_3$ (FLÜCKIGER & FLÜCK) as the main salts. Here it is possible to perceive the stabilizing influence of the salt composition of the sea-water on the heart rate in connection with the animals' power of survival at dangerous points within their life.

There are data given in literature about the influence of several ions on *Daphnidae*. HOLM-JENSEN (1948) pointed out that the exchange of ions is very fast in *Daphnia magna*. The speed of the Na-exchange depends on the amount of Ca-ions available in the solution. In a former publication HOLM-JENSEN (1944) discussed the time of survival in this species when kept in several concentrations of diluted sea-water, especially in those, in which salinity was too high. FLÜCKIGER (1952) found that K-ions, as well as Mg-ions, accelerate the heart rate. On the other hand Ca-ions inhibit this and show the effect of an antagonist of K and Mg.

Unfortunately, all these investigations were made at different or unknown points within the life of an animal. It is, however, clear that several ions influence the heart rate in the one sense or the other. There may be antagonistic effects occurring mainly between the above mentioned ions and Ca. Further investigations which are coordinated in time at fixed points in an instar are possible, however, and very desirable.

The same should be emphasized with respect to investigations concerning resistance, especially after shocks. In such cases the time of survival is calculated from the shock to the death of the animal and the measure of resistance is determined by the value obtained. Having in mind the position within an instar or within the life is highly useful in these matters, since the effect of treatment may be different, it depending on whether an animal has just moulted or not.

In order to facilitate orientation within the life we have worked out a scheme of time in *Daphnia magna*. Since length of life depends to a large extent on the speed of life, which is influenced by the rate of metabolism (FRITSCH, 1962), we find very different data on the duration of life in literature. Factors of the environment, such as temperature (MCARTHUR & BAILLIE, 1929) or feeding (INGLE, WOOD & BANTA, 1937; DUNHAM, 1938), which influence the rate of metabolism, influence the speed and so the duration of life. In order to get a standard measure we looked for a measurement connected with metabolism. Most suitable for these purposes were heart beats. The duration of the longest recorded life span (BODENHEIMER, 1938, 1958) was expressed in millions of heart beats (FRITSCH, 1958b; MEIJERING, 1958).

Certain physiological events can be calculated in heart beats very exactly within the life span. In female *Daphnia magna, Daphnia pulex* DE GEER and *Macrothrix hirsuticornis* NORMAN & BRADY the number of heart beats was constant in juvenile, adolescent and adult instars (MEIJERING, 1958; MEIJERING & V. REDEN (in press) HUCHZERMEYER, 1963). The number of heart beats between moulting and the following laying of parthenogenetic eggs was also constant under constant conditions of the environment (diluted sea-water as mentioned above) (MEIJERING, 1958). This also holds for *Simocephalus vetulus* O.F.M. (MEIJERING, 1960). On the other hand FRITSCH (1958b) noted that the number of heart beats per instar increases in male *Daphnia magna* throughout life. FRITSCH (1962) gathered all these figures into one scheme.

It was found that mortality in a population is avoidable within the first 40% of the life span, after which mortality sets in catastrophically at the so-called death-limit (MEIJERING, 1958; V. REDEN, 1960; ENGEL, 1961). Then mortality continues at a constant rate. Thus the life of a *Daphnia*-population can be divided into two phases, the vital and the debile phase, the latter setting in at about 18 million heart beats (MEIJERING & REDFERN, 1962).

Now it seems possible to have a good orientation within the life so that we can begin to characterize further effects of diluted sea-water on *Cladocera*. Some information is already available, which we have obtained by using the methods of working mentioned above.

According to BODENHEIMER we can distinguish between the ecological longevity and the physiological longevity of a population, the latter being an ideal figure which does not appear in nature because of the inconstancy of factors of the population's surroundings. Fig. 1 shows survival curves of female *Daphnia magna* kept in diluted sea-water, taken from MEIJERING (1958), which we now consider to be almost physiological. This was discussed by MEIJE-

RING & REDFERN on the following criteria: there is almost no mortality in the vital phase; there is a constant rate of mortality in the debile phase; and the maximal life span was attained. Until now a survival curve has assumed this form only when a population of female *Daphnia magna* was kept in diluted sea-water. It was not even possible for pantothenic acid, which prolongs the life of *Cladocera* considerably (FRITSCH, 1953), to improve the environment any more.

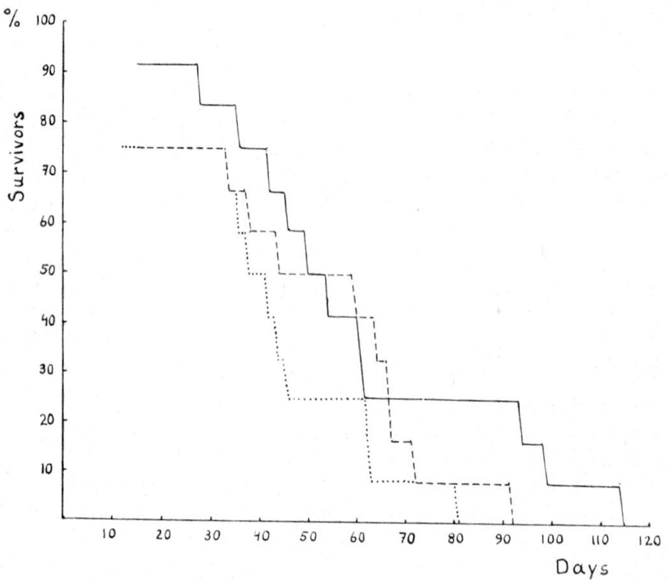


Fig. 1. Survival curves of female *Daphnia magna* STRAUS. – · – · – · = Survival curve based on units of physical time. ——— = The corresponding curve based on heart beats. The animals performed 0.726 million heart beats on the first day of life. The heart beat curve was based therefore on the scale 0.726 million/1 day. Temperature 18° C. (After MEIJERING & REDFERN, 1962).

Different types of effects on the survival curve are noticeable when the medium deviates from this optimum. Mortality may remain small in the vital phase, but then nearly all the animals die at the end of it, at the so-called death-limit. This kind of curve is typical of animals living in media which contain a large amount of Ca-ions (FRITSCH, 1956; MEIJERING, 1958). Another kind of survival curve is that exhibiting the mortality of the animals within the vital phase, especially in the region of the adolescent instar. This was found by FRITSCH (1953) in a population of *Daphnia pulex* DE GEER kept in the medium after FLÜCKIGER & FLÜCK, and by the same author (1958b) in male *Daphnia magna* kept in diluted sea-water containing 2.3‰ salts (β-oligohaline water). The latter, however, does not seem to be optimal for males, as it is for females of the same species. ENGEL (1961) showed mortality within the vital phase of female *Daphnia magna* when working with stronger solutions of diluted sea-water (S-content 3.2‰ and 4.6‰, α-oligohaline waters), and so did HUCHZERMEYER in a population of female *Macrothrix hirsuticornis* when using a salt content of 2.3‰.

Since the latter figure mentioned by HUCHZERMEYER was not optimal for *Macrothrix hirsuticornis*, investigations were started with other concentrations of diluted sea-water by MEIJERING & v. REDEN (in press). Unfortunately the survival curves are only available in measures of astronomical chronometry so that the line between the vital and debile phase is not so well marked as in fig. 1. A comparison, however, shows that for *Macrothrix hirsuticornis* this line lies within the region from the 60th to the 70th day of life. The best curve is that of females reared in the diluted sea-water containing 2.9‰ salts (β-oligohaline water). Juvenile mortality in the animals kept in 2.3‰ (β-oligohaline water) and 4.0‰ (α-oligohaline water) dilutions is higher and, above all in the case of those living in the dilution with the lower S-content, mortality increases within the vital phase (Fig. 2). The survival curve of the population kept in 2.9‰ is the best hitherto recorded for *Macrothrix*

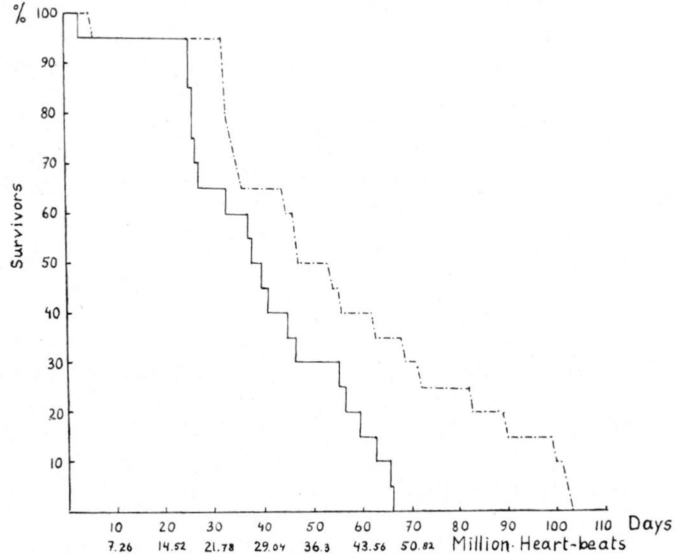


Fig. 2. Survival curves of female *Macrothrix hirsuticornis* NORMAN & BRADY. Animals kept in diluted sea-water containing · · · · · = 2.3‰, ——— = 2.9‰ and – – – – – = 4.0‰ sea-salt. (After MEIJERING & v. REDEN, in press).

*hirsuticornis* and contains the longest recorded life span of this species (115 days). This high figure depends on a retardation of the animals caused by a scarcity of food. But the shape of this survival curve cannot deviate greatly from that exhibiting the physiological longevity of this species, when we compare it according to the principles of MEIJERING & REDFERN with the curves of female *Daphnia magna* (Fig. 1).

In nature, i.e. outside the laboratory, similar survival curves cannot be recorded for populations, as the time of good living conditions is much shorter (BERG, 1933; OLOFSSON, 1918; MEIJERING, 1961). The very good survival curves of figs. 1 and 2 are a result of the constancy of temperature, salt content (being optimal), feeding, and living-space.

Now let us consider some results concerned with growth and reproduction, insofar as they give information about the productiveness of a population. They should, however, be seen in connection with the population's strength of survival.

From GREEN (1956) we know that in *Daphnia magna* the number of parthenogenetic eggs increases during the first adult instars. After passing a maximum figure this number decreases steadily. Since the maximum figure occurs in the region of the 12th instar, we can consider it as being a further mark between the vital and debile phase. HUCHZERMEYER found the same reproduction curve in his population of *Macrothrix hirsuticornis* kept in diluted sea-water. The maximum number of offspring was liberated in the 12th instar (the mean number of young being 41.6 in this instar). The maximum number of young produced by one individual in this instar was 52. These numbers are more than twice as high as any known for this species until now. In nature most specimens have not more than about 10 eggs in their brood chambers and usually even less. From HUCHZERMEYER's results we can see that this species of *Cladocera* is not inferior to many others as far as reproduction is concerned. *Macrothrix hirsuticornis* appears to be very much retarded in its natural habitats, from which, therefore, its optimum of life cannot be derived.

HUCHZERMEYER had similar results when he observed the growth of his animals. The length of 1.35 mm was considered to be the maximum size (BERG, 1933; MEIJERING, 1961). Other authors mention figures as low as 1 mm (SCOURFIELD & HARDING, 1958). HUCHZERMEYER's females, however, reached a maximum length of 1.65 mm in diluted sea-water.

These examples may show that it is possible to raise a species, which is retarded in nature, to its full potential in size, reproduction and longevity, when diluted sea-water is used as a medium.

Growth depends to a large extent on the salt content of the medium (KINNE, 1952). It may be possible, therefore, to find even larger animals in *Macrothrix hirsuticornis,* when the optimal salt content for this species is used.

Some data have been published on the connection between growth and the beginning of puberty in *Cladocera.* When the initial sizes are different, female (GREEN) and male (FRITSCH, 1958a) *Daphnia magna* can mature in different instars and at different sizes. Male *Daphnia magna* can mature at 1.7 mm, when growth is retarded during the juvenile instars (GREEN) or at 1.85 mm when living in an outdoor pond (MEIJERING, 1962b). Males of this species which were kept in diluted sea-water under constant conditions, however, reached a length of 2.15 mm at the end of the juvenile phase (MEIJERING, 1962b). These results show that the size of the animal in the adolescent instar depends largely on the quality of the medium.

Salinity has an important influence on the duration of the juvenile phase. This was pointed out by ENGEL, who caused female *Daphnia magna* to mature after 4, 5 or 6 instars by using different concentrations of diluted sea-water. In these experiments the juvenile phase was prolonged by the higher concentrations.

Data are available about the mechanism of the influences of the salt content on growth and maturation. It was shown by KINNE (1952) and by SUOMALAINEN (1956) that the total salt content influences the $O_2$-consumption of some species of *Gammarus* found in the Baltic Sea. KINNE (1952) also found relations between the salt content and heart rate in *Gammarus duebeni* LILLJEBORG. It seems to be clear, then, that the influence of the salinity acts via the animal's metabolism.

Since diluted sea-water proved to be such a suitable medium for freshwater *Cladocera,* we were encouraged to undertake further investigations concerning the influence of environmental factors which act on the rate of reproduction via metabolism. It was proved mainly by MORTIMER (1936) that overcrowding in a culture of *Cladocera* forces the females to produce males. Scarcity of food as a consequence of overcrowding in nature, however, causes the production of gamogenetic eggs and ephippia. MORTIMER also found a tendency in *Daphnia magna* to produce males, when temperature was lowered markedly. He obtained the highest percentage of males and gamogenetic females by combining all the active factors. These factors of the environment are of very different types (BANTA). Some of them such as crowding or underfeeding are very complex so that it is difficult to investigate them.

We have tried to get some more light on these matters by counting heart beats in the instars of animals, which lived under suboptimal conditions. Females of *Daphnia magna* kept under crowding-conditions in diluted sea-water always show a low heart rate. Females were isolated from such a mass-culture in order to find the number of heart beats performed during an instar, when gamogenetic eggs were laid. This number was much lower than in normal adult instars (1.1 million to 1.8 million). The space of time between moulting and egg-laying, however, was prolonged extremely to 300.000 heart beats (8000 normal) (MEIJERING, 1962a). Negative factors, such as crowding, do not only alter the rate of metabolism, but also the scheme of time in the animal. Since "normal" figures were observed in animals living under optimal conditions, we can speak of optimal numbers of heart beats between certain physiological events in *Cladocera*, or we can speak of an optimal scheme of time. If animals are not living under optimal conditions, it could be possible to determine the deviation from the optimum by scoring it in numbers of heart beats.

In his work on *Macrothrix hirsuticornis*, HUCHZERMEYER compared the numbers of heart beats performed by this species with those of *Daphnia magna*. He found the normal length of adult instars to be greater than in *Daphnia magna*. Since *Macrothrix hirsuticornis* is distributed mainly over colder regions (MEIJERING, 1961), HUCHZERMEYER altered the temperature from 13° C to 18° C in order to study the animal's reaction under these suboptimal temperature-conditions. He noticed the surprising fact that the effect was a lower heart rate. The number of heart beats per instar decreased markedly, and males and gamogenetic females appeared instantly. In *Daphnia magna* MORTIMER did not find males at a higher temperature, but, as already mentioned, only when he lowered the temperature. Here we have similar effects by using opposite stimuli.

It is quite clear that unspecific stimuli from the surroundings, which lower metabolism and alter the scheme of time, alter the modus of reproduction in *Cladocera*.

Relations between salt content and temperature were found in several species (STEINER, 1935; PRECHT, 1949; KINNE, 1954; GRESENS, 1928; SCHLIEPER, 1958 and BROEKEMA, 1942). Further investigations might well deal with the optima in salt content and temperature. In *Cladocera*, above all, living conditions can be evaluated by measuring the numbers of heart beats; in other words, it seems possible to find environmental optima systematically.

## Experimental Studies on the Influence of Salinity on Growth and Morphology of Agropyron junceum boreoatlanticum

The following part of this article will deal with some botanical results which have not been published before. We thought that it would be expedient to see all matters concerned with diluted seawater on a common biological basis, that is, to gather experience both in the zoological as well as the botanical field. The research on *Agropyron junceum boreoatlanticum* SIMON. & GUIN. was started in 1960 in cooperation with Mr. D. M. WERNER and continued in 1961 in cooperation with Mr. W. STEUDE. Mr. WERNER worked out the methods of irrigation and is responsible for all the measurements of the shoots in 1960, Mr. STEUDE for doing the same work in 1961. I am very grateful to both gentlemen for giving me permission to publish their results in this context. The author is responsible for observations on the root system, some ecological remarks, literature, and the general interpretation of the results.

*Agropyron junceum* was chosen for our research work for several reasons: we wanted to study sea-water influences over a longer period of time. Since *Agropyron junceum* is a perennial grass, it served well in this respect. We wanted to use a grass of economic value (Literature see BOYKO & BOYKO, 1959). We expected to succeed in growing a species of the sea-border in strong concentrations of diluted sea-water in order to obtain marked physiological effects of irrigation. Some data were already available in the literature and these helped us to avoid mistakes and made some comparison possible.

Late in July 1959, seeds of *Agropyron junceum boreoatlanticum* were taken from plants growing on the small dunes of the large eastern sands of Spiekeroog (WIEMANN* & DOMKE, 1959). This material was kept indoors until March 1st, 1960, when it was sown in Mitscherlich-jars, containing nearly 6 l of pure quartz sand from our dunes. The mean diameter of the grains of sand here is 0.33 mm (HABERMANN & HABERMANN, 1931). We used only the upper part of the jars which were dug in outside in a dune-area completely void of vegetation. There was a good connection with the ground underneath through a hole of 8 cm in diameter at the bottom of the jars, which allowed the irrigation water to flow out. 20 seeds were sown in each jar, and 4 jars were taken for each concentration of diluted sea-water. Three of these jars were observed during 1960, four during 1961, so all the data concerning 1960 hold for 60 seeds per concentration and for 80 seeds per concentration in 1961.

These cultures were irrigated with sea-water from the North Sea, which contains about 32‰ salts. This was diluted with unchlorinated tapwater. Pure sea-water was regarded as a 100% dilution, pure tapwater as 0% dilution, being euhaline and fresh water according to the "Venice System". Between these figures we used 25% diluted sea-water (S-content 8‰, β-mesohaline water), 50%

* I am grateful to Dr. P. WIEMANN (Hamburg) for providing some articles on the ecology of *Agropyron junceum* on the Frisian islands.

diluted sea-water (S-content 16‰, α-mesohaline water) and 75% diluted sea-water (S-content 24‰, polyhaline water).

The first period of irrigation started on April 27th, 1960. Every evening 0.5 l of irrigation water was poured onto each jar. This quantity corresponds to 16 mm rainfall per day. The irrigation was stopped on July 17th, 1960. The second period of irrigation began on June 4th, 1961 and was carried out in the same way as that described for 1960. This irrigation lasted until August 2nd, 1961. For the rest of the year the cultures were left without tendance in their places. During all this time they were under the influence of our climatic conditions. They were not protected against either rainfall or sunshine, against high or low temperatures or the influence of the wind. Rainfall was measured 250 m to the west of the cultures, temperature taken to coincide with that on Norderney in 1960 and that on Langeoog in 1961.

The days on which the plants were measured can be seen from figs. 3 and 4. Apart from one, all the measurements were taken within the periods of irrigation.

Not all the seeds germinated during 1960. In 1961, however, new shoots had to be reckoned with. So all the heights of the plants irrigated by the respective dilutions have been added together severally in order to get comparable measurements of the yield. The mean height of the plants was taken at the end of each irrigation period as well.

Unfortunately the site of the cultures was flooded by the sea in February 1962, so no further consideration was possible, as all the groups were "irrigated" with 100% sea-water. This flood was the highest in this region since 1825, and broke all the dykes on Spiekeroog, leaving only a small strip of dunes out of the water.

On June 12th, 1962 one jar containing plants irrigated with each concentration was dug out so that the root system could be examined, the other jars still remaining in their former positions.

Some information about temperature during the two periods of irrigation should be given before the growth of our plants is discussed. The differences between the daily maxima and minima did not exceed 7° C. During May 1960, the mean temperature rose steadily from 8° C to 15° C. The latter temperature held through June and July to the date when irrigation ceased. In 1961 there was the same temperature throughout the whole period of irrigation. The mean temperature was a little higher (17.7° C), however, during the days between the measurements taken on June 23rd and July 3rd. As temperature conditions were quite similar during both irrigation periods, rainfall can be looked upon as the main factor to cause fluctuations of growth, which are reflected in the curves of figs. 3 and 4.

A first look at fig. 3 gives one the impression that growth is notably retarded by higher salt concentrations and that a 100% dilution is much more harmful than that of 75%. When we compare the state of all the groups from 0% to 50% as attained on June 18th, it is apparent that they are nearly equal. Until this date there was very little rainfall. The last month of irrigation, however, altered the picture very much. The amount of rainfall increased steadily, and until the end of irrigation the 25% and above all the 50% dilution forced the plants to grow faster than those irrigated

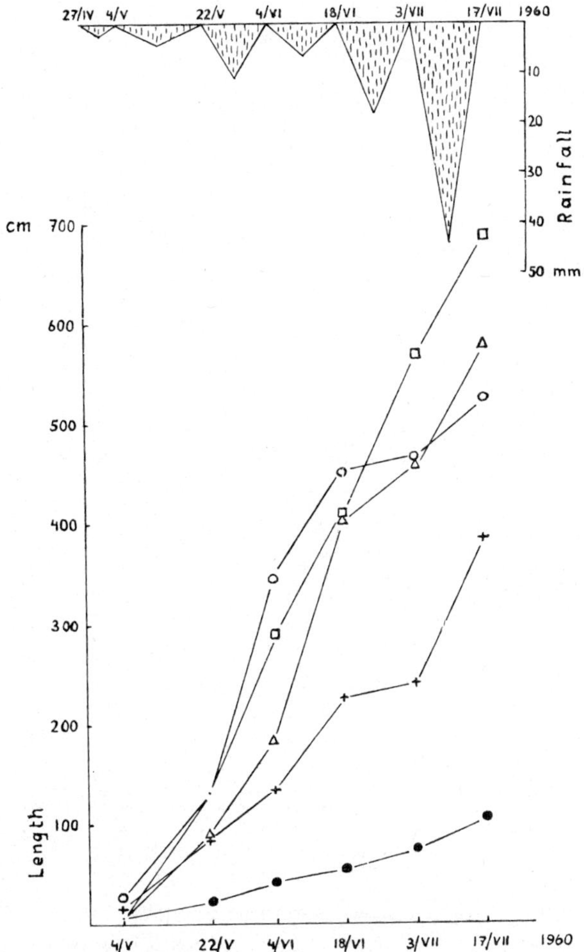


Fig. 3. Growth of *Agropyron junceum boreoatlanticum* SIMON & GUIN. under sea-water irrigation and rainfall. Irrigation water containing ○ = none, △ = 8‰, □ = 16‰, + = 24‰, and ● = 32‰ sea-salt.

with tapwater, which had seemed to be the best at the very beginning. On July 17th the 50% dilution had produced the best yield, and, on the whole, the growth curve of these plants was rising most steadily.

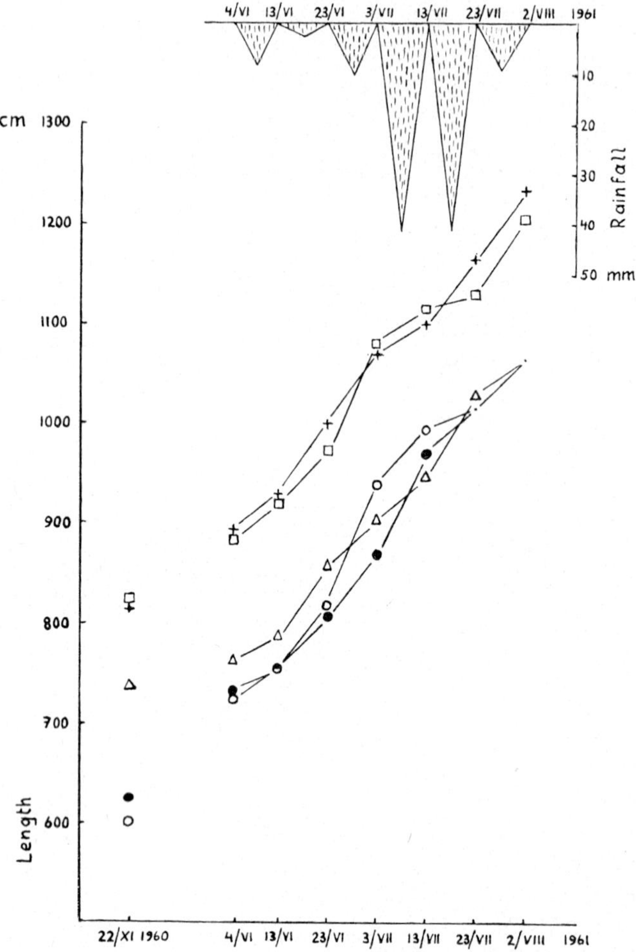


Fig. 4: Continuation of the growth curves of fig. 3.

Measurements taken on November 22nd, 1960 by Mr. STEUDE indicate that the effect of four months' rainfall, with no additional irrigation was very impressive (Fig. 4). The five rows were then divided into three groups, the best containing the 50% and 75% rows, the second the 25% row, and the worst consisted of the 100%

and 0% rows. These facts show that the stronger solutions caused the plants to accelerate growth as soon as the salt content of the medium was lowered.

At the beginning of the second irrigation period two groups were noted, the better containing plants of the 50% and 75% rows, and the worse consisting of the other concentrations.

Careful comparison of all the curves makes it clear that, whenever rainfall was higher, the growth of the plants in the higher concentrations was accelerated. At the end of the watering period, the row irrigated with the 75% dilution was the best, followed immediately by the 50% group. In these two rows Mr. STEUDE observed the only 6 plants which flowered in 1961. Also in 1962 the author observed that only plants in these groups flowered, mainly in the 75% row. Higher and lower concentrations of irrigation water eventually retarded growth and even prohibited fruiting up to the summer of 1962.

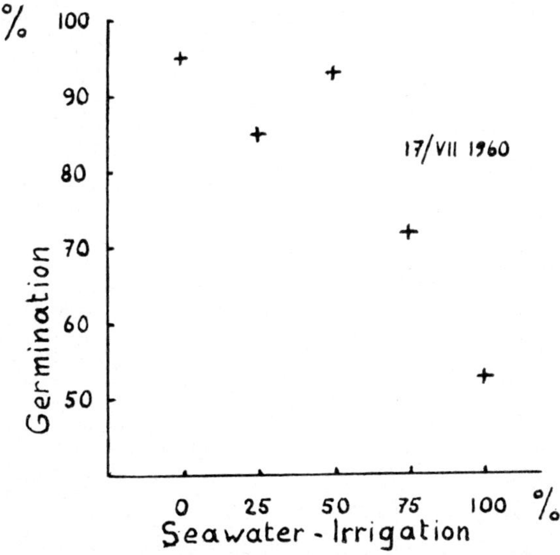


Fig. 5. The effect of sea water irrigation on the germination of *Agropyron junceum boreoatlanticum* SIMON. & GUIN.

Fig. 5 gives an impression of the effect of irrigation on germination, for which lower concentrations are better than higher, including even the 75% row. Reasons for this will be discussed later in connection with literature.

Fig. 6 contains the values of the average height of all the plants. It gives the same impression as that described for growth. It also makes clear that the harmful, or, more precisely, the retarding influences of higher concentrations were compensated later by accelerated growth.

Finally fig. 7 shows the same effect with respect to the general development of the plants as indicated by the mean number of leaves per plant.

In order to get an impression of the root system, one jar of each row was taken up by the author on June 12th, 1962. The following findings hold for the descendants of 20 seeds per concentration.

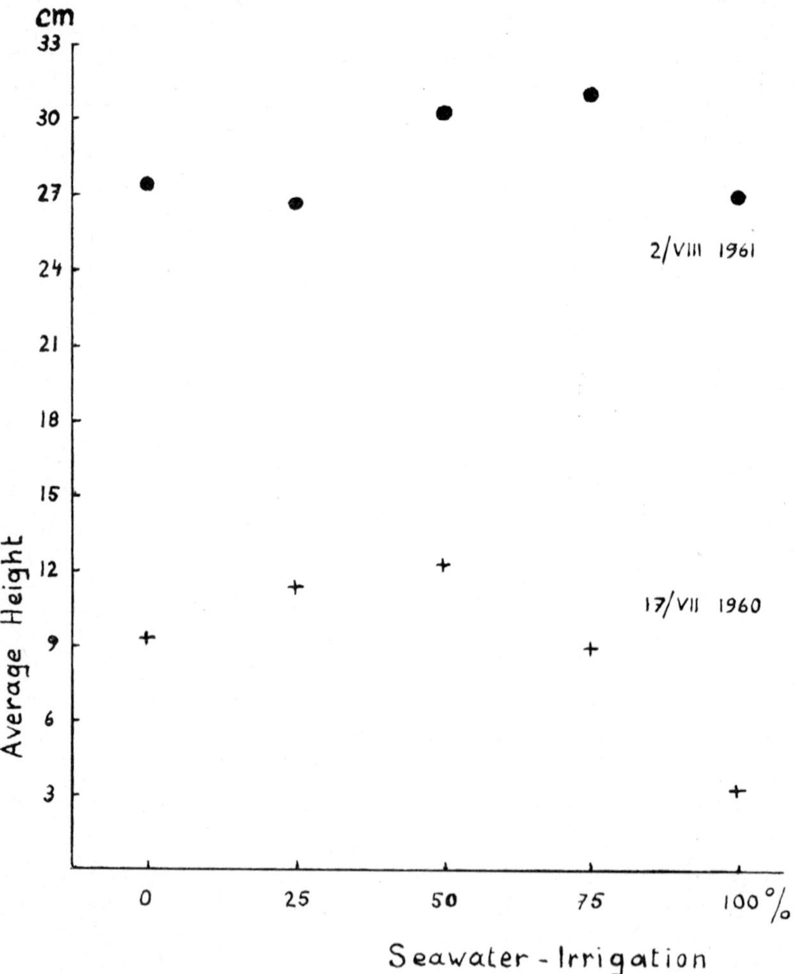


Fig. 6. Average height of *Agropyron junceum boreoatlanticum* SIMON. & GUIN. under sea-water irrigation, when measured on July 17th, 1960 (+) and August 2nd, 1961 (●).

The jars irrigated with 25%, 50% and 75% were very similar as regards the development of the root system. The two latter were only slightly better than the first. The roots were of the type shown on the right of the drawing (fig. 8). Each jar contained two or three rhizomes winding round and following the shape of the jars. Only

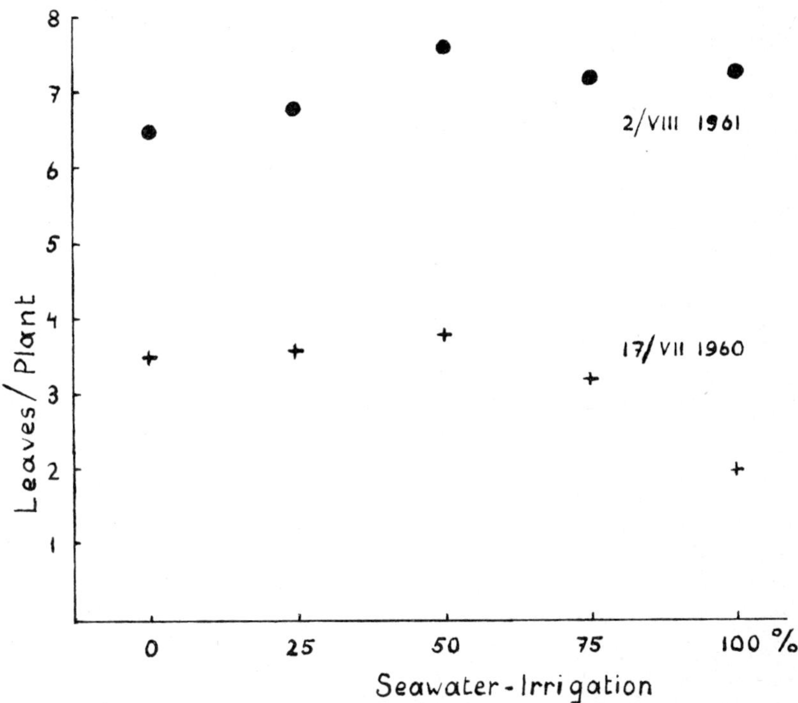


Fig. 7. The effect of sea-water irrigation on the development of the leaves of *Agropyron junceum boreoatlanticum* SIMON. & GUIN. in 1960 (+) and 1961 (●).

the plants in the jar of the 0% dilution contained a much poorer root system with just one rhizome. The 100%-roots were a little more branched than those of the 75% and 50% rows, but there were no rhizomes. It was from this jar that the plant shown in fig. 8 was taken. The other drawing on the left is of a plant from the natural habitat on Spiekeroog island given for comparison. It was taken from a dune of about 1.5 m in height. This type is very different. Long rootstocks are found among the plants of this area, bearing small roots, which spring from the internodia. These roots are irregularly covered with root-hairs. In our cultures these hairs are found mainly within the first 5 cm layer of sand just below the shoot.

We see reasons for these different types of root development in the fact that our cultures were not covered regularly with new sand, an occurrence which is characteristic of new dunes at the sea-side, the natural habitat of *Agropyron junceum*. For further information on this type of root development caused by shifting sands see MEIJERING (1964). The sketches of fig. 8 are only given here, since they throw some light on the findings of other authors. These data will be compared now with our results.

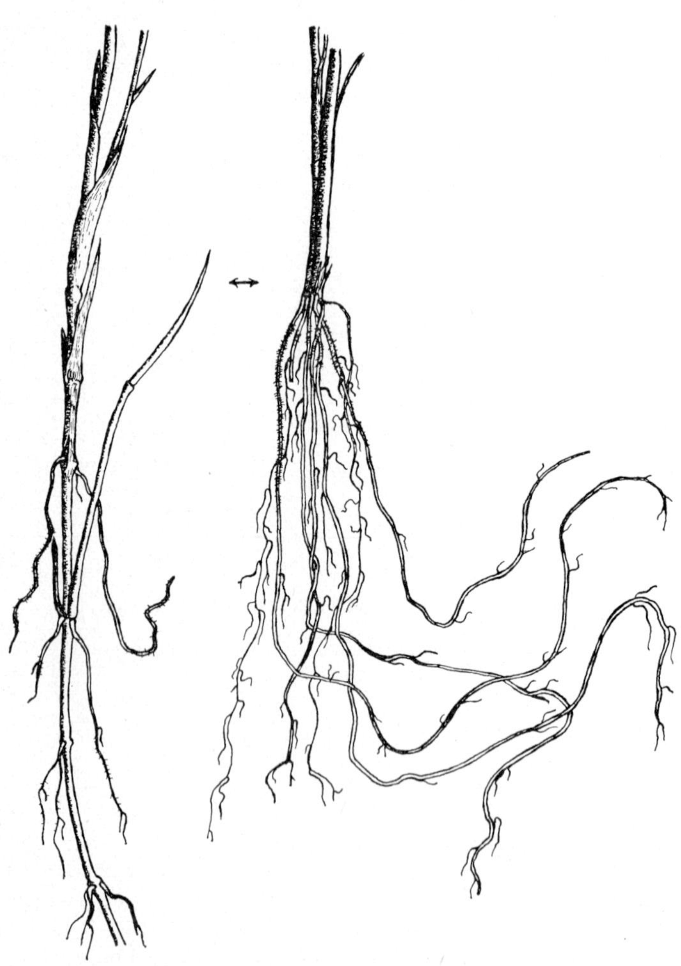

Fig. 8. Part of the shoot and root system of *Agropyron junceum boreoatlanticum* SIMON. & GUIN. Left: Plant from shifting dunes on Spiekeroog island. Right: Plant which was irrigated with sea-water (S-content 32‰). ←→ = Soil surface.

*Agropyron junceum boreoatlanticum* does best when irrigation water contains a certain amount of sea-salt, ranging from 16—24 g/l (α-mesohaline to polyhaline water). So it is clear that this species is a halophyte, as was already stated by REINKE (1909), who investigated the ecology of the species* on the East Frisian islands. He found *Agropyron junceum* only in the immediate neighbourhood of the sea, it being the first plant pioneer on the sandy west, north and east coasts of these islands, just as *Salicornia herbacea* L. is the pioneer along the more muddy south coast opposite the mainland. STOCKER (1924) counted *Agropyron junceum* among "obligate Strandpflanzen".

Some investigators measured the salt concentration of the subsoil water in the *Agropyron*-region. SCHRATZ (1936) found the isohaline 20—25 g/l to be the limit of the main region of *Agropyron junceum*. According to ARNOLD & BENECKE (1935) the species does well at 18 g/l, but can be found even in places, where this figure reaches 80—90 g/l, though this is an exception. In such environmental conditions the plants are very poorly developed. On the island of Wangerooge, the next eastward from Spiekeroog, ARNOLD (1936) found *Agropyron junceum* within the isohalines of fresh water up to 45 g/l.

Unfortunately there is only little information given on the plants themselves, the main point of investigation always being to describe the extent of the territory covered with *Agropyron junceum*. The descriptions of the plants are vague, such as "kümmerlich" or "Hauptvorkommen". This second term appears, however, in connection with the isohalines of 18 or 20 g/l, which is really the optimal region for *Agropyron junceum*.

The only author to deal with irrigations of *Agropyron junceum boreoatlanticum* was BENECKE (1930). He cultivated shoots in Mitscherlich-jars. His material was collected on Spiekeroog island just as our plants were. BENECKE, however, wanted to find the highest concentration that *Agropyron junceum* can bear, so he gradually increased the salt content of his irrigation water. His plants died at the concentration of 62 g/l. At concentrations approaching this figure, growth was reduced markedly.

According to SCHRATZ seeds of *Agropyron junceum* contain no salts. ARNOLD found that plants of *Agropyron junceum* growing in their natural habitats may contain up to 10‰ NaCl, a figure which did not rise further even when the salt concentration of the surroundings still increased. In the Mediterranean subspecies *Agropyron junceum mediterraneum* SIMON. & GUIN. BOYKO & BOYKO (1959) also found high osmotic values, which did increase, however, with the increasing salt concentration of the irrigation

* *Agropyron junceum* was formerly known as *Triticum junceum* L.

water. According to REPP (1958) the concentration of salts in most halophytic plant cells increases during life. Older plants, therefore, are generally more resistent to higher concentrations than young shoots (REPP).

BICKENBACH (1932) found that the transpiration of *Agropyron junceum* was reduced when he kept the plants in higher concentrations. But his findings are not quite comparable with our results, since he irrigated very young plants with various concentrations of van der Crone-solution, which might not have quite the same effect as sea-water solutions. He found concentrations of 26 and 29 g/l reduce transpiration. *Agropyron junceum* can presumably bear higher concentrations of sea-salts before transpiration is retarded.

Keeping in mind all these facts we can now start to explain our findings on *Agropyron junceum*.

Since the osmotic pressure of seeds is very low, it is evident that the intake of water is faster when the concentration of the irrigation water is lower. Tapwater is the best solution for germination (figs. 3 and 5). The advantage of tapwater, however, lasts only for several weeks. The intake of sea-salt seems to be necessary during the first four or six weeks of life. During these weeks the salt content of the medium should not exceed 16‰ and should not be lower than 8‰. In older plants optimal living-conditions range within the limits of 16 and 24‰. Higher and lower salt contents reduce growth and development more or less (figs. 4 and 7). Since the sand which we used contained no salts, our rows irrigated with low concentrations may be considered as hunger cultures. On the other hand we may expect that transpiration in the plants irrigated by the strongest concentration was reduced through difficulties occurring in connection with the intake of water.

Since we noticed, however, that the differences between all the rows were not too large, *Agropyron junceum* can be considered as being euryhaline.

It would be interesting to obtain more data about the mechanism of regulation, by which *Agropyron junceum* manages to bear fluctuations of the medium's salt content. In some Danish halophytes REPP observed a "limit of permeability" ("Permeabilitätsschranke") against higher concentrations. In *Agropyron junceum boreoatlanticum* the same limit may exist (ARNOLD). We see a most important hint in the fact that the lowering of the concentration by rainfall, as happened between our two irrigation periods as well as during July 1960 and 1961, forced plants in the higher concentrations to speed up their growth considerably. This can be seen as a reaction depending upon the increasing difference of ion concentration inside and outside the roots. The opposite case can be regulated too. Plants in lower concentrations grew faster during the dry weeks of June 1961 (fig. 4). An explanation will be given later in connection with literature.

It is quite obvious that *Agropyron junceum* is able to "wait" for the most favourable conditions as regards the salt content of the medium. Suboptimal conditions reduce metabolism more or less and, within certain limits, changes of salt conditions towards the optimum force the plant to speed up metabolism as has been indicated already in growth.

Finally the root system of our plants should be explained. Since the sand was not disturbed throughout the investigations, our *Agropyron junceum* did not develop such large rootstock systems as they do in nature (fig. 8). The root-hairs were distributed mainly within the region of the first 5 cm below the sand surface, so it is certain that they took water and salts chiefly from this uppermost sand layer. According to BOYKO & BOYKO (1959) and BOYKO (1952), pure sand or similar light soil does not accumulate salts. Our plants, therefore, absorbed water from that solution, which was afforded them by irrigation or rainfall.

In nature *Agropyron junceum* lives on shifting primary dunes. On Spiekeroog the height of these dunes ranges from 20 cm to 300 cm, the very low dunes being reached by sea-water nearly every spring tide, that is approximately every fortnight. Even those dunes which are roughly 100 or 200 cm high, however, are usually flooded once or twice a year. These higher dunes on Spiekeroog bear rich *Agropyron junceum* vegetation, the plants being 50 to 70 cm high. They fruit every year. On very low dunes on the other hand, the plants are normally not more than 20 cm high and seldom fruit. In dunes the root system of *Agropyron junceum* stretches down 100 cm or more. It is clear that the salt content of the soil water in higher dunes varies markedly in different layers. This was already pointed out by SCHRATZ, who assumed that *Agropyron junceum* takes water from that layer, in which the soil water contains less salts. This would mean, however, that it has to be absorbed from the uppermost layer, which rainfall for the most part rids of salts. In reality plants in nature take in water from that layer, which contains the optimal dilution of sea-water. This layer may be situated higher or lower, everything depending on whether heavy rains have fallen or whether the region has just been flooded by the sea. Dunes of approximately 1 or 2 m in height will always contain such a layer of optimal salt conditions and so the large root system is always able to reach it. Further investigations would be useful to bring more light on these matters. Dunes, which are no longer reached by the sea or are flooded too often have a bad vegetation of our species.

Since tapwater is the best solution for germination, it is obvious that in nature, where the seeds lie mostly near the surface, rainfall will help to make them germinate quickly. When young shoots have developed a root system, they are able to reach more and more layers where the salt content is higher. These layers are necessary for their further development, as has been pointed out above.

Since the distribution of certain isohalines is so complicated and variable in nature, it is quite clear that the halophytic character of *Agropyron junceum* can only be proved by physiological investigations on plants grown under culture conditions. The isohalines of the subsoil water, as mainly used by ecologists, do not quite suffice in this case. Both the vertical isohalines of the soil water, which are extremely variable, and the possibilities of the plant to take in water by higher or lower roots, just as the salt content of the soil water demands it, must be considered as well. Poor plants appear only when the scale of salt contents is limited, i.e. either too high or too low (see also MEIJERING, 1964).

It would be useful now to review some other papers dealing mainly with agricultural plants in order to see whether they are comparable with the results on *Agropyron junceum.* Deserving attention in the first place is the work of ZIJLSTRA (1946), who investigated ten agricultural species in vitro under the influence of diluted sea-water. As this publication is written in Dutch with only a brief English summary appended, it would serve a useful purpose to give one or two details from his results. This author found that a small amount of sea-salts stimulates growth in *Triticum vulgare* VILL., *Lolium perenne* L. and *Pisum sativum* L. The stimulation was noticeable within the region of 3.27 g/l, containing 2 g/l van der Crone-salts and 1.27 g/l sea-salts. A similar effect was observed by SCHARRER & SCHROPP (1949), who used NaCl for their investigations. ZIJLSTRA worked mainly with seeds, young shoots and plants, so his results are comparable with our figs. 3 and 5. An important finding of ZIJLSTRA was that a decrease in the salt content of the medium forced seeds of *Lolium perenne* L. which had lain in sea-water for 45 days to germinate within 5 days, whereas seeds which remained in sea-water needed 75 days for comparable germination. A similar principle is involved here among plants which are able to wait for favourable conditions and which do better with a certain amount of sea-water. This is especially true of *Triticum vulgare,* though the optimal amount is lower than in *Agropyron junceum,* since *Triticum vulgare* is not a real halophyte.

The fact that certain amounts of sea-salts may increase the yield of agricultural plants was already evident from the results of AHI & POWERS (1938), though these authors did not stress this phenomenon. This holds above all in the case of *Trifolium fragiferum* when the temperature is rather low. It has been shown that high temperatures are much more harmful when salt contents are high. BOYKO & BOYKO also found lower concentrations better for irrigating *Juncus arabicus* and *Agropyron junceum mediterraneum* than higher concentrations of Mediterranean sea-water. This can be looked upon as an effect of temperature being high in Israel. When our figures for the optimal salt content of *Agropyron junceum* are being considered, therefore, it must be borne in mind that these hold for regions, where the mean temperature is relatively low. In a warmer climate it would be necessary to dilute sea-water more than we did in our work.

Acceptance of this point of view enables us to explain why in our cultures plants irrigated with low concentrations speeded up growth during the last week of June 1961, when the temperature was higher (fig. 4). When temperature is high, it is easier for the plants to take water in from lower concentrations, while metabolism is higher as an effect of temperature. Here the close connection between temperature and salinity is once more apparent.

In all these examples of halophytic and non-halophytic plants sea-water was used for irrigation. As soon as non-balanced salt solutions were taken, plants were much more harmed. Many articles deal with waters containing non-marine salts. Here we can only single out those of AHI & POWERS, who worked with sea-water as well as with other salt solutions; or KELLER (1925), working on *Salicornia herbacea* L. According to HEIMANN (1961) it may be possible to raise the fertility of brackish waters of various qualities by adding certain antagonists. As far as we can see now from our little knowledge of diluted sea-water, this would not be necessary for sea-water, as it seems to be optimally balanced for either animals or plants. It is quite obvious that questions concerning diluted sea-water as a medium for organisms are questions of the quantities and not of the qualities of salts. In searching for optimal conditions we have to investigate salinity in connection with other environmental factors, mainly with temperature, as KINNE (1956) pointed out.

For economic purposes *Agropyron junceum* is worth knowing better, since it has been proved that this plant possesses a very large scale of adaptation to various salt conditions and shifting soil.

## REFERENCES


AHI, S. M. & POWERS W. L., 1938. Salt tolerance of plants at various temperatures. *Plant Physiol. U.S.A.* **13,** *767—789.*

ARNOLD, A., 1936. Beiträge zur ökologischen und chemischen Analyse des Halophytenproblems. *Jb. wiss. Bot.* **83,** *105—132.*

ARNOLD, A. & BENECKE W., 1935. Zur Biologie der Strand- und Dünenflora auf Borkum, Juist und dem Memmert. *Planta* **23,** *662—691.*

BANTA, A. M., 1939. Studies on the Physiology, Genetics, and Evolution of some *Cladocera.* Pap. Dep. Genet. Carneg. Inst. **39,** 285 pp., Washington, D.C.

BENECKE, W., 1930. Zur Biologie der Strand- und Dünenflora. 1. Vergleichende Versuche über die Salztoleranz von *Ammophila arenaria* Link,

*Elymus arenarius* L. und *Agropyrum junceum* L. *Ber. dtsch. bot. Ges.* **48.**

BERG, K., 1933. Note on *Macrothrix hirsuticornis* Norman & Brady, with description of the male. *Vidensk. Medd. Dansk. naturh. Foren.* **97,** *11—24.*

BICKENBACH, K., 1932. Zur Anatomie und Physiologie einiger Strand- und Dünenpflanzen. *Beitr. Biol. Pfl.* **19,** *334—370.*

BODENHEIMER, F. S., 1938. Problems of animal ecology. Oxford University Press, Oxford, England.

BODENHEIMER, F. S., 1958. Animal Ecology Today. Mon. Biol. VI. Dr. W. Junk Publishers, The Hague.

BOYKO, E., 1952. The Building of a Desert Garden. *J. Roy. hort. Soc.*, **76,** *4.*

BOYKO, H. & BOYKO, E., 1959. Seawater Irrigation. A new line of Research on a Bioclimatological Plant-Soil Complex. *Int. J. Bioclimatol. Biometeorol.* **III,** II, B 1, *1—24.*

BROEKEMA, M. M. M., 1942. Seasonal movements and the osmotic behaviour of the shrimp, *Crangon crangon* L. *Arch. néerl. Zool.* **6,** *1—100.*

DUNHAM, H. H., 1938. Abundant feeding followed by restricted feeding and longevity in *Daphnia. Physiol. Zool.* **11,** *399—407.*

ENGEL, E. K. H. R., 1961. Die Abhängigkeit der Sterblichkeitsmaxima von der Pubertät bei Weibchen von *Daphnia magna* Straus. *Z. wiss. Zool.* **165,** *422—427.*

FLÜCKIGER, E., 1952. Beiträge zur Verwendung von *Daphnia* als Pharmakologisches Testobjekt. These Nr. 2090, Zürich.

FLÜCKIGER, E. & FLÜCK, H., 1949. Ein künstliches Milieu für das Züchten von Daphnien im Laboratorium. *Experientia* **5,** *486.*

FRITSCH, R. H., 1953. Die Lebensdauer von *Daphnia* spec. bei verschiedener Ernährung, besonders bei Zugabe von Pantothensäure. *Z. wiss. Zool.* **157,** *35—56.*

FRITSCH, R. H., 1956. Drei Orthoklone von *Daphnia magna* Straus ohne Lansing-Effekt. *Pubbl. Staz. Zool.* **XXVIII,** *214—224.*

FRITSCH, R. H., 1958a. Längenwachstum und Häutungen der Männchen von *Daphnia magna. Z. Morph. Ökol. Tiere* **47,** *193—200.*

FRITSCH, R. H., 1958b. Herzfrequenz, Häutungsstadien und Lebensdauer bei Männchen von *Daphnia magna* Straus. *Z. wiss. Zool.* **161,** *266—276.*

FRITSCH, R. H., 1962. Measures and orders of time in the life of *Daphnia magna* Straus. *Proc. Fifth Congr. Int. Assoc. Geront.* **3,** *4—7.* Columbia Univ. Press, New York and London.

FRITSCH, R. H. & MEIJERING, M. P. D., 1958. Die Herzfrequenzkurve von *Daphnia magna* Straus innerhalb einzelner Häutungsstadien. *Naturwiss.* **45,** *346—347.*

FRITSCHE, H., 1917. Studien über Schwankungen des osmotischen Druckes der Körperflüssigkeit bei *Daphnia magna. Int. Rev. Hydrobiol.* **8,** *22—80* and *125—203.*

GIBITZ, A., 1922. Verbreitung und Abstammung mariner Cladoceren. *Verh. zool.-bot. Ges. Wien* 1.

GREEN, J., 1956. Growth, size and reproduction in *Daphnia (Crustacea: Cladocera). Proc. zool. Soc. Lond.* **126,** *173—204.*

GRESENS, J., 1928. Versuche über die Widerstandsfähigkeit einiger Süßwassertiere gegenüber Salzlösungen. *Z. Morph. Ökol. Tiere* **12,** *707—800.*

HABERMANN, G. & HABERMANN, H., 1931. Über den Sand. Messungen und Betrachtungen. (Unpublished, archives of Hermann Lietz-Schule, Spiekeroog).

HAYWARD, H. G. & WADLEIGH, C. H., 1949. Plant growth on saline and alkali soils. *Adv. Agron.* **1,** *1—38.*

HEIMANN, H., 1961. Salt water farming. *New Scientist* **9,** *410—411.*

Holm-Jensen, I., 1944. *Daphnia magna* som Giftindikator. *Lunds Univ. Årsskr. N. F.* Avd. 2, **40,** Nr. 5.

Holm-Jensen, I., 1948. Osmotic regulation in *Daphnia magna* under physiological conditions and in the presence of heavy metals. *Kgl. Danske Vidensk. Selsk. Biol. Medd.* **20**.

Huchzermeyer, E. W., 1963. Herzfrequenz und Lebensablauf von *Macrothrix hirsuticornis* Norman & Brady. *Z. wiss. Zool.* **168,** *119—132.*

Ingle, L., Wood, T. R. & Banta, A. M., 1937. A study of longevity, growth, reproduction and heart rate in *Daphnia longispina* as influenced by limitations in quantity of food. *J. exp. Zool.* **76,** *325—352.*

Keller, B., 1925. Halophyten and Xerophyten-Studien. *J. Ecol.,* **13,** *224—261.*

Kinne, O., 1952. Zur Biologie und Physiologie von *Gammarus duebeni* Lillj., V: Untersuchungen über Blutkonzentration, Herzfrequenz und Atmung. *Kieler Meeresforsch.* **9,** *134—150.*

Kinne, 1954. Experimentelle Untersuchungen über den Einfluß des Salzgehaltes auf die Hitzeresistenz von Brackwassertieren. *Zool. Anz.* **152,** *10—16.*

Kinne, O., 1956. Über Temperatur und Salzgehalt und ihre physiologisch-biologische Bedeutung. *Biol. Zbl.* **75,** *314—327.*

Lagerspetz, K., 1955. Physiological studies on the brackish water tolerance of some species of *Daphnia. Arch. Soc. zool.-bot. fenn.* **9,** *138—143.*

McArthur, J. W. & Baillie, W. H. T., 1929. Metabolic activity and duration of life. *J. exp. Zool.* **53,** *221—268.*

Megyeri, J., 1959. Vergleichende Untersuchungen der Natrongewässer der ungarischen Tiefebene (Alföld). Különlenyomat a Szegedi Pedagógiai Föiskola Evkönyvéböl, Szeged, *91—170.*

Meijering, M. P. D., 1958. Herzfrequenz und Lebensablauf von *Daphnia magna* Straus. *Z. wiss. Zool.* **161,** *239—265.*

Meijering, M. P. D., 1960. Herzfrequenz und Herzschlagzahlen zwischen Häutung und Eiablage bei Cladoceren. *Z. wiss. Zool.* **164,** *127—142.*

Meijering, M. P. D., 1961. Zur Verbreitung von *Macrothrix hirsuticornis* Norman und Brady in Europa. *Zool. Anz.* **167,** *334—341.*

Meijering, M. P. D., 1962a. Häutung und Eiablage bei miktischen Weibchen von *Daphnia magna* Straus. *Z. wiss. Zool.* **167,** *103—113.*

Meijering, M. P. D., 1962b. Längenwachstum und Geschlechtsreife bei Männchen von *Daphnia magna* Straus. *Z. wiss. Zool.* **167,** *114—119.*

Meijering, M. P. D., 1964. Der Strandweizen in seinem ausser-gewöhnlichen Lebensraum. *Natur und Museum* **94,** *319—324.*

Meijering, M. P. D. & v. Reden, H. K. J. (in press). Vergleichende Untersuchungen an Zeitplänen dreier Cladoceren-arten. *Z. wiss. Zool.*

Meijering, M. P. D. & Redfern J. B., 1962. Heart-beats as a measure of biotic time in the life of *Daphniidae. Amer. Nat.* **96,** *61—64.*

Mortimer, Cl. H., 1936. Experimentelle und cytologische Untersuchungen über den Generationswechsel bei Cladoceren. *Zool. Jb.* Abt. allg. Zool. u. Physiol. **56,** *323—388.*

Olofsson, O., 1918. Studien über die Süßwasserfauna Spitzbergens. *Zool. Bidrag Uppsala* **6,** *183—646.*

Passowicz, K., 1935. Studien über das Verhalten des Wasserflohes *Daphnia pulex* de Geer in Zuchtlösungen von verschiedenen Wasserstoffionenkonzentrationen. *Bull. int. Acad. polon. Sci.,* Cl. Sci. math. et natur., S.B. **II,** 3/5 *59—86.*

Precht, H., 1949. Die Temperaturabhängigkeit von Lebensprozessen. *Z. Naturf.* **46,** *26—35.*

Redeke, H. C., 1933. Über den jetzigen Stand unserer Kenntnisse der Flora und Fauna des Brackwassers. *Verh. int. Limnol.* **6,** *46—61* (cit. after Schlieper).

REDEN, K. A. v., 1960. Sterblichkeitsmaxima bei *Daphnia magna* Straus. *Z. wiss. Zool.* **164,** *119—126.*

REINKE, J., 1909. Die ostfriesischen Inseln. *Wiss. Meeresunters. Kiel* NF **10**.

REMANE, A., 1950. Das Vordringen limnischer Tierarten in das Meeresgebiet der Nord- und Ostsee. *Kieler Meeresforsch.* **7,** *5—23.*

REMANE, A., 1958. Ökologie des Brackwassers. (From: Die Biologie des Brackwassers). E. Schweizerbart'sche Verlagsbuchhandlung, Stuttgart *1—216.*

REPP, G., 1958. Die Salztoleranz der Pflanzen I. *Österr. bot. Z.* **104,** *454—490.*

SCHARPER, K. & W. SCHROPP, 1949. Untersuchungen über die Wirkung von Chlor und Brom auf die Keimung und Jugendentwicklung einiger Kulturpflanzen. *Z. Pflanzenern., Düngung u. Bodenk.* **46,** *88* (cit. after REPP).

SCHLIEPER, C., 1958. Physiologie des Brackwassers. (From: Die Biologie des Brackwassers). E. Schweizerbart'sche Verlagsbuchhandlung, Stuttgart *217—348.*

SCHRATZ, E., 1936. Beiträge zur Biologie der Halophyten. *Jb. wiss. Bot.* **83,** *133—189.*

SCHWARTZKOPFF, J., 1955. Vergleichende Untersuchungen der Herzfrequenz bei Krebsen. *Biol. Zbl.* **74,** *480—497.*

SCOURFIELD, D. J. & HARDING, J. P., 1958. A Key to the British Species of Freshwater *Cladocera.* Freshwater Biol. Ass. **5,** 2nd Ed.

SIMONET, M. & GUINOCHET, M., 1938. Observations sur quelques espèces et hybrides d'*Agropyrum.* II. Sur la répartition géographique des races caryologique de l'*Agropyrum junceum* (L.) P.B. *Soc. Bot. France* **85,** *175—179.*

STEINER, G., 1935. Der Einfluß der Salzkonzentration auf die Temperaturabhängigkeit verschiedener Lebensvorgänge. *Z. vergl. Physiol.* **21,** *666—679.*

STOCKER, O., 1924. Beiträge zum Halophytenproblem. *Z. Bot.* **16,** *289—330.*

SUOMALAINEN, P., 1956. Sauerstoffverbrauch finnischer *Gammarus*-Arten. *Verh. int. Ver. Limnol.*, (cit. after SCHLIEPER).

"VENICE SYSTEM", 1958. The Venice System for the Classification of Marine Waters according to Salinity. Symposium on the Classification of Brackish Waters. Venice.

WIEMANN, P. & DOMKE, W., 1959. Vegetationsübersicht der ostfriesischen Insel Spiekeroog. Vermessungsamt d. Baubehörde Hamburg.

ZIJLSTRA, K., 1946. Over de gevoeligheid van enige landbouwgewassen voor zeewater. Rijksuitgeverij, Den Haag, No. 52 (2) B, 27 pp.

# SALINITY PROBLEMS

## SUMMARY OF THE UNESCO — WAAS — ITALY SYMPOSIUM

BY

HUGO OSVALD, Sweden

During the International Symposium on *Irrigation with highly Saline or Sea-water with or without Desalination* held in Rome September 5th—9th, 1965, a great number of lectures were given, dealing with different aspects on the salinity in soils and its effects upon plants as well as with the effect of irrigation with saline water. The desalination techniques were also considered.

In the introductory lecture The President of the National Academy of Agriculture, Senator GIUSEPPE MEDICI, gave a general survey of the water problem and of the use of water for different purposes: household, industry and agriculture. Water for household and industry must of course be fresh water, containing less salt than 0.3 g/l, and with the rapidly growing population of the world there is, therefore, a steadily increasing demand for fresh water. The water for irrigation represents about 3/4 of the water used by mankind nowadays.

The salinity problems are mainly confined to plant growing, in the first place of course to agricultural production.

The salt in the soils is of varying origin. In arid regions, where soils are irrigated with fresh water, the salt originates from the weathering of minerals — in the first place sodium bearing minerals — in the soil profile and the rise of the water-table within the capillary fringe. Along the sea coasts salt may also come directly from the sea, tidal water and so on, or it is carried to the fields with saline water used for irrigation. Finally, salt is also carried over rather long distances by rain. That is what is referred to by Dr. BOYKO as the "Global Salt Circulation".

Several salts are necessary for the plants, for instance potassium salts, but others are detrimental not to say toxic, for instance sodium chloride, NaCl. Most of the salinity problems are closely related to the NaCl content.

The salt content in the soils varies within very wide limits. It may even be high enough to prevent plant growth.

In his lecture on "Vegetation and Salinity" Professor V. J. CHAPMAN, New Zealand, emphasized that in relation to salt concentration plants can be classified on various bases, and in the future progress must be directed towards this end. The characteristics of plants should be related to the relative proportions of Cl and $SO_4$ ions. The tolerance of vegetation towards salinity depends not only on the saline ions (Cl, $SO_4$, $CO_3$) but also upon the plant species or variety and its stage of development. — Seed germination of all plants takes place most readily under non-saline conditions.

The effect of salts on plants can be due either to toxic influence or to the reduction of the water uptake by plants due to increased osmotic pressure of the soil solution. More attention should be paid to the toxic effect as well as to the mechanism for the absorption of different ions. The toxic effect of excess chloride on some plants seems to be associated with a disturbance of the normal nitrogen metabolism. Finally, if we are to make effective use of saline lands, much more work has to be carried out concerning the effect of salinity upon photosynthesis and respiration.

Prof. H. HEIMANN, Israel, talked about the physiological balance between the ions in the soil solutions also. In spite of high water salinity good crops can be obtained, if potassium is present in appreciable concentrations in relation to sodium. The antagonism between these two ions has been overlooked in plant physiology. If potassium is not present in a proper ratio to sodium, it must be added in order to readjust the balance. A balanced ionic environment is much more important than the osmotic pressure in the soil water system. All factors stimulating root growth, like an efficient organic manure, assist the plant in resisting salinity. A good soil aeration and the presence of all essential trace elements are also important.

The detrimental effect of for instance NaCl can be counteracted also by lime and salts containing calcium, for instance gypsum.

In Italy a lot of experimental work has been carried out in order to study the biological effects on plants on saline soils. Dr. G. LOPEZ for instance has found that the critical period of the plants is the germination and the early growth period. These experiments were carried out on "red earth", irrigated with water containing 32 g/l. A great number of cultivated plants and several salt concentrations were studied. *Hordeum sativum* turned out to be the most resistant species, its capacity of germination dropped only to 93% compared with the control. Both *Triticum durum* and *T. vulgare* lost about half of their capacities of germination. *Vicia sativa* proved to be the most resistant of the leguminous plants studied. *Medicago sativa* and *Trifolium alexandrinum* had only a scarce resistance.

*Solanum lycopersicum*, although rather resistant to irrigation with brackish water, was deeply affected during the germination and outgrowth periods. Also several vegetables were deeply influenced.

In many regions there is no fresh water for irrigation and, therefore, it has since long been studied, whether it might be possible to use saline water, brackish water or sea-water for irrigation. According to eng. I. ESTEBAN-GOMEZ, Spain, who has made experiments in plant growing with ocean water, man has ever since

Aristotle pondered over the possibilities to use sea-water for irrigation, but it is not until recent time that positive results have been obtained. ESTEBAN-GOMEZ gave an interesting description of the positive results obtained in the last six years from the experiments on the Orinon beach.

Dr. Manuel MENDIZABAL, Spain, also gave an interesting account of successful experiments carried out in the south of Spain along the coast. A great number of plants had been grown.

Several experiments with saline water have also been carried out in Italy. In an interesting lecture Prof. L. CAVAZZA described the use of saline water from different sources, mainly aquifers. He emphasized that in the last 15 years good results have been obtained in certain Italian regions to reduce the salinization of irrigation water on underground hydrology.

Irrigation trials with saline water gave discouraging results on clay soils in Sicily. In Apulia on the other hand, where irrigation with brackish water has a very ancient tradition, the results were good because of very favourable soil conditions (high calcium content, good permeability, etc.). Small amounts of water, 250 $m^3$/ha, were applied with an interval of 6—7 days, so that leaching of a large proportion of the salts was permitted during the irrigation period.

The saline water/plant relationship was also studied, and it was observed that when moderately salinized water, 3 g/l, was used, an increase in fruit number but decrease in fruit size on pepper took place. With the same water both yield and quality were raised, but with increasing water salinity both yield and quality decreased. Much more research on the agronomic aspects on the use of saline water is strongly needed.

Another group of Italian experiments were carried out by Dr. FICCO. His intention was to test the resistance of the cultivated plants, when they were irrigated with saline water containing 4—9 g/l, and to find out which method was the most suitable. In the experiments the following methods were tried:

a) irrigation with large quantities and long intervals.
b) irrigation with small quantities and short intervals.

In both cases the plots were fertilized with a) only mineral fertilizers, b) only manure, and c) a combination of both.

A great number of cultivated plants were tested and the best result was obtained by small quantities of water and short intervals and a combined fertilization.

In another group of experiments FICCO studied the result of irrigation and mineral fertilization of salty soils. He found that the irrigation had a positive influence in leaching the soil and that the fertilization did not have any influence on the salinity of the soil.

According to Dr. J. W. VAN HOORN, Tunis, irrigation with brackish water has been studied in Tunis since 1935. It has been found that in order to counteract salinization it is necessary to drain the soils, so that the accumulated salts can be washed out.

India, where there are about 8 million hectares of saline soils (mainly as coastal sandy soils along a 3.500 miles long coastline), affords good possibilities for utilizing sea water for irrigation.

Prof. P. C. RAHEJA said in his lecture that the salinity has mainly developed from canals and on inundated land along rivers, by tidal encroachment and by irrigation with highly saline water. Tidal saline lands on the west coast of India develop salt encrustations after the monsoon season.

The provision of drainage followed by the growing of crops requiring much water, such as rice and sugar-cane, ameliorates the salinity. Green manuring is also beneficial. Gypsum application is needed on sodic soils with high clay content.

Saline water irrigation is feasible, when it is practized in order to maintain an optimum salt balance in the root zone, with due regard to the properties of the soil, cropping technique and aridity factors. Salt tolerant varieties of millet, sorghum and wheat have been selected for such areas.

Drs. E. IYENGAR and T. KURIAN reported, in a paper delivered and distributed but not read, on the utilization of sea-water and coastal sandy belts for crop growing. For this purpose plants which can withstand the fluctuating salinity were selected. Several varieties of cereals, pulses, millet, oil seeds and fodder plants were examined. It was found that many species developed quite well in water containing 10—20 g/l salt. The growth was remarkably improved, if the soils were treated with potassium and calcium salts.

A most important contribution to our knowledge concerning salinity problems was delivered by Drs. H. and E. BOYKO. It was a good survey of the growth regulating factors characteristic of the ecosystem in saline soils.

Induced by ecological studies and observations of desert-, steppe-, and saline habitats in various countries Dr. H. BOYKO started extensive series of experiments with various plant species grown on dune sand and irrigated with natural sea-water in concentrations from 10 to 50 g/l. These experiments have, indeed, been very successful. On sand the percolation is quick and the aeration very good and the most detrimental salts, NaCl and $MgCl_2$, are easily soluble.

The studies were based on a number of new principles found by him and explained by slides: the "Partial root contact", the "Viscosity", "Subterranean Dew", "Raised Vitality", "Biological Desalination" and "Global Salt Circulation". "Biological Desalination" means that many plants accumulate salts in their tissues, and when the crop is harvested a lot of salt is removed.

The results of the experiments have led to the following general rules of application:

1. If salt concentration is constant then a) the higher and/or more frequent the rainfalls, the higher can be the clay content; b) the higher the temperature, the lower must be the clay content.

2. If clay content is constant or lacking at a given concentration from oceanic concentrations downwards, then a) the more frequent or evenly distributed are effective rainfalls the higher is the number of plant species that can be grown;

b) the higher the saturation deficit, the smaller is the number of plant species that can be grown.

3. In semi-arid regions annual crops with a growing period during the rainfall season will give better economic results than others and need in many cases only additional saline or sea-water irrigation, the latter having at the same time a potential fertilizing effect.

4. Seeds from plants grown with saline water in general grow better than those from plants grown with fresh water or water with only a low salt content.

5. Accumulation of NaCl or $MgCl_2$ is not to be feared in sand or gravel in the layer, where the root system is developed, if at least once a year an effective rainfall occurs.

6. Agrotechnical details must of course be worked out for each species separately.

These principles and results can be applied to desert areas of the total size seven times as large as the agricultural area of the United States, areas which are now completely uninhabitable and which may be converted into highly productive land without great technical difficulties. Many food and fodder crops as well as industrial raw materials can be grown on the basis of these results.

In another lecture Dr. H. and Dr. E. Boyko reported about the results of experiments carried out in the period 1957—1963 on dune sand using natural sea-water of four types:

1. East Mediterranean water,
2. Oceanic concentration,
3. North Sea type,
4. Caspian Sea type.

Fresh water was used for control.

Ten species of economic value were used for the experiments, among them barley and sugar-beet.

All the plants showed a much greater salt tolerance than plants grown on normal agricultural soil.

In all species the heliotropical growth was more or less retarded by sea-water with a salt concentration higher than 10 g/l, but horizontal and geotropical growth was not significantly influenced.

Vitality as expressed by drought resistance increased with the salt concentration; it was lowest with freshwater irrigation.

Analyses showed that no salt accumulation is to be feared if the sand is deep enough to allow an adequate continuous drainage.

If the results of the experiments and the experiences now briefly summarized be applied on a large scale, it would be possible to increase the food production quite considerably. Therefore, — as Dr. H. BOYKO has expressed it — in view of its global impact on Human Welfare, the World Academy of Art and Science has established a Working group on this subject in order to organize close international cooperation for this new line of research, and it is to be hoped that the experiments will soon be enlarged on a global scale with the help of the Agencies of U.N.

Irrigation with saline water can be employed not only for agricultural production but also for creating a not only inhabitable but also an attractive environment in desolate desert-places as has been shown by Dr. E. BOYKO. At the harbour-town of Eilat a desert garden was planted hy her 15 years ago. A large variety of ornamental plants, mainly trees, altogether 180 heat-, drought-, wind- and saltresistant species, producing a maximum of shadow and of ornamental value, were planted on completely vegetationless gravel hills and irrigated with saline water with a salt content of 2—6 g/l and more.

Now the place looks green and adorned. Gradually, this now makes Eilat to an outstanding recreation place all the year round and particularly during the winter months.

In a contribution, which was not read at the Symposium, Prof. IGNATYUK, U.S.S.R., gave an account of theoretical principles of designing, construction and operation of drainage on saline lands under irrigation in the Soviet Union.

The last day of the Symposium was mainly devoted to Desalination problems in lectures given by prof. R. DI MENZA, Italy, on The Use of Nuclear Energy for Sea-water Desalination, and Prof. G. NEBBIA, Italy, on Economics of the Conversion of Saline Waters into Fresh water for Irrigation.

The lectures dealt with the technical and economic problems of desalination with different methods. Although these problems are not salinity problems in its proper sense, they are nevertheless closely related with these. The cost of desalination is still too high for the use of desalinated water for agricultural purposes. And even if the cost of desalination will be reduced in the near future, the use of salt water will be much cheaper. Nevertheless, it seems probable that desalinated water will be used for irrigation to some extent. In my opinion the situation is this:

Sea-water and highly saline water can only be used for a restricted number of cultivated plants — on the other hand desalted water can only be used for a small number of very valuable crop plants. In many cases the solution to the problem might, therefore, be to mix highly saline water with desalinated water. By such a method it might be possible to increase the number of plants which can be irrigated in an area, where the available saline water has a salt concentration too high for most of the plants.

In addition to these lectures Prof. J. STERNBERG, Canada, gave a lecture on the use of Radio-Isotopes and nuclear Energy in Saline Irrigation research.

Finally, The Resolution committee presented three resolutions concerning the support of research on saline and sea-water irrigation and salinity problems in general. They were unanimously accepted.

---

Editor's remark: Most of the lectures mentioned in this summary are being published in full in Vol. IV of the WAAS-Series.

# PART III:

# STUDIES ON PLANT AND ANIMAL LIFE IN A BRINE

# THE FLORA AND FAUNA OF THE GREAT SALT LAKE REGION, UTAH

BY

SEVILLE FLOWERS* and FREDERICK R. EVANS**

(with 5 figs.)

## Origin of Great Salt Lake

Great Salt Lake is the remnant of Lake Bonneville, a much larger Pleistocene lake which formerly occupied the eastern part of the Great Basin Province of western North America. Evidence of this ancient lake is clearly shown by a series of wave-wrought terraces and faceted headlands on the lower slopes of the surrounding mountains where clean sands and gravels are capped with loam of later development. The lake rose to a level of about 1,000 feet (310 metres) above the present level of Great Salt Lake. At that time it had a long period of constant level, as shown by the highest terrace, and was extremely irregular in outline. The maximum dimensions were 346 miles (586 kilometres) long and 145 miles (233 kilometres) wide, with a maximum depth of 1,050 feet (320 metres) and an area of 19,750 square miles (31,786 square kilometres). It occupied a large part of western Utah and extended slightly into eastern Nevada and southern Idaho. Glacial conditions existed at that time, as shown by the topography in the mountains and by fossil remains of trees embedded in the lake sediments that are found today only high in the mountains.

An increase in precipitation caused the lake level to rise, and it finally overflowed at a low point at the northern end of the lake, since called Red Rock Pass, Idaho. The bed of this outlet was composed of soil, and a large river soon formed. Evidence shows that the bed was eroded at a terrific rate, and it is estimated that it cut through 375 feet (114 metres) of unconsolidated soil and rocks in about 25 years when a bed of hard limestone was encountered. The area of the lake was reduced by one-third and a more-or-less steady rate of drainage established. The lake remained at a relatively constant level for the longest period of its existence. Eventually the climate changed, precipitation declined and the rate of evaporation increased. The lake level decreased slowly with relatively long periods of constant levels, as shown by numerous small terraces. When the level reached about 200 feet (61 metres) above the present level of Great Salt Lake, salt had concentrated to a notable degree and began to be deposited in the sediments. The time interval of

* Professor of Botany. ** Professor of Zoology and Entomology, University of Utah.

the decline of the lake has been variously estimated from 30,000 to 12,000 years, during which it became divided into several smaller ones, all of which dried up except Great Salt Lake.

## Geography and Geology of the Region

The general configuration of the land presents a series of parallel mountain ranges and broad valleys extending in a north-south direction and connected by diagonal passes. To the east of the lake the high Wasatch mountains rise to nearly 8,000 feet (2,438 metres) above the valley floor and extend for 200 miles. To the west the lake is bordered by several short, low basin ranges. The Promontory Mountains extend into the lake from the north and form a peninsula while the islands in the lake are isolated salients of the Oquirrh and Stansbury Mountains to the south. The greatest extent of the lake shore is bordered by lowlands composed of lacustrine sediments dominated by clays and loams with some local sandy areas. The general topography of the valley floor shows a flat or slightly undulating plain interrupted here and there by playas. The Salt Lake Desert lies west of the lake and is a vast playa of sinuous outline, 100 miles (160.9 kilometres) long and 50 miles (80.46 kilometres) wide. Most of the surface is encrusted with white salt and for the most part barren. Isolated mountains rise in its midst like islands in a sea of lake sediments.

In many places long alluvial slopes gradually rise from the plains to the foothills, the soils changing upwards from clays to loam, while gravel constitutes the ancient lake terraces on the lower slopes of the mountains. In some places the mountains rise abruptly from the valley floor, and in a few places steep headlands extend to the lake shore.

Most of the water draining into Great Salt Lake comes from the mountains to the east. The Bear, Weber and Provo rivers arise in the Uintah mountains far to the eastward and pass through deep canyons in the Wasatch range. The latter river empties into Utah Lake, from which the Jordan River flows northward into Great Salt Lake. The smaller basin ranges to the west have only a few permanent streams, none of which reaches the lake. Several large saline springs at the bases of the mountains formerly drained into the lake.

The valleys of this region are structural in origin, having been formed by normal faults. The mountains are composed of a wide variety of igneous, sedimentary and metamorphic rocks ranging from Pre-Cambrian to Cretaceous, with some more recent conglomerates.

## Climate

Climatic data from several stations surrounding Great Salt Lake show that the average annual rainfall in the western portion of the region is about 6 inches (16 cm), while that of the eastern portion is about 16—17 inches (42 cm), a difference of about 10 inches (26 cm). The high Wasatch mountains to the east of the lake account for this difference. The average temperature from several stations is about 50° F, with slight variation. The relative humidity measured at Salt Lake City, 14 miles east of the lake, shows wide daily variation during the summer months, and on the average shows about 46% at 6:00 A.M. and 26% at 6:00 P.M. In January the amounts are 75% and 70%, respectively. Occasional measurements made on the plains near the lake in mid-afternoon during August show the humidity as low as 10%, while amounts of 15—22% commonly occur from late June to September.

The average annual evaporation (open pans of pure water) at Salt Lake City is 63.4 inches (164.04 cm), and the average velocity of the wind is 4.3 miles (7 km) per hour. At Midlake, a station on a railroad trestle crossing the lake, the evaporation is 71.5 inches (181.61 cm) of water, and the average wind movement is 10 miles (16 km) per hour.

## Great Salt Lake

Great Salt Lake is roughly mitten-shaped but the exact outline and dimensions vary with fluctuations of the level. The long axis extends in a slightly northwest-southeast direction. In 1906, when the lake level was rather high, a survey showed it to be 75 miles (120 km) long and 35 miles (56 km) wide. This is the shape shown on most older maps and is marked by distinct beach bars and bluffs. Most of the beaches are flat and broad, while abrupt headlands occur on the west sides of the Promontory mountains and Antelope and Stansbury islands. The deepest point in the lake is located in a trough between Antelope and Stansbury islands; at the highest recorded level it was 40 feet (14.9 m) deep, while the estimated average depth was 19 feet (5.19 m). During the great recession of 1903, measurements in similar places showed only 36 feet (10.97 m), a difference of 16 feet (4.75 m), which corresponds to the surface measurements.

Oscillations in the lake level have been measured over a period of 111 years. A gauge was established with the zero point at 4,194.8 feet above sea level. Seasonal fluctuations show an average difference of about 2 feet (0.6 m) between the high point about June 1st and a low point about November 1st. Chronologically a maximum difference of 17 feet (5.18 m) has been recorded. In 1865 the level rose to plus 14 feet (4.26 m) on the gauge, and at that time the bars connecting Stansbury and Antelope islands with the mainland at their southern ends were submerged under three and 10 feet (3 m) of water, respectively.

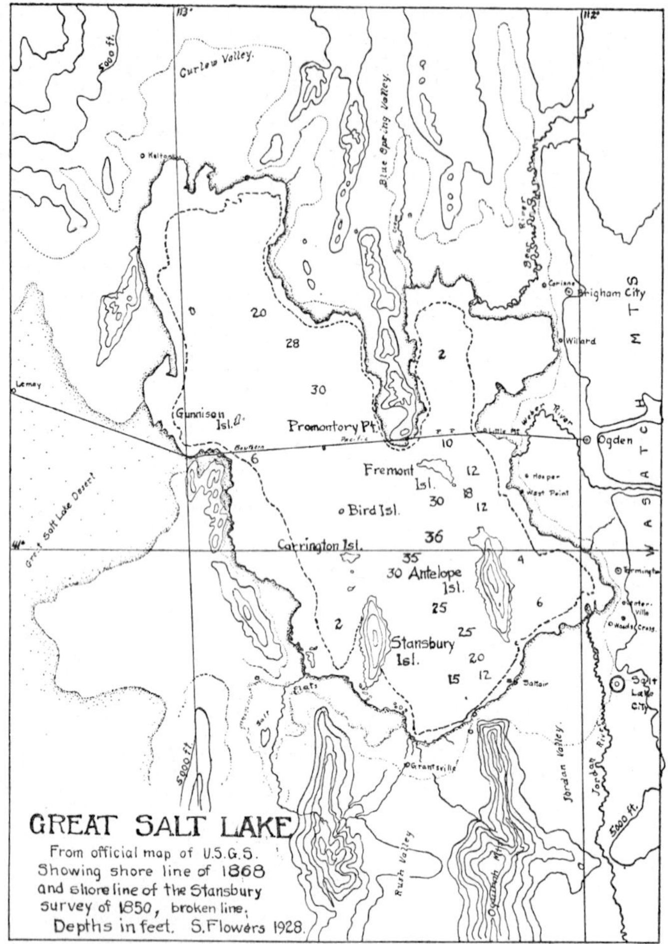


GREAT SALT LAKE
From official map of U.S.G.S.
Showing shore line of 1868
and shore line of the Stansbury
survey of 1850, broken line.
Depths in feet. S. Flowers 1928.

The fluctuations roughly correspond to the average annual rainfall with about a one-year lag following wet cycles. Since 1900 progressively larger amounts of water have been diverted from the main rivers for agricultural and culinary uses, and as a result there is less correlation between rainfall and the lake level at present.

The salinity of the lake water varies inversely with the rise and fall of the level, there being about 1% increase in salt concentration for each foot decrease in the surface until saturation is reached at zero on the gauge. Since records have been kept, the salt content has varied from 13.7% when the level stood at plus 14 feet (4.26 m) on the gauge and 27.7% (saturation) when the level stood at zero or less on the gauge. Analyses of the brine are shown in Table I.

**Table I.**

Theoretical percentages of salts in the water of Great Salt Lake, dry weight.

| Observer | Allen A | Bassett B | Waller C | Talmage D | Talmage E | Morton Salt Co. F |
|---|---|---|---|---|---|---|
| Date | 1869 | 1873 | 1892 | 1885 | 1889 | 1940 |
| Lake level on gauge | +13 | +14 | +6 | +10 | +7 | —2 |
| Salinity of lake | 14.99% | 13.42% | 23.80% | 16.71% | 19.55% | 26.40% |
| NaCl | 79.12 | 65.95 | 81.02 | 81.27 | 80.51 | 87.908 |
| $Na_2SO_4$ | 6.204 | 8.122 | none | 8.504 | 5.249 | none |
| $MgCl_2$ | 9.940 | 8.867 | 6.317 | 6.756 | 10.520 | 3.524 |
| $MgSO_4$ | | | 2.191 | | | 7.942 |
| $CaSO_4$ | .600 | 1.49 | 3.461 | .859 | 1.427 | .095 |
| $CaCO_3$ | | | | | | .530 |
| $K_2SO_4$ | 4.451 | | 3.940 | 2.597 | 2.424 | |
| KCl | | 1.408 | | | | |
| $Li_2SO_4$ | trace | | .0697 | | | |
| $SiO_2$ | | | .006 | | | |
| $Fe_2O_3Al_2O_3$ | | | .00168 | | | trace |
| Excess Cl | | 1.49 | | | | |
| Excess $SO_3$ | | | .0214 | | | |
| Totals | 100.3 | 87.327 | 97.01 | 99.98 | 100.11 | 100.00 |

When the lake level is low and during the winter months when the temperature of the water falls to 35° F or lower, sodium sulphate is precipitated as Glauber salt ($Na_2SO_4 \cdot 10H_2O$), mirabilite of the mineralogists, and is often deposited on some beaches in parallel ridges by wave action. A solid layer of mirabilite ranging upwards to eight feet (2.43 m) thick occurs offshore along the southeastern part of the lake. Sodium chloride is also deposited on the lake bed in coarse square or rectangular crystals, often becoming interlocked in masses which may become as much as a foot (20.5 cm) thick. These salts dissolve again as the temperature of the water rises during the following spring and summer.

Carbonates occur in the water but these usually do not appear in ordinary analyses using undiluted brine because it is saturated with sodium ions which prevent the ionization of the weak carbonate radical. If the brine is diluted with 6—8 times the volume of distilled water, the carbonate and bicarbonate radicals then become dissociated and can be detected. Likewise, the pH of the brine is normally 7.4 when the concentration is high, but when progressively diluted with 6—8 times the volume with distilled water it ranges to 8.4.

## Algae of the Lake

The only plants growing and reproducing in the main body of Great Salt Lake include two species of Cyanophyta, *Coccochloris elabens* Drouet & Daily* and *Entophysalis rivularis* (Kütz.) Drouet** (?) and two undescribed species of *Chlamydomonas* of the Chlorophyta. *Coccochloris elabens* is by far the most abundant alga and consists of single small oval or shortly oblong cells, great numbers of which are rather closely disposed in gelatinous colonies. Young colonies are minute bluish-green granules of firm texture, and as they increase in size the color changes to brown and finally pink or yellowish, while the gelatinous matrix becomes soft and flaccid. When the lake level is high and the salt concentration less than 20%, the spherical colonies enlarge to about 1 cm in diameter, become hollow and eventually break open, forming more-or-less expanded undulate gelatinous laminae reaching 6—12 cm long. Great masses of the alga frequently become locally aggregated and on bright days appear at a distance as brown areas in the surrounding blue water. Storm waves frequently deposit large masses of the alga on some of the eastern beaches, sometimes 10—25 cm deep.

*Entophysalis rivularis* consists of aggregates of single subspherical cells, each with a rather thick, firm gelatinous sheath and disposed in short irregular filaments which sometimes become compacted side by side in mats. The mass as a whole forms bluish-green, yellowish or dark brown mats less than 1 mm thick on rocks and wood in shallow water.

*Chlamydomonas* are free-swimming biflagellate green algae, and the two forms, one ovoid and the other oblong, occur widely distributed in all parts of the lake. They seem to adjust readily to gradual changes in salt concentrations, although they reach their maximum development when the water contains 13—15% salt.

In Bear River Bay and East Bay of Great Salt Lake the water is very shallow and the salt content is greatly reduced by the inflow of fresh water from Bear River and Jordan River, respectively. In these places *Cladophora fracta* (Dill.) Kütz. often becomes so abundant as to appear like a green meadow at a distance. It becomes adapted to a gradual increase in salt content up to as much as 3 to 4%, whence it ceases to grow but apparently is not all killed. *Rhizoclonium hieroglyphicum* (Ag.) Kütz. and *R. crispum* Kütz. are less tolerant of salt but will persist where the water contains as much as 0.8% salt. Strong winds from the west often blow the brine

* Formerly known variously as *Polycystis packardii* Farlow, *Microcystis packardii* (Farl.) Setchell and *Aphanothece utahensis* Tilden.

** Formerly referred tentatively to *Pleurocapsa entophysaloides* Setch. & Gard. by Dr. Francis Drouet. This name has since gone into synonomy. See comments in Eardley, 1938, p. 1333.

from the main body of the lake into these bays, killing great quantities of algae.

## Bacteria

FREDERICK (1924) reported eleven species of bacteria inhabiting the waters of the lake. These organisms cause decay in dead algae, animals and organic wastes entering the lake by way of rivers and wind, and at times during the summer months foul odors pollute the air in the vicinity of the lake.

The bacteria are as follows: *Micrococcus subflavus* BUMM.; non-motile spheres. *Bacillus cohaerens* GOTTHEIL; motile rods, single or in pairs. *Bacillus freudenreichii (Urobacillus freudenreichii* MIGUEL); motile rods, single or in chains. *Bacillus mycoides* FLUGGE *(B. ramosus* EISENB., *B. radicosus* ZIMM.); rods in chains with spores. *Achromobacter solitarium (Bacillus solitarius);* slender motile rods, non-motile in cultures. *Achromobacter album (Bacillus albus);* non-motile rods. *Achromobacter hartlebii (Bacillus hartlebii);* motile single rods. *Flavobacterium arborescens (Bacillus arborescens);* non-motile rods in pairs or chains. *Bacterioides rigidus (Bacterium rigidum* DESTOSO); motile slender rods, single or in pairs. *Serratia salinaria (Pseudomonas salinaria);* single motile rods, non-motile in Great Salt Lake. *Cellulomonas subcreta (Pseudomonas subcreta);* single motile rods, non-motile in Great Salt Lake.

## Fauna of the Lake

According to published reports, animals belonging to only two phyla, Arthropoda and Protozoa, have been found in the waters of Great Salt Lake. The arthropods may be found there in vast numbers in spring, summer and fall, but only rarely during the winter months. Most conspicuous is the brine shrimp, *Artemia salina.* Eggs of two types are formed: thin-walled summer eggs which hatch internally (viviparous) and thick-walled winter eggs which seem to require dessication prior to hatching. It has been observed that males are scarce and that parthenogenetic development is common. Winter eggs numbered in the billions collect in windrows along the leeward beaches and hatch in the spring into nauplii. About three weeks later, having passed through twelve or more instars, adults appear. It is said that hatching and development is affected by concentration of salts, by specific ions, by temperature, and by light. Winter eggs more than 25 years old have maintained their viability. Development is better in diluted lake water than in concentrated brine, indicating that the brine shrimp is not fully adapted.

Observers have noted a tremendous decline in brine shrimp during 1961; this may be the result of low level of the lake (new all-time record) and accumulation of industrial and organic wastes.

Larvae and pupae of the brine flies *(Ephydra cinerea (gracilis))* are abundant in summer. They are found in open waters; *E. hians* are found near shore. The female adults lay eggs on the surface of the water, the eggs hatching into long cylindrical larvae which respire through "tracheal gills" found in long, forked anal tubes. The pupal coat consists of the last larval skin. Pupation occurs in the water, the larvae shrinking, leaving a gas-filled space. Coming to the surface, the adult can fly away.

Several species of protozoa have been found in Great Salt Lake, some of them not identified. Thus far, none has been thoroughly investigated and much work is required to work out details of morphology, taxonomy and physiology. The ciliates comprise *Uroleptus packii, Chilophrya utahensis* and species of *Podophrya* (a Suctorian), *Euplotes, Cyclidium, Pseudocohnilembus* and *Cothurnia.* Other undetermined species have been seen. With the exception of *Podophrya,* which feeds upon *Euplotes* and *Pseudocohnilembus,* all are bacterial feeders. It has been shown that all will grow at optimum rates in salt concentrations of 2% or 3% and up to 18% or 20%. Little if any growth occurs at saturation (27.7%). Cysts of *Pseudocohnilembus* will survive complete dessication in the brine, however. Two amoebae have been seen in large numbers and several species of Flagellates, including *Tetramitus, Oikomonas,* and at least two others have been seen in large numbers in cultures taken from the lake and the briny and brackish water ponds nearby. Additional studies should result in finding many more forms of protozoa, and if reports from casual observers are correct, it is likely that larvae of several insects may be found and identified.

## Affects of Organisms on Sedimentation

The formation of calcareous deposits in the lake, either with or without the influence of living organisms, has been studied by several investigators (ROTHPLETZ, 1892; MATHEWS, 1930; FLOWERS, 1934; EARDLEY, 1938). The studies of sedimentation of Great Salt Lake by EARDLEY (1938) summarize the previous work and add the most significant conclusions to date.

Oolites. Oolitic sand is formed in two ways. The commonest and most abundant type is formed by the deposition of calcium and magnesium carbonates around very small mineral particles of various kinds, but principally fine siliceous clay particles. The grains of oolite formed in this manner are subspherical to irregular in shape and range from 0.1 to 1.5 mm across. They follow the general contour or angularity of the nuclei but as the minerals accumulate on the outside the angles become rounded, while wave and wind action further rounds them off.

The second type of oolite is formed by the deposition of carbonates in and around the faecal pellets of *Artemia*. The action begins while the faeces are in the intestine of the animal and then further accumulations continue after the pellets have been discharged. The grains thus formed are short to rather long cylindrical rods about 0.1 mm in diameter and range from 0.05 to 1 mm long. Cleavage of the faeces into various lengths by the muscular action of the intestine of *Artemia* and subsequent breakage of long sections after they have been discharged accounts for the wide range of lengths.

The grains of both types of oolite may be smooth or roughened, mostly pearly white, but often with a faint brownish tint, sometimes finely dotted with darker spots. EARDLEY (1938) gives an extended description of the structure and mineral composition of the oolites together with their genesis and distribution. As to composition, he gives the following summary:

"The concentric layers are white and quite opaque, except in exceptionally thin sections. Many, but by no means all, of the white opaque layers are intersected and traversed normally by radiating rays of colorless, transparent mineral, ordinarily calcite but probably in part aragonite in some oolites. The rays may be confined to certain layers, commonly the inside ones, leaving others, ordinarily the outer, entirely dense."

"The concentric layers consist of about 84% of $CaCO_3$, 5.5% of $2MgCO_3$. $CaCO_3$ and 5.6% of very fine clay. The dense carbonate mineral is cryptocrystalline aragonite. The small amount of $MgCO_3$ probably exists in the state of dolomite in mechanical admixture with the aragonite because the magnesium atom is too large to fit into the space lattice of either calcite or aragonite. The composition of the dolomite in oolites is not known but is assumed to be the same as in clay, viz., $2MgCO_3 \cdot CaCO_3$."

The oolites predominate at the southeastern end of the lake, on the western shores of the islands and the promontory and along the entire western side of the lake. Wind has piled up dunes of oolitic sand in some places as much as twenty feet (6 m) high.

Sediments of algal origin. Accumulations of *Coccochloris* on firm hard substrata beneath relatively quiet shallow water become attached by the gelatinous sheaths and gently wave back and forth in the slightly agitated water. The solubility of calcium and magnesium carbonates in the brine depends upon the amount of dissolved carbon dioxide which is about one-half the amount that dissolves in pure water. Photosynthesis by the algal cells reduces the amount of dissolved $CO_2$ and thereby reduces the solubility of the carbonates. In this instance great numbers of separate cells dispersed in the gelatinous matrix assimilate the $CO_2$, and it would seem logical that the dissolved calcium carbonate would be precipitated within the matrix. However, there is no evidence that this takes place throughout the algal mass. Instead, the calcium and magnesium carbonates are deposited on the firm substratum under the algal mass, practically cementing it down. The exact mechanism by which this is brought about is still not fully explained. Localized masses of algae resting on rocks, hard clay or any other firm substratum build up mounds of calcareous deposit of various sizes ranging from 6—36 inches (15—90 cm) across. By coalescence these mounds may become upwards to 30 feet (9 m) across with irregular V-shaped troughs between them.

The crusts formed by *Coccochloris* are nodular and irregular on the surface; sections show no definite stratification.

*Pleurocapsa* is much less frequent and is more or less scattered here and there in local masses among the more prevalent *Coccochloris*. By nature it is always attached in the form of thin gelatinous mats of somewhat firmer texture on hard substrata. By a similar process it gives rise to a smooth, lamellar calcareous deposit of tabular form. The mounds are generally smaller than those formed by *Coccochloris* and occur isolated or in groups 4 to 6 inches (10—15 cm) across.

Fig. 1. Biscuit-like masses of *Coccochloris* incorporated with sand drying on the eastern beach of Great Salt Lake.

Incorporated in both kinds of deposits are oolites, sand, clay, remains of *Artemia*, and larva cases of *Ephydra*, Figs. 1 and 2. The crusts are usually gray to light yellowish-white and when dried much paler, rather porous and friable; some may be broken with the fingers, while others are rather hard. EARDLEY (1938, p. 1398) gives the following chemical analyses: the nodular crusts formed by *Aphanothece* consists of 77% calcium and magnesium carbonates in a ratio of about 11:1, 20% insoluble material consisting of clay, angular silicate minerals, as silt and sand, and faecal pellets of *Artemia*, and 2% organic matter.

The lamellar portion of the deposits laid down under *Entophysalis* is much more compact and harder and shows 93% carbonates and 6.3% clay but no silt or sand particles.

Along the eastern shores from the area around the Jordan river delta northward to Bear River Bay the lake is quite shallow, the deepest points being about 10—12 feet deep, while the bed is composed of clay and silt, oolites being largely lacking. During periods of high lake level *Coccochloris* accumulates on some of the beaches in a similar manner described above and becomes incorporated with the clay and silt together with organic matter from sewage entering the lake. As the lake recedes during the summer the mixture is left exposed, the surface frosted with white salt and apparently solid (but really soft and gelatinous beneath) 6—10 inches (15—25 cm) deep, and jet black. As the hot sun plays upon it, it puffs up and emits a strong sulphurous briny odor. Crusty deposits of silt and clay heavily impregnated with organic matter resemble those described above and give rise to compact clays mottled or laminated with black layers.

### The Strand

The term "strand" as used in this paper includes any ground left exposed between the margin of the water and the highest level attained by the lake since records have been kept. The latter level is well marked by old beach bars and bluffs. The lateral extent may range from less than 440 yards (400 m) near headlands to 3—4 miles (1.5—2.5 km) on some flat beaches. Seasonal variations in lake level alter the extent of the beaches. At low lake level some beaches may be left exposed for as long as thirty years, as indicated to date.

For the most part the strand is barren, Fig. 2. Trees are absent and the pioneer plants invading the beaches are mostly restricted to the extreme margins but occasionally widely scattered or in isolated groups on the otherwise barren flats, especially after prolonged periods of low lake level. Fleshy chenopodiaceous plants invade the salty soils, their seeds germinating on the surface when ample fresh water from snow and spring rains is present. The roots quickly become adapted to increasing salt concentration of the soil

Fig. 2. A general view of the strand, showing pioneer *Salicornia rubra* and *Suaeda erecta* in the foreground and drying algal masses beyond.

solution as the water soaks into the soil. Seeds of the most salt-tolerant plants of this region have been found to germinate best in fresh water or very mildly saline solutions.

*Salicornia rubra* A. NELS., *S. utahensis* TIEDS. and *Allenrolfea occidentalis* (S. WATS.) KUNTZ. are the foremost species invading the beaches. These plants have strongly jointed, fleshy stems and very small scale-like leaves. *S. rubra* is an annual and becomes established in various ways according to the physical character of the soil and amount of salt and water present. Much depends upon the abundance of the seeds. On soils containing 3—6.5% salt an abundant crop of seeds may produce plants in very dense sodlike colonies 3—10 cm tall. Dense colonies persist better in sandy soil than in clay, where competition for water becomes acute as it decreases in amount during the summer. In some clay soils the plants wilt and die while there is still as much as 10% water present due to the high adsorptive properties of the colloids. In soil containing 0.2—0.5% sodium chloride or in soils containing upward to 3% sodium sulphate, which is much less toxic than chlorides, the plants may become 20—50 cm tall, diffusely branched and robust. In the autumn the plants acquire a reddish color. (Figs. 2—3).

*Salicornia utahensis* is a coarser perennial species of caespitose habit, less widely distributed but locally abundant. It tends to form low hummocks of wind-blown soil.

*Allenrolfea occidentalis* is a larger, more woody perennial with alternate branches and is much more widely distributed, not only on the beaches but far beyond the lake on saline plains (Fig. 4). On hard surfaced soil it reaches only 15—30 cm tall and persists in clay-loams when the water content becomes very low during summer. In some sandy areas the bushes become large and form hummocks of wind-blown soil which may become as much as 90 cm tall and 2 metres across.

At one time it was thought that these three plants had different limits of salt toleration, but as more analyses of the salt content of various soils accumulated, it became evident that they all have about the same degree of toleration, up to about 6.5% where sodium chloride is the dominant salt, and that zonation and distribution are due to local physical and chemical conditions. All of them may invade the barren salt flats singly or in combinations.

*Suaeda erecta* (S. WATS.) A. NELS. is another prominent plant invading the strand. It is a freely branched annual with fleshy linear leaves about 2 cm long. It tolerates salt in the soil up to about 3%, which limits its growth to about 15 cm tall, but in soil

Fig. 3. A view of the strand with bordering bluff showing dense growths of *Salicornia rubra* invading the barren strand and followed by *Suaeda erecta* and *Distichlis stricta*. *Sarcobatus vermiculatus* on the bluff.

containing about 0.3% it may reach 30—35 cm tall. It often grows in a rather distinct zone following *Salicornia* and *Allenrolfea*, and commonly occurs abundantly along the bases of beach bars and on broad beaches where the salt is not too concentrated. *Suaeda torreyana* S. WATS. is a taller perennial with a woody base, darker in color and more bushy in habit. It often follows the annual species in the succession and is frequent in some playas and on saline plains where the soil is drier most of the year.

*Distichlis stricta* (L.) GREENE, salt grass, is abundant throughout the region and is conspicuous in the beach succession. Long rhizomes with aerial shoots in straight rows grow outward on the beaches toward the saltier soils. It frequently forms an irregular zone following *Suaeda erecta*, although the two plants commonly grow together. In many places it becomes densely aggregated and eliminates many associated plants. It depends upon a constant supply of soil moisture and tolerates salt up to about 2.5% and thrives in lower amounts.

*Atriplex hastata* L. is an annual of great plasticity. It commonly grows in a distinct zone following *Suaeda erecta* or *Distichlis* along the outer limits of the beaches where the soil is less salty and drier earlier in the season. In such places it assumes a rather slender habit with thin hastate leaves and a coating of white scurfy scales. It becomes upwards to 60 cm tall and seldom grows in soil containing more than 1% salt. In wetter places where the salt is greatly reduced it shows a progressive change in appearance. It becomes as much as 1 metre tall, the leaves become much larger and thicker and they shed the white scales very early, becoming dark green. The contrast is so striking that the latter form at one time was thought to be a separate species. *Atriplex argentea* NUTT. is a more diffusely branched annual with smaller, thinner, undulate leaves with shiny, silvery

**Table II.**

Analyses of soil samples from the Saltair area; calculated to dry weight of soil in terms of sodium salts. July 12, 1929 .

| Location | Water Content | pH | NaCl | $Na_2CO_3$ | $NaHCO_3$ | $Na_2SO_4$ | Total |
|---|---|---|---|---|---|---|---|
| Invading *Salicornia rubra* | 14.10 | 8.4 | 4.08 | 0.052 | 0.022 | 0.853 | 4.99 |
| *Suaeda erecta* zone, surface | 9.4 | 8.8 | 2.68 | 0.084 | trace | 1.421 | 4.80 |
| Same, 2 feet (60 cm) deep | 18.20 | 8.4 | 1.51 | 0.084 | trace | 0.433 | 2.01 |
| *Distichlis stricta* zone, surface | 3.59 | 8.6 | 0.07 | 0.057 | trace | 0.497 | 0.62 |
| Same, 2 feet (60 cm) deep | 10.81 | 8.6 | 0.06 | 0.10 | 0.012 | 0.384 | 0.54 |

scurfy scales. It often follows *A. hastata* in the beach succession.

The salt content of soils in the root region of plants across several zones of invasion were investigated, and from these studies two transects showing the maximum toleration of salt by the pioneer plants are given in Tables II and III.

In this first area the *Distichlis* zone was along the base of low sand dunes from which precipitation water had leached considerable salt.

**Table III.**

Analyses of soil samples from the Little Mountain area; calculated to dry weight of soil in terms of sodium salts. July 20, 1929.

| Location | Water Content | pH | NaCl | $Na_2CO_3$ | $NaHCO_3$ | $Na_2SO_4$ | Total |
|---|---|---|---|---|---|---|---|
| Barren salt flat 25 feet (8 m) from plants | 20.84 | 8.9 | 17.68 | 0.339 | trace | 4.572 | 22.59 |
| 2 feet (60 cm) from 1st *Salicornia* plants | 15.94 | 8.8 | 5.89 | 0.005 | 0.05 | 0.852 | 6.72 |
| *Salicornia rubra* pioneers | 15.81 | 8.8 | 5.30 | 0.005 | 0.05 | 0.401 | 6.26 |
| *Suaeda erecta* zone | 14.80 | 9.8 | 2.60 | 0.005 | 0.672 | trace | 3.31 |
| *Distichlis stricta* zone | 8.22 | 8.8 | 2.21 | 0.053 | 0.141 | trace | 2.60 |
| *Cressa depressa* zone | 8.20 | 9.0 | 0.651 | 0.055 | 0.605 | trace | 1.30 |

## Playas

Playas are low flat depressions in the valley floor that were formed by bottom currents of water in Lake Bonneville in its last stages of recession. They range in size from about 100 metres to 2 kilometres across, rather regular to sinuous in outline and 1—5 metres below the terracelike surrounding plains. Some of them are continuous with the beaches of the lake while others are closed basins. The Great Salt Lake desert west of the lake is a vast playa embracing numerous local depressions and irregular bars. In the spring some of the playas accumulate shallow water while others are merely wetted on the surface. In either instance the water leaches soluble salts from the underlying soils and deposits a thin crust of white crystals on the surface when it dries out later in the season.

The soils are dominantly clay and the depth of the water-table is 1—2 metres below the surface in most places. Usually the salt content decreases with depth but varies in amount in different playas. The following is a generalized condition:

| Depth of water-table | Percentage salt in soil |
|---|---|
| 1 metre or less | 1.5—3% or more |
| 2 metres | 1—3% |
| 3 metres or more | 0.25—1% or less. |

The pioneer plants invading barren playas from the outer margins are the same species found on the beaches of Great Salt Lake. Frequently they show more distinct zonation due to the increasing gradient of salt concentration toward the center. There are several variations in the order of invasion but the usual sequence is as follows: Zone 1. *Salicornia rubra* or *S. utahensis;* Zone 2. *Suaeda erecta;* Zone 3. *Allenrolfea occidentalis;* Zone 4. *Distichlis stricta;* and Zone 5. *Suaeda torreyana,* sometimes with mixed annual plants.

A typical small playa of clay soil was studied in detail in 1929 by means of a transect 120 m across the short diameter and 2 m wide. This was a natural undisturbed area in which the plant succession had extended for 9 to 10 m toward a broad barren central area thinly coated with white salt. A dense growth of *Salicornia rubra,* about one metre wide, formed the first invasion zone. A second zone of scattered *Salicornia,* about four metres wide, was being invaded by *Suaeda erecta* followed by a few young bushes of *Allenrolfea* in an overlapping zone about two metres wide. From the base of the surrounding terrace a narrow zone of *Distichlis* sent out a few long rhizomes into a sparsely populated zone about 4 m wide.

Across the transect, soil samples were taken from the barren central area and from the root region of the plants in each zone and the total salt content determined. The results are shown in Table IV.

The Great Salt Lake Desert. This vast playa was the last extensive basin vacated by Great Salt Lake in its decline. Most of the surface

**Table IV.**

Soil transect across a pioneer playa succession showing the percentage total salt, dry weight, clay soil, water-table about 1 m. June 25, 1929.

| Zone | Depth in centimetres | | |
|---|---|---|---|
| | 5 | 60 | 90 |
| Barren area, center | 7.06 | 6.55 | 10.9 |
| Barren area 10 m from plants | 3.55 | 6.60 | 5.30 |
| *Salicornia rubra,* east side | 2.42 | — | — |
| *Salicornia rubra,* west side | 2.00 | 4.15 | 4.11 |
| *Suaeda erecta,* east side | 1.47 | 2.03 | 1.98 |
| *Suaeda erecta,* west side | 1.48 | 1.68 | 1.70 |
| *Allenrolfea,* east side | 0.98 | 1.00 | 1.06 |
| *Allenrolfea,* west side | 1.54 | 1.61 | 1.60 |
| *Distichlis* zone, east side | 0.80 | — | — |
| *Distichlis* zone, west side | 0.74 | — | — |

is white with a crust of crystalline salt covering the yellowish or gray clays or forming a fluffy mixture of loose clay and crystals. At wide intervals the flat surface is interrupted by sinuous wave-wrought terraces upwards to two metres high. During the summer, heat waves shimmer over the desert and mirages are a constant feature. The area is subject to wind erosion and large amounts of salt are removed during storms. The wind has also built non-calcareous sand dunes at some points around the margins and on some of the larger terraces in mid-desert. The average annual rainfall over the area is about 11 cm. By summer the surface becomes very dry, although in most places the water-table is 1—2 m below the surface. Some places are entirely barren, without even an occasional plant to interrupt the monotony of the white surface. However, the area as a whole is far from being entirely barren, as *Allenrolfea* has become established on higher ground around the margins (Fig. 4) and on isolated hummocks of wind blown soil formed around tumbleweeds and other objects far out on the flats. The usual hummocks also form round the branches of the shrubs and in time the salt leaches from them, permitting an occasional plant of *Suaeda torreyana, Sarcobatus vermiculatus* or *Atriplex nuttallii* to grow on the hummock also.

Fig. 4. *Allenrolfea occidentalis* invading a barren playa. Note ancient lake terraces on skyline left.

Analyses of the soil are given in Table V.

**Table V.**

Analyses of the soil from the root region of *Allenrolfea occidentalis*, Great Salt Lake desert. August 12, 1955.

| Depth, cm | pH | Moisture % | Salts, % dry weight of sample | | | | |
|---|---|---|---|---|---|---|---|
| | | | NaCl | $Na_2SO_4$ | $Na_2CO_3$ | $NaHCO_3$ | Total |
| 5 | 8.4 | 4.4 | 20.80 | .842 | .0 | .018 | 21.600 |
| 30 | 8.4 | 21.7 | 4.90 | .204 | .021 | .087 | 5.300 |
| 60 | 8.4 | 22.1 | 4.05 | .275 | .037 | .081 | 4.240 |
| 90 | 8.6 | 8.7 | 5.40 | .204 | .023 | .070 | 5.470 |

The rate at which plants invade playas until the area is completely occupied depends upon the size of the area, the rate at which salt is removed from the soil, depth of the water-table and the average precipitation in the area. Salt is reduced in the soil in two general ways: by wind and by drainage. In large playas this is a very slow process since drainage is often negligible. In small closed basins the salt tends to become concentrated toward the center but in areas where drainage occurs the process is accelerated. The effects on the rate of succession through human influence was vividly shown by the construction of drainage ditches across several small playas in 1932, including the one described above. By 1950, eighteen years later, the entire central area of most of them became occupied with scattered growths of *Salicornia rubra*, *Suaeda erecta* and *Allenrolfea*, while *Distichlis stricta* had followed half way or more from the perimeter of the playas. In the playa mentioned above the plants advanced about 50 metres in eighteen years, and in other playas the distance was greater.

In some playas the gradient of salt in the soil is very gradual and of relatively low concentration so that the succession proceeds faster, in some instances reaching an intermediate stage when the entire area is populated with plants of variable density and composition, some mixed and others more or less homogeneous.

Mixed playa successions. Following the pioneer species of *Salicornia*, *Suaeda*, and sometimes *Allenrolfea* and *Distichlis*, the following plants may become established in various combinations:

*Triglochin maritima* L.
*Puccinellia airoides* (Nutt.) Wats. & Coult.
*Atriplex argentea* L.
*A. hastata* L.
*Suaeda intermedia* S. Wats.
*S. torreyana* S. Wats.

*Salsola kali* var. *tenuifolia* TAUSCH.
*Bassia hyssopifolia* (PARL.) KUNTZE
*Sessuvium sessile* PERS.

A second group often following these plants includes:
*Deschampsia danthonioides* MONRO
*Myosurus apetalus* GRAY
*Lepidium perfoliatum* L.
*L. dictyotum* GRAY
*Capsella procumbens* (L.) FERIES
*Allocarya nitens* GREENE
*Plantago elongata* PURSH.
*Plantago purshii* ROEM. & SCHULT.
*Psilocarphus globiferus* NUTT.

The salt content of the latter community ranges from 0.6—0.8%.

Reclaimed homogenous playas. It is assumed that a playa is reclaimed when it is occupied by plants in such numbers as to create sharp competition among them. In many instances a single species may invade an area almost completely except for an occasional plant of other species here and there. Uniform conditions throughout the area partly explain this situation, but the more apparent reason is that the one species got there first in the greatest numbers. Where playas are reclaimed by a single species there is usually a robust growth showing the plant at its optimum. Soil samples taken from selected communities of this sort were analyzed and the salt contents are shown in Table VI.

**Table VI.**

Percentage salt in soils of reclaimed playa communities in terms of dry weight; the water-table about 1 m in each instance. June 28, 1929.

| Community | Depth in Centimetres | | |
|---|---|---|---|
| | 5 | 60 | 90 |
| *Salicornia utahensis* | 2.50 | — | — |
| *Allenrolfea occidentalis* | 1.41 | 1.12 | 1.16 |
| *Allenrolfea occidentalis* | 0.98 | 1.00 | 1.03 |
| *Allenrolfea occidentalis* | 1.54 | 1.61 | 1.60 |
| *Suaeda torreyana* | 0.81 | 0.80 | 0.74 |
| *Suaeda torreyana* | 1.03 | 0.95 | 0.99 |
| *Distichlis stricta* | 0.96 | 1.01 | 1.12 |
| *Distichlis stricta* | 0.96 | 1.45 | 1.22 |
| *Distichlis stricta* | 0.82 | 0.84 | 1.00 |
| *Sporobolus airoides* | 0.55 | 0.89 | 1.07 |
| *Sporobolus airoides* | 0.54 | 0.83 | 0.97 |

## Saline Plains

Saline plains extend beyond the strand and playas to the bases of the mountains, where the junction with steeper slopes of headlands is abrupt, but in most of the wider valleys the plains gradually rise to long alluvial slopes. At lower levels clay and clay-loams predominate, with a rather high water-table and a relatively high salt content, but on the gradually rising alluvial slopes, loams and gravelly soils predominate, while the water-table becomes progressively more remote and the salt content decreases.

The flora is diverse according to the topography and physical condition and is disposed in rather well-defined communities.

*Sarcobatus vermiculatus* communities. *Sarcobatus vermiculatus* (HOOK.) TORR. is an erect shrub with short, widely spreading, spiny branches of pale color and bears fleshy, linear, green leaves 2—3 cm long. The older stems are irregularly furrowed and have gray bark and green wood. The long tap root may penetrate the soil to a depth of 3 m or more; the plant is an indicator of groundwater. It grows on the lower plains and along drainage ways on higher ground where the soil holds ample moisture. The texture of the soil appears to limit the size of the bushes. In heavy clays, even where the water-table is only about two metres deep, the plant is generally less than 1 m tall, but in sandy-clay and clay-loam it commonly becomes 1—2 m tall, while in some sandy soils of low salt content it may reach a maximum of 4 m.

Fairly robust plants tolerate a maximum of about 1.5% sodium chloride but where sodium sulphate is the dominant salt it may tolerate as much as 14% although the exact maximum limits have not been ascertained. Sodium carbonate is the most toxic salt, and while some analyses of soil from the root region of *Sarcobatus* run as high as 0.3% and occasionally as high as 0.5%, it is generally much lower. Table VII gives the analyses of soil from two communities of fairly robust plants.

**Table VII.**

Analyses of soils from communities of *Sarcobatus vermiculatus*. August 10, 1957.

| Depth, cm | pH | Moisture % | Salts, % dry weight of samples | | | | |
|---|---|---|---|---|---|---|---|
| | | | NaCl | $Na_2SO_4$ | $Na_2CO_3$ | $NaHCO_3$ | Total |
| Area A: 3 | 8.6 | 1.1 | .063 | .009 | 0 | .092 | .165 |
| 30 | 9.4 | 6.2 | 1.164 | .080 | 0 | .164 | 1.400 |
| 60 | 9.2 | 6.9 | 1.478 | .097 | trace | .111 | 1.684 |
| 90 | 9.0 | 7.3 | 1.605 | .097 | 0 | .067 | 1.771 |
| Area B: 3 | 9.6 | 2.66 | .040 | .026 | 0 | .106 | .180 |
| 30 | 9.6 | 5.9 | .330 | .168 | .042 | .080 | .612 |
| 60 | 9.2 | 11.75 | .965 | .532 | .043 | .140 | 1.74 |
| 90 | 9.0 | 13.17 | 1.210 | .942 | .020 | .081 | 1.940 |

The bushes are generally quite evenly spaced, about 1—3 metres apart with considerable bare soil between them, the density varying from about 30—60% with a few denser stands here and there.

The botanical composition among different communities includes herbs and smaller shrubs, the number of species and their frequency depending largely on the character of the surface soil and the rainfall. In the areas receiving only 15 cm annual average precipitation, the clay soils harbor a very limited number of frequent species, such as *Suaeda torreyana*, the only shrub, and a few annual herbs such as *Bromus tectorum* L., *Hordeum gussonianum* PARL., *Salsola kali* var. *tenuifolia* TAUSCH., *Bassia hyssopifolia* (PALL.) KUNTZE and *Lepidium perfoliatum* L., *Tortula ruralis* (HEDW.) SMITH is a frequent moss while the soil lichens, *Synechoblastus coccophorus* (TUCK.) FINK and *Lecidea crenata* (TAYL.) STIZENB. are often exceedingly abundant, giving rise to numerous small crests of lichen-bound soil. Slight depressions often accumulate water in the spring and harbor a thin coating of *Oscillatoria angustissima* W. & W. on the clay.

In areas of loamy soil where the annual rainfall is about 45 cm there is a much greater variety of associated shrubs, annual and perennial herbs, and several species of moss. Some of the more conspicuous ones include:

Shrubs:
- *Suaeda torreyana* S. WATS
- *Kochia vestita* (S. WATS.) RYDB.
- *Atriplex nuttallii* S. WATS.
- *Atriplex confertifolia* (TORR.) S. WATS.

Grasses:
- *Distichlis stricta* (TORR.) RYDB.
- *Bromus tectorum* L.
- *Sitanion hystrix* (NUTT.) J. G. SMITH
- *Sporobolus airoides* TORR.

Herbs:
- *Bassia hyssopifolia* (PALL.) KUNTZE
- *Salsola kali* var. *tenuifolia* TAUSCH.
- *Lepidium perfoliatum* L.
- *Capsella procumbens* (L.) FRIES
- *Draba nemorosa* L.
- *Draba cuneifolia* NUTT.
- *Aster leucanthemifolius* GREENE

Mosses:
- *Pterigoneurum ovatum* (HEDW.) DIXON
- *Pterigoneurum subsessile* (BRID.) JUR.
- *Pterigoneurum lamellatum* (LINDB.) JUR.
- *Tortula brevipes* (LESQ.) BROTH.
- *Tortula bistratosa* FLOWERS

*Crossidium aberrans* HOLZ. & BART.
*Funaria hygrometrica* HEDW.
*Bryum caespiticium* (L.) HEDW.

Locally *Pterigoneurum ovatum* and *Tortula brevipes* form dense and often extensive tufts.

*Atriplex confertifolia* communities. The most extensive basin areas vacated by Lake Bonneville are occupied by communities of *Atriplex confertifolia,* locally called shadscale. It is a silvery-gray, round-topped shrub 60—90 cm tall with spreading irregular, spiny branches bearing rather thick ovate to orbicular leaves heavily covered with scurfy scales. It has rather shallow, widely spreading roots and it feeds almost entirely upon precipitation water. From the level plains it extends high up on surrounding slopes. It grows best in clay-loam and loam and gravelly soils, and, in areas of higher rainfall, reaches its maximum size.

Sodium chloride is the dominant salt in the soils of the lower plains around Great Salt Lake and the desert to the west, but higher on the alluvial slopes sodium sulphate becomes dominant. Elsewhere in the Bonneville basin and in the Colorado river basin sulphates are consistently higher than chlorides in both plains and slopes. *Atriplex confertifolia* seldom grows well in chloride soils exceeding 0.8% in the root region, although higher amounts may be present in the lower layers. Its maximum toleration is reached at only 2.3% but the plants are small, stunted and with scanty small leaves. In sulphate soils the maximum has not been determined, but amounts in the root region may reach as high as 2.3% without showing any marked effect on the vigor of the plants. Table VIII shows the analyses of soils from an average community on a lower slope of clay-loam.

**Table VIII.**

Analyses of soil from an *Atriplex confertifolia* community.

| | | NaCl salts, percent dry weight | | | | |
|---|---|---|---|---|---|---|
| Depth, cm | pH | NaCl | $Na_2SO_4$ | $Na_2SO_3$ | $NaHCO_3$ | Total |
| 30 | 7.6 | .060 | .120 | trace | .100 | .283 |
| 60 | 7.8 | .130 | .130 | .031 | .100 | .396 |
| 90 | 7.8 | .623 | .134 | .022 | .101 | .855 |

Botanical composition. On the lower plains the most frequent associated shrubs include *Sarcobatus vermiculatus, Suaeda torreyana, Atriplex nuttallii, Eurotia lanata, Kochia vestita, Tetradymia spinosa* H. & T., *T. nuttallii* T. & G., *Chrysothamnus pumilus* NUTT., *C. nauseosus* (PALL.) BRIT. and *Gutierrezia microcephala* GRAY. Among the frequent herbs are *Festuca octoflora* WALT., *Bromus tectorum* L., *Poa secunda* PRESL., *Agropyron trachycaulum* (LINK) MATLE, *Agropyron desertorum* (FISCH.) SCHULT., *Eriogonum cernuum* NUTT., *Salsola kali* var. *tenuifolia* TAUSCH., *Lepidium perfoliatum* L., *Draba nemorosa* NUTT., *D. cuneifolia* NUTT., *Capsella procumbens* (L). FRIES., *Arabis holboellii* HORNEM., *Malcolmia africana* (WILLD.) R. BR., *Astragalus utahensis* T. & G., *Sphaeralcea munroana* (DOUGL.) SPACH., *Opuntia fragilis* (NUTT.) HAW., *Cleome lutea* HOOK., *Chrysopsis foliosa* NUTT., *Aster leucanthemifolius* GREENE, *A. leucelene* BLAKE, and *Erigeron divergens* T. & G.

Several communities of less extent are dominated by two shrubs in about equal numbers. They are as follows:

*Sarcobatus-Atriplex confertifolia* communities. These shrubs grow together on the lower plains where the water-table is about 1 m below the surface and the salt content of the soil shows an average range of about 0.5—0.8%, Fig. 5.

*Atriplex-Kochia* communities. On some broad plains in dry areas adjacent to the southern extremity of the Great Salt Lake Desert, these shrubs form local communities which survive on a minimum of rainfall. *Kochia vestita* (S. WATS.) A. NELS., locally called gray molly, is a lowgray shrub 15—40 cm tall, slender, erect or spreading branches bearing erect, linear, fleshy leaves 6—20 mm long and

Fig. 5. Saline plain showing the greasewood-shadscale community (*Sarcobatus vermiculatus* and *Atriplex confertifolia*).

densely covered with white villous hairs. It is frequent or occasional in many other communities.

*Atriplex-Eurotia* communities are local in many areas. *Eurotia lanata* (PURSH) MOQ., locally called winter fat, is a small pale gray shrub producing several tall, slender, flowering stalks 40—70 cm tall which bear numerous fruits covered with copious long silky hairs which give the plant a decidedly white appearance late in the season. The leaves are linear to narrowly lanceolate, 18—36 mm long, with the margins revolute and the surface covered with dense stellate hairs. It is widely distributed throughout the western United States and is highly palatable to sheep and cattle, for which reason it has become greatly reduced in numbers. It also grows in minor communities by itself.

*Atriplex-Artemisia spinescens* communities occur mainly on alluvial slopes. *Artemisia spinescens* D. C. EATON, locally called bud sagebrush, is a spiny undershrub 10—50 cm tall with spreading branches bearing pedately parted leaves, the segments again divided into linear lobes. Dense white hairs give the plants a whitish color which stands out conspicuously among the silvery gray *Atriplex* bushes, especially in the spring.

## Sand Dunes

Locally, sand dunes are formed along some of the eastern shores of Great Salt Lake and on the plains and foothills bordering the salt desert. Those near the lake are composed of whitish calcareous oolite while those farther removed are tawny colored and composed of mixed silaceous and ferro-magnesium grains. The latter are often markedly localized and formed by peculiarities of air currents. The origin of the sand is obscure. The salt content ranges from 0.1—0.3%.

The botanical composition varies somewhat in different dunes but in general includes the following plants:

Dominant shrubs:

*Atriplex canescens* (PURSH.) NUTT.
*Chrysothamnus viscidiflorus pumilus* (NUTT.) T. & G.
*C. puberulus* (EATON) GREENE
*Tetradymia glabrata* A. GRAY

Frequent shrubs:

*Grayia spinosa* (HOOK.) MOQ.
*Sarcobatus vermiculatus* (HOOK.) TORR.
*Chrysothamnus viscidiflorus* (GRAY) GREENE
*C. viscidiflorus tortifolius* (GRAY) GREENE
*C. stenophyllus* (GRAY) GREENE
*C. nauseosus speciosus* (NUTT.) H. & A.
*Gutierrezia sarothrae* (PURSH.) B. & R.
*Tetradymia spinosa* H. & A.

Grasses:
*Sporobolus airoides* TORR.
*Oryzopsis hymenoides* (R. & S.) RICKER
*Spartina gracilis* TRIN.

Herbs:
*Eriogonum cernuum* NUTT.
*E. ovalifolius* NUTT.
*E. dubium* STOKES (endemic here)
*Rumex venosa* PURSH
*Abronia elliptica* A. NELS
*A. micrantha* TORR.
*A. salsa* RYDB.
*Psoralea lanceolata* PURSH
*Euphorbia arenicola* PARISH
*E. flagelliformis* ENGELM.
*Oenothera pallida* LINDL.
*Sphaerostigma alyssoides* (H. & A.) WALP.
*S. tortum* (L. EVL.) A. NELS
*S. utahensis* SMALL
*Chylisma parryi* (WATS.) SMALL
*Leptodactylon pungens* (TORR.) NUTT.
*Ptiloria exigua* (NUTT.) GREEN
*Lygodesmia grandiflora* T. & G.

On dunes formed high on hillsides, *Juniperus osteosperma* (TORR.) LITTLE marks the closest approach of a native tree to the lake and salt desert. Here also *Ephedra nevadensis* S. WATS. enters.

## Algae of the Strand and Inland Salterns

At times when the lake level is low, small depressions along the higher parts of the beaches (beyond the reach of the lake water), accumulations of fresh water from rain and snow or from littoral springs, and seepage areas create a habitat for certain algae and protozoa. The water becomes variably salty, the amount of salt in solution varying according to how long the water stands over the salty beach sediments. Analyses show as little as 0.2% and as much as 8 or 10% sodium chloride with smaller amounts of other salts. The depressions are formed in a variety of ways, such as by old wave-formed ridges, automobile tracks and hoof marks of cattle. Thus the habitats may contain as little as 200—330 ml of water and range upwards to small ponds of 40—50 metres' extent.

During January of 1942 a particularly rich algal flora occurred on the beach at Black Rock, about 200 m from the lake shore, in which 10—15 forms were distinguished. Some of these were restricted to a single population occupying a very limited habitat, such as a single hoof-mark in the sand and the water scarcely exceeding

300 ml in volume. Others occurred in larger puddles where the populations were mixed. Cultures of some of these forms were maintained in the laboratory for about two months and efforts made to observe the life cycles, but they died out before any could be completely followed. Most of these forms appeared to be undescribed species. At least eight of them were identified as *Chlamydomonas*, one *Carteria* and several uncertain. During subsequent years most of these forms disappeared and only a few in small populations have been found since.

In some strand puddles and particularly in the evaporating ponds of commercial salt manufacturing companies, *Dunaliella salina* becomes abundant in certain seasons. This organism is a unicellular, biflagellate form having a red pigment which masks the chlorophyll, and it often occurs in such dense populations as to give the water a bright red color. It also tints dried salt piled up for future processing.

In offshore marshes and drainage ditches traversing saline plains the water is variably salty, usually less than 0.2%, but occasionally as much as 0.5%. In these places *Rhizoclonium hieroglyphycum* (AG.) KÜTZ. and *Cladophora fracta* KÜTZ. frequently grow prolifically while *Rhizoclonium crispum* KÜTZ. and *Chaetomorpha aurea* (DILL.) KÜTZ. are less common. *Enteromorpha inestinalis* (L.) GREV. is locally abundant while *E. plumosa* KÜTZ. and *E. prolifera* (FL. DAN.) AG. are occasional. *Ulva lactuca* L. occurs in the outflow of a large spring of mildly saline water at Timpe, an unusual disjunct habitat for this marine alga. Numerous other species of green algae are less frequent or occasional. Diatoms are especially abundant in shallow sloughs of brackish water.

*Batrachospermum moniliforme* ROTH and *Audouinella violacea* (KÜTZ.) HAMEL are Rhodophyta frequently growing in rocks and wood in flowing water from cold saline sulphur springs and wells.

A large number of Cyanophyta occurs in many situations but particularly in and around hot sulphur springs which carry from 0.2—3.4% salt. The most frequent species include *Oscillatoria amphibia* AG., *O. animalis* AG., *O. angustissima* W. & W., *O. brevis* KÜTZ., *O. chalybea* MERT., *O. janthopora* KÜTZ., *O. lemmermannii* KÜTZ., *O. limosa* AG., *O. nigro-viridis* THW., *O. princeps* VAUCH., *O. sancta* KÜTZ., *O. subtilissima* KÜTZ., *O. terebriformis* AG., and *O. tenuis* AG., *Phormidium angustissimum* W. & W., *Ph. foveolarum* (MONT.) GOM., *Ph. inundatum* KÜTZ., *Ph. ramosum* BOYE. and *Ph. subfuscum* KÜTZ., *Spirulina major* KÜTZ. and *Tolypothrix tenuis* KÜTZ.

## REFERENCES*

ALDRICH, J. M., 1912. The Biology of Some Western Species of the Dipteron Genus Ephydra. *J. New York ent. Soc.* **20**: *77—99. Ephydra hians* SAY.

ALLEE, W. C., 1926. Some Interesting Animal Communities of Northern Utah. *Sci. Monthly* **23**: *481—495.* 1926. Three species of *Ephydra* larvae and pupae in their cases. *Artemia gracilis.*

DAINES, L. L., 1917. Notes on the Flora of Great Salt Lake. *Amer. Nat.* **51**: *499. Aphanothece packardii, Chlamydomonas, Navicula, Cymbella.*

EARDLEY, A. J., 1938. Sediments of Great Salt Lake, Utah. *Bull. Amer. Ass. Petroleum Geologists.* **22**: *1305—1411. Pleurocapsa entophysaloides* SETCH. & GARD. forma. (p. 1333). Flagellates: *Trachelomonas* or *Peridinium* or both (?) 1333.

FLOWERS, S., 1934. Vegetation of the Great Salt Lake Region. *Bot. Gaz.* **95**: *353—418.*

FREDERICK, ELFREDA, 1924. Bacterial Flora of Great Salt Lake. Thesis Univ. of U. 11 bacilli and 1 coccus.

JENSEN, A. C., 1918. Some Observations on *Artemia gracilis*, etc. *Biol. Bull.* **34**: *18—25.*

KIRKPATRICK, RUTH, 1934. Life of the Great Salt Lake with Special Reference to the Algae. Thesis, Univ. of U.

PACK, DEAN A., 1919. Two Ciliates of Great Salt Lake. *Biol. Bull.* **36**: *273. Porodon utahensis* Pack and *Uroleptus packii* Calkins. (Provisional.)

PACKARD, JR., A. S., 1871. On Insects Inhabiting Salt Water. *Amer. J. Sci.* **1**: *100—110.* 1871.

PACKARD, JR., A. S., 1879. The Seaweeds of Great Salt Lake. *Amer. Nat.* **13**: *701—703.* Descriptions by W. G. FARLOW. 1879. *Polycystis packardii, Ulva marginata, Rhizoclonium salinum.*

PATRICK, RUTH, 1936. Some Diatoms of Great Salt Lake. *Bull. Torrey Bot. Club.* **63**: *157—166.*

ROTHPLETZ, A., 1892. On formation of oolite. *Bot. Cbl.* **51**: *265—268.* Also in *Amer. Geologist* **10**: *279—282.* 1892. On *Gloeothece* and *Gloeocapsa.*

SCHWARTZ, E. A., 1891. Prelim. Remarks on the Insect Fauna of Great Salt Lake. *Can. Ent.* **23**: *235—241. Ephydra gracilis.*

TILDEN, J., 1898. *Aphanothece utahensis, Polycystis packardii, Dichothrix utahensis, Enteromorpha tubulosa, Chara contraria.* Amer. Algae, Cent. III, No. 298. Also cited in Minnesota Algae.

VERRILL, A. E., 1869. Twelfth Annual Report, U.S. Geological and Geographical Survey of the Territories of Wyoming and Idaho, Part I. p. 130. 1878. *Ephydra fertilis* (gracilis).

VORHIES, C. J., 1917. Notes on the Fauna of Great Salt Lake. *Amer. Nat.* **51**: *494.* 1917. *Artemia gracilis, Ephydra gracilis, Amoeba* of 2—3 varieties, *Euglena* (seen once), Ciliates (1 resembling *Uroleptus*).

* With indication of species.

# INDEX OF PLANT NAMES, ANIMAL NAMES AND AGRICULTURAL PRODUCTS

# GENERAL INDEX